FORBIDDEN KNOWLEDGE

FORBIDDEN KNOWLEDGE

Medicine, Science, and Censorship
in Early Modern Italy

For Robert —

With happy memories of tea +
cookies at the IAS. ☺

HANNAH MARCUS

THE UNIVERSITY OF CHICAGO PRESS

CHICAGO AND LONDON

The University of Chicago Press, Chicago 60637
The University of Chicago Press, Ltd., London
© 2020 by The University of Chicago
All rights reserved. No part of this book may be used or reproduced in any manner
whatsoever without written permission, except in the case of brief quotations
in critical articles and reviews. For more information, contact the University of
Chicago Press, 1427 E. 60th St., Chicago, IL 60637.
Published 2020
Printed in the United States of America

29 28 27 26 25 24 23 22 21 20   1 2 3 4 5

ISBN-13: 978-0-226-73658-7 (cloth)
ISBN-13: 978-0-226-73661-7 (e-book)
DOI: https://doi.org/10.7208/chicago/9780226736617.001.0001

Library of Congress Cataloging-in-Publication Data

Names: Marcus, Hannah (Historian of science), author.
Title: Forbidden knowledge : medicine, science, and censorship in early modern
    Italy / Hannah Marcus.
Description: Chicago : University of Chicago Press, 2020. | Includes bibliographical
    references and index.
Identifiers: LCCN 2020008837 | ISBN 9780226736587 (cloth) | ISBN 9780226736617
    (e-book)
Subjects: LCSH: Science—Censorship—Italy. | Medicine—Censorship—Italy. |
    Libraries—Censorship—Italy. | Censorship—Italy—History.
Classification: LCC Z658.I8 M373 2020 | DDC 363.310945—dc23
LC record available at https://lccn.loc.gov/2020008837

♾ This paper meets the requirements of ANSI/NISO Z39.48-1992 (Permanence
of Paper).

This book is dedicated to Morgan MacLeod
with my deepest gratitude.

CONTENTS

FIGURES

The Paradox of Censorship

When the Italian doctor Francesco Redi wrote to Leopoldo de' Medici in 1670, he rhapsodized about the intense allure of books in biblical terms: "I believe that my soul will certainly be lost to perdition on account of prohibited books. If instead of creating Adam God had created me in Eden, and if instead of prohibiting me from eating that fig and that apple he had prohibited me from reading books, I am so weak that I surely would have done worse than Adam."[1] Redi's insatiable bibliophilia, described with a substantial degree of humor and self-deprecation, deftly raises several important themes about prohibited books and early modern physicians that are central to this study.

First, Redi was quite open with his patron, de' Medici, about reading prohibited volumes. While prohibited religious texts remained off-limits, prohibited professional books in medicine, law, and astronomy could still be used and read through a vast and visible Catholic censorship system that involved petitions, expurgations, and licenses. Indeed, Redi had a license issued by the Roman Inquisition to keep many prohibited books in his library.[2] A professional like Redi would have had to substantially circumscribe his medical practice if he could not access prohibited books in support of his work as a physician. Prohibited texts were part of physicians' personal libraries, on the shelves of public libraries, and circulating in secondhand book markets. The practice of reading proscribed texts was far more widespread among learned and elite society in early modern Italy than we previously understood. In this book, I tell the story of how prohibited medical texts came to be such an open and integrated part of Catholic society by the end of the seventeenth century.

Redi's reflection also presents prohibited books as fundamentally irresistible. Surely Redi's inability to refrain from reading books may have been due in part to his personal curiosity, but he was also part of a professional community that had demonstrated a need for prohibited texts. Over the previous century, physicians had developed a broad, flexible discourse about the utility of prohibited medical books. Church officials and physicians explicitly debated which prohibited medical books were useful and which parts needed to be removed to render them safe for circulation in the eyes of Catholic authorities. The context of censorship forced physicians to articulate the utility of these books to their profession and the utility of their profession to Catholic society. The compromise of expurgation resulted in libraries with shelves of medical books that bear physical signs of censorship and generations of readers, like Redi, who were licensed to use these books openly. In the following pages, I describe the process by which physicians living a century before Redi came to read banned medical texts, and I explore the motivations that led them to do so. These early modern physicians repeatedly justified this practice through claims that prohibited books contained useful knowledge that was necessary to their work.

In this book I take the concept of scientific utility as a distinct historical subject and detail the essential and shifting valences of the important category of medical utility in relation to Catholic censorship. Take for example the copy of Paschal Le Coq's *Bibliotheca medica* (*Medical Library*) printed in 1590 and pictured here (figure I.1). Although Redi did not own Le Coq's book, his library did include the competing volume published by Israel Spach in 1591.[3] This page from Le Coq's bibliography is an evocative example of the ways that prohibited books circulated after expurgation and a telling piece of evidence about how the concept of scientific and medical utility was shaped by physicians' encounters with ecclesiastical censorship. Most of the Catholic censors described in the following pages focused primarily on evaluating the content of Leonhart Fuchs's texts and then on removing Fuchs's name from books like this one, damning the memory of the important Protestant physician. However, in this case, the censor was drawn to a different problem. Following the instructions for expurgation given in the Spanish Index of 1612, he crossed out the word *useful* from Le Coq's entry which read, "Leonhart Fuchs wrote in his *Paradoxes* book one, many things in a useful fashion about medicinal simples and about errors concerning those medicines."[4] The censor struck the word *useful* because the idea of utility in early modern Catholic Europe had become a declaration of piety in addition to a general, positive remark about the ends of knowledge.

In the second half of the sixteenth century, censors and doctors dis-

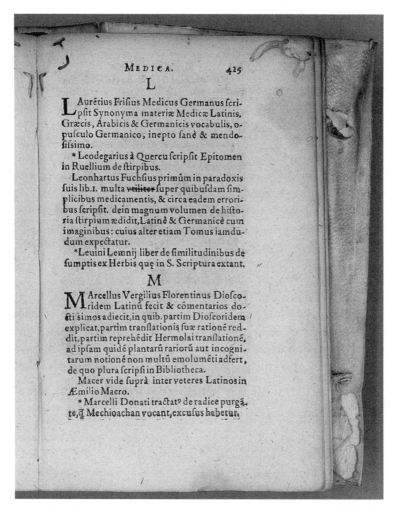

Fig. I.1. Paschal Le Coq, *Bibliotheca medica* (Basel, 1590), 425, showing the word *utiliter* (useful) crossed out from Leonhart Fuchs's entry by a Catholic censor. BH MED 96, Biblioteca Histórica de la Universidad Complutense de Madrid.

cussed utility in relation to the contribution of medicine and physicians to the health of society. However, by the seventeenth century, when the pictured copy of Le Coq's bibliography was expurgated, the concept of utility itself had taken on a religious meaning as well. Following the rules for expurgation, painstakingly formalized in the sixteenth century, this censor, who may even have been a physician himself, dutifully crossed out the praise describing the Protestant Fuchs's books as "useful," because what was useful was pious and Catholic, and Fuchs was neither by the standards of the theologians who wrote the 1612 *Index of Prohibited and Expurgated Books*. Over the course of the sixteenth and seventeenth centuries, physicians confronted an extensive system of ecclesiastical censorship by repeatedly, explicitly, and creatively defining what it meant for medical books and the medical profession to be useful. While Redi, in the second half of the seventeenth century, was too weak to resist the pull of prohibited books, he also did not have to. Over the past one hundred years physicians had labored within and against Catholic censorship regimes to create a place for these useful books in their libraries, and Redi could take advantage of these systems.

Finally, Redi's letter to Leopoldo de' Medici shows him to have been delightfully aware of the timelessness of the problem of forbidden knowledge. Since Eden, people have been reaching for knowledge that authorities sought to deny them, whether on the branches of trees or the shelves of libraries. I examine the age-old theme of forbidden knowledge through the censorship of medical texts in early modern Italy, a story that culminates in the society in which Redi wrote. The history of the ecclesiastical censorship of science in Italy has traditionally been told as a Galilean struggle between faith and science, destined to end in conflict.[5] However, the Catholic censorship of medical books that I describe took place largely in the years before the Galileo affair, with important—and different— emphases and consequences.

THE PARADOX OF CENSORSHIP

Censorship was a ubiquitous fact of intellectual life in early modern Europe, and it took many forms.[6] Systems of prepublication censorship and review sprang up across Europe in an attempt to control the spread of political and religious ideas in the new age of mass media facilitated by the invention of moveable type. Early modern authors even exercised varying degrees of self-censorship before a book reached the stage of licensure to be printed, so as to ensure that their works would be published.[7] This book deals primarily with the restrictions placed on books after they had already been printed and were in circulation in Europe. As we shall see, postpublication Catholic censorship in the early modern period was different in several fundamental ways from the modern systems of state-sponsored censorship that seek to remove ideas completely from public view.[8]

In the aftermath of the Reformation, various Catholic communities published Indexes of Prohibited Books—lists of books that Catholics were not permitted to read. These lists were read aloud from pulpits, nailed to the doors of cathedrals, and distributed to booksellers. The first was the Paris Index of 1544, which was soon followed by the Indexes of Louvain, Venice, Spain, Portugal, and finally Rome in 1559. While the lists differed in particulars, they all initially banned religious works by Protestant leaders such as Martin Luther and John Calvin, books written anonymously, and even books written by Protestants that did not discuss religion. In so doing, the Indexes of Prohibited Books tied Protestant authors' religion to their scientific works, and the fate of those works to the intricacies of Catholic censorship.

Censorship was but one of the Catholic Church's responses to the Protestant Reformation, which also included councils, inquisitions, and pastoral reforms.[9] Scholars have increasingly come to understand the Catholic Church and its sixteenth- and seventeenth-century agents as functioning as a network of actors with distinct goals and motivations for controlling thought and behavior.[10] The same view holds true for our understanding of censorship. While censorship prohibited many books from being read in Italy, the normative prescriptions for book control diverged significantly from the experiences of readers.[11] Furthermore, Catholic rules about censorship were constantly being revised and would eventually allow for the accommodation of texts through expurgation and licensing rather than total prohibition. An examination of the full range of Catholic censors' activities, not just the burnings of books and authors, is essential for understanding the goals of the Counter-Reformation Church with respect to regulating both people and knowledge. As the Congregation of the Index of Prohibited Books reiterated in letters sent throughout the Mediterranean world, Catholic censorship was the business of "prohibiting, permitting, correcting, and printing books."[12]

From the Middle Ages onward, the Catholic Church relied on the pope, universities, and local inquisitorial tribunals to censor books. In the early modern period in Rome, responsibility for censorship was divided between the Master of the Sacred Palace (the pope's theologian), the Roman Inquisition, and, beginning in 1572, the newly formed Congregation of the Index of Prohibited Books. The functioning and effects of these regulatory bodies on Italian society are vast subjects that have been a source of intense scholarly debate for many years. From Adriano Prosperi's tribunals of conscience to Carlo Ginzburg's story of the burning of the miller Menocchio and Gigliola Fragnito's accounts of the burning of Italian Bibles, we are well aware of the insidious and violent acts of the Italian Inquisitions.[13] Paul Grendler's account of how censorship affected Venetian booksellers presents us with a different image of Catholic attempts at intellectual control in the cosmopolitan city on the lagoon, where it seemed banned books were always hidden under a bookseller's bench or being surreptitiously tossed into a canal moments before an inquisitor's arrival.[14] There is truth to all of these accounts, and since the opening of the archives of the Roman Inquisition and the Congregation of the Index to researchers in 1998, we are in a better position than ever before to assess the early modern Church's mechanisms of social and intellectual control.

The flood of new sources available since the opening of the Roman archives has recently led to a revisionist and conflicted historiography about ecclesiastical censorship in Italy.[15] Much of this literature is framed by a debate about the "effectiveness" of Catholic censorship and looks to case studies of individual books or subjects as evidence.[16] I reframe the question about the effectiveness of censorship to ask instead: What were the cultural and scientific products of censorship?

Through analysis of archival sources from across Italy and the Vatican, I argue that the effects of ecclesiastical censorship were both material and discursive. The Catholic censorship of medical books was a form of promulgation, albeit limited, and a Catholic endorsement of a discourse about the utility of scientific knowledge. The Indexes of Prohibited Books catalyzed a conversation about medical texts and, paradoxically, convened a learned forum in which physicians and ecclesiastics discussed and analyzed prohibited texts and recorded and archived their opinions about these works. Indeed, this is the paradox of censorship in early modern Europe. Catholic censorship succeeded in repressing the circulation of some texts while simultaneously creating a structured arena for discussion and debate about scientific knowledge. Both projects drew on the professional expertise of ecclesiastics and lay scholars and created a select, elite readership of prohibited medical books on the Italian peninsula. This account compliments Gigliola Fragnito's conclusion that the systematic prohibition of devotional texts written in the vernacular had the effect of creating two separate "registers of communication," an elite register for those with a classical education and a lower register for those ignorant of Latin.[17] However, in the case of medical texts, elite physicians were also afforded avenues to continue to engage with prohibited materials. Catholic censorship thus sought to delineate particular forms of texts, to authorize the terms of readership, and to articulate legitimate contexts for prohibited books, in addition to keeping them out of the hands and beyond the understanding of large portions of European society.

Focusing on the stark paradox of censorship reveals the complex reception histories of many prohibited texts in Catholic Italy. These histories are usually described as either nonexistent, on the basis that the books were removed from circulation, or heterodox, on the basis that readers continued to access these books through the black market and with nonorthodox intentions. By considering ecclesiastical censorship as a limited form of promulgation and giving equal attention to processes of expurgation and licensing as to prohibition, I trace Catholic reception histories of medical books that were prohibited in Counter-Reformation Italy. Examining the process by which books by authors such as Girolamo Cardano, Conrad Gessner, and Leonhart Fuchs remained crucial parts of Catholic libraries reveals how early modern physicians evaluated the utility of these works and facilitated their continued circulation with the oversight of Catholic authorities. The censorship of medical knowledge is a particularly interesting subset of the larger history of censorship because physicians were so vocal about the utility of their profession and the utility of their books for maintaining a healthy Catholic society. The scientific and religious stakes of medicine were inextricably connected through this highly developed discourse of utility, to which the Catholic Church and Galileo would both turn in the conflict over Copernicanism in the seventeenth century.

PHYSICIANS AND THEIR USEFUL BOOKS

During the period of censorship discussed in this book, physicians became increasingly professionalized. Although university-trained practitioners represented only part of early modern medical practice and healing, over the course of the sixteenth century physicians were actively consolidating their position as social and intellectual elites.[18] From the elevated status of the learned expert, physicians debated the boundaries of prohibited knowledge with ecclesiastics. Physicians' engagement with the Catholic Church through censorship contributed to the recognition of medical knowledge as an independent realm of professional expertise.[19] This expertise placed the authority of physicians over that of other medical practitioners and separate from, though not equal to, that of ecclesiastics. The legal scholar Frederick Schauer has theorized that the very ontology of censorship is that of expertise, professionalism, and separation of authorities.[20] Significantly then, my consideration of the conflict between religion and science proceeds with the recognition that religious regulation played a role in establishing the professional credentials of practitioners of early modern science.

Learned medicine in the sixteenth century was also an international enterprise. Italian universities, in particular at Padua and Bologna, were arguably the two most important sites of medical learning in sixteenth-century Europe. Swiss, German, English, and French physicians flocked to these cities on a *peregrinatio medica*, or medical travel, where they studied with Italian scholars and met physicians from across Europe.[21] William Harvey, the English physician who came to Padua to study under Girolamo Fabrici d'Acquapendente and then went on to describe the circulation of blood through the body, was but one of the famous medical travelers on the Italian peninsula. In addition to matriculated students like Harvey, medical travel also included scholars' short-term visits to places and people within the European world of medical learning. These personal connections and the shared culture of Latin scholarship formed the basis for what has been described as the medical republic of letters.[22] Sixteenth-century Catholic censorship, however, would drive a wedge into a community that was at once personal and professional, and entangled a multiconfessional, transnational community of scholars in an era of religious conflict.

When Catholic censorship and medical scholarship first came to loggerheads in Italy following the publication of the Pauline Index of Prohibited Books in 1559, the formal systems of expurgation and licensed reading had yet to be established. With its Index of Prohibited Books, the Catholic Church created its own ideal universe of proper Catholic learning in which Protestant scholars played no part.[23] However, many Catholic physicians throughout Italy complained to local ecclesiastical officials that the prohibitions interfered with their work. Physicians then became involved in a negotiation with the Church that came to define the boundaries of what was important and which authors and professionals were at the center of useful knowledge. Debate about the censorship of medical knowledge was a struggle between the inexpertly dictated regulations of the Index and the necessity of scientific knowledge to Catholic society.

In both the universities and the medical republic of letters, the study of medical texts in ancient Greek and Latin (medical humanism) was particularly widespread.[24] Many of the best new editions were edited or translated in the first half of the sixteenth century by Protestant, humanist physicians in Northern Europe including Leonhart Fuchs, Conrad Gessner, and Janus Cornarius. When the Catholic Index of Prohibited Books banned works written or even edited by Protestants, it inadvertently denied physicians licit access to the best translations of ancient texts. This was an extremely problematic outcome for humanist scholars in Catholic Europe who put a premium on accurate textual editing and precise translation.

Complicating matters still further, in the sixteenth century scholars of varying religious beliefs, some of which were incompatible with the Counter-Reformation Church, had produced a unique and highly useful body of literature, almost all of which would be prohibited in Catholic Europe over the course of the century. Books describing plants and pharmacology written by Protestant physicians—including Otto Brunfels, Gessner, and Fuchs—were considered the best references available based on the knowledge they gathered and the precision of their images. Paracelsus's iconoclastic forays into chemical medicine earned him first a great deal of scorn and later interest from both Protestant and Catholic scholars. The seven hundred medical cases collected by the Portuguese physician and crypto-Jew Amatus Lusitanus were indispensable references, even though the author was vocal about his skepticism regarding clerical celibacy. Catholic censorship condemned all of these texts and, in so doing, forced Catholic physicians to confront the religious contexts of these authors in addition to the content of their works.

In the context of the Counter-Reformation, discussion of the religious status of medical knowledge was fundamentally confessionalized. My research is part of a larger historiographical conversation that is reconsidering learned medicine in an explicitly religious context.[25] Much of this literature has focused on reassessing the impact of the Reformation on medical learning and practice with particular attention to heterodox thought.[26] One of the goals of this book is to turn our focus from cases of heterodoxy to better understand the world of Catholic physicians.[27] Catholic physicians devised strategies, both intellectual and logistical, to navigate the culture of censorship in which they lived and worked. In the following pages, we will meet Catholic physicians involved in book smuggling, self-censorship, and both pious and devious expurgation. Throughout, I will draw attention to these doctors' attempts to justify and explain their engagement with prohibited books, as Redi's quote exemplifies in the opening lines of this introduction.

One such strategy for validating and obtaining access to prohibited books was a persistent and explicit discourse about the utility of medical knowledge, which emerges repeatedly from the wealth of archival evidence documenting physicians' interactions with Catholic censorship. While discussions about the utility of medical and scientific knowledge were hardly new, the concept of utility came to take on new meanings in Counter-Reformation Italy. We are familiar with a traditional, cynical concept of medical utility that points to how physicians expected to make money from their practice. As Petrarch ferociously held forth in his 1355 *Invectives against a Physician*, "Your medicine has money as its goal, is subordinate to it, and exists for its sake. Draw the conclusion, O dialectician: Therefore, medicine is the servant of money."[28] From this remark Petrarch moves on seamlessly to a critique of the skills and subjects "useful and necessary" for medicine. Katherine Park has shown how fifteenth-century Florentine physicians drew on the term *utility* to refer to their ability to make money in their chosen profession.[29] For Renaissance Florentines, the concept of medical utility had long been tied to less than high-minded goals.

Other fields of knowledge also laid claim to utilitarian justifications, though perhaps more high-mindedly than fourteenth- and fifteenth-century physicians. Mathematics and its allied discipline of astronomy were described in terms of their utility.[30] Encomiums written by Renaissance scholars of astronomy detailed the many applications of astronomy for the calendar, medicine, agriculture, navigation, and pedagogy.[31] The technical arts, in general, had a special claim to the direct application of knowledge, a discourse which became central to the value of experience in early modern natural philosophy.[32] Even within the realm of literature, Horace had described the best poetry as pleasant and useful. This description of good writing as useful prompted the seventeenth-century Spanish censor Juan Caramuel to instruct censors to not only correct errors in faith in texts, but also to fix errors in grammar, mathematics, and historical fact.[33] Situated firmly at the intersection of theoretical, practical, and literary knowledge, early modern medicine had a claim to each of these utilitarian traditions.

Additionally, as the medical humanists of the Renaissance read, translated, edited, and commented upon Galen, they encountered an explanation of their craft as both utilitarian (healing the sick and preserving health) and fundamentally philosophical and theological. In his *De usu partium* (*On the Usefulness of the Parts of the Body*), Galen drew connections between the actions that specific parts of the body performed and their underlying utility. For the anatomist of Pergamon, it was not sufficient to understand the way the hand worked in order to heal it; one also needed to grasp the functions that the hand performed. The usefulness of each part of the body was "related to the soul," as Galen understood it, since "the body is the instrument of the soul."[34] Indeed, Galen laid out the theological and philosophical implications of the study of anatomy in the final, seventeenth book, the "Epode," which he named specifically to liken it to the closing section of a hymn of praise to the gods:

> Then a work on the usefulness of the parts, which at first seemed to him a thing of scant importance, will be reckoned truly to be the source of a perfect theology, which is a thing far greater and far nobler than all of medicine. Hence such a work is serviceable not only for the physician, but much more so for the philosopher who is eager to gain an understanding of the whole of Nature.[35]

Caring for and closely studying the body could be an act of piety, as even the ancient Greek Galen described it.

In Counter-Reformation Italy, physicians and ecclesiastics alike repeatedly invoked the utility of medicine, most often as a justification for making books available selectively to certain readers, rather than burning them in their entirety. This justification operated on two levels. Physicians described medicine as a useful discipline and described their books as necessary to that endeavor. They also extended this reasoning to define themselves as part of a profession that was fundamentally useful to Christian society. The work of physicians included the theory of medicine, the practice of the medical arts, and the pious act of understanding and caring for the human body. Calling attention to the utility of a prohibited book or describing the medical profession as useful rationalized the contribution of scientific studies to Catholic society.[36]

Utility has long been considered an important discourse of the Scientific Revolution and Enlightenment, drawing inspiration in particular from the works of Francis Bacon.[37] This book takes the concept of medical utility as a central focus of historical study and argues that we must work harder to understand the many valences of this touchstone concept in its specific historical contexts. I confront this broad, flexible category of utility in each chapter to trace how medical professionals and ecclesiastical officials explained the particular roles of medicine and physicians in their society. Ecclesiastical censorship, which necessitated justifications for the continued use of prohibited medical knowledge despite Catholic bans, had the effect of amplifying and confessionalizing a discourse of medical utility.[38] As utility became central to European conversations about the value of medical and scientific knowledge, we should remain attentive to the surprising ways that the Catholic Church participated in shaping this discourse through censorship.

Finally, my research positions the history of medicine as integral to understanding the cultural forces shaping the so-called Scientific Revolution in astronomy and mechanics.[39] Medicine is a particularly well-documented realm of early modern censorship that ultimately had great influence on the more famous encounter between science and religion in the seventeenth-century debates about Copernicanism. From censorship, professionalization, and utility in the history of medicine, I will move in the epilogue to reveal how these powerful labels were leveraged in the Catholic Church's ban on Copernicanism in 1616 and in Galileo's responses. The censorship of medical books was an especially well-articulated part of a broader contemporary discourse about the social, political, and economic stakes of scientific knowledge.

DEFINING MEDICINE IN EARLY MODERN EUROPE

This book examines the censorship of medical books in Italy by the Catholic Church in the sixteenth and seventeenth centuries. But what exactly did learned medicine encompass in this period? Throughout this study, I adopt an early modern understanding of medicine and medical knowledge that includes many texts that we would be hard-pressed today to describe as pertaining to medicine. In addition to books about anatomy, surgery, therapeutics, or materia medica, in the sixteenth and seventeenth centuries, the field of medicine also encompassed astrology, botany, natural history, and chemical medicine, reflecting the broad interests and studies of early modern doctors.

The breadth of early modern medicine can best be appreciated by examining the interests and publications of some of its leading practitioners. The life and work of Girolamo Cardano—the physician, mathematician, humanist, astrologer, philosopher, and occult enthusiast whose prohibited works were among the most popular and widely requested in Italy—encapsulates much that is intriguing and inherently complicated about the field of medicine in the sixteenth century. Reflecting on his career in November 1575, Cardano reckoned that he had probably made in total about five thousand suggestions for medical treatments, solved or investigated forty thousand problems, and composed another two hundred thousand minutiae! Based on numbers alone, he counted himself worthy of the title that the Italian jurist and humanist Andrea Alciati had bestowed upon him: "The Man of Discoveries."[40] Cardano's career was unique, but his accounting reflects the sense that knowledge, including medical knowledge, was increasing in leaps and bounds and physicians were contributing to and learning from this explosion in related and unrelated fields. By one calculation, the number of medical titles circulating in print increased by a factor of more than one hundred over the course of the sixteenth century.[41]

While Cardano was uniquely productive and perhaps uniquely self-reflective, he presents a typical problem for defining the field of learned medicine in early modern Europe: How should scholars separate the medical from the nonmedical in the career and work of physicians? Nancy Siraisi has suggested that we "reconceptualize the view of Renaissance medical learning to include elements that have hitherto seemed extraneous to either the social or the scientific history of medicine."[42] I build on Siraisi's study of history and medicine and Ian Maclean's studies of medical publishing and scholarship to define the amorphous category of the early modern learned medical book from three angles: based on readership, authorship, and early modern bibliographical categorizations.[43] Defining medicine over the course of the sixteenth century is actually a shifting task because the field was rapidly, and constantly, changing throughout this period.[44] However, each of these three approaches expands and delimits the categories of the medical in ways that early modern physicians would have found familiar. Ultimately, my definition of the medical book is capacious, encompassing what physicians read and wrote, as well as the social and professional capital that libraries provided. This approach attends to readership, authorship, and contemporary classifications to ground definitions within the realities of sixteenth-century physicians and their professional world.

One of the defining characteristics of early modern learned medicine was the enthusiasm of physicians for collecting knowledge, an undertaking that often resulted in large, varied, and widely appreciated libraries.[45] Ulisse Aldrovandi, the Bolognese physician, naturalist, and botanist, had an extensive library (in addition to his museum of plant samples and other curiosities) that he collected over the course of his life and donated at his death to the city of Bologna.[46] Achilles Pirmin Gasser was a physician and astrologer who had a remarkable book collection of his own and who also worked as an agent collecting books for the Fugger family of Augsburg.[47] Physicians were not only men of letters; they were bibliophiles and accomplished collectors who understood their libraries to be essential tools for their medical practice and teaching. Leonhart Fuchs justified his decision to turn down a position offered by Albrecht Margrave of Brandenburg in 1538 in part because it would have been inconvenient to move his children and his pregnant wife and in part because it would have been "impossible" to transport his books such a great distance. It would have been equally impossible, in Fuchs's view, to leave the books behind "since I have to read medicine and give the public my services."[48]

Despite Fuchs's protestations, other physicians did choose to move with their libraries. In Italy after 1559, a physician's library often contained prohibited books that required special permissions to be transported. In 1595, when the physician and medical professor Girolamo Mercuriale moved from Bologna to Pisa to take up new positions teaching medicine at the university and working as a court physician for the Medici, he had to secure a license for his library to travel with him. Mercuriale wrote to Cardinal Giulio Santini in Rome, "Working in the profession that I do, in order to read it behooves me to have many books, and especially those I have studied."[49] Mercuriale's declaration of the necessity of his books indicated the importance, for a physician, not only of having a library but also of having one's own volumes available, perhaps to take advantage of manuscript annotations and corrections in the margins.[50] Physicians' libraries were repositories of books, notes, and notes in books, which serve now to document the intellectual work of these early modern practitioners of medicine.[51]

In addition to the medical texts in his library, Mercuriale believed that reading in classical literature was also vital for physicians. He advised medical students to read such authors as Homer, Lucretius, Virgil, Horace, Juvenal, Herodotus, and Strabo. "And do not be surprised that I propose poets and historians to you," Mercuriale admonished his students, citing Galen's precedent in turning to these unlikely sources to "shed no small authority and light on medical science."[52] Medical and nonmedical texts had much to offer aspiring physicians. Similarly, according to Mercuriale, the approach to reading medical and literary authors should be fundamentally the same: careful reading combined with excerpting passages into notebooks.[53] Cardano noted that reading history, philosophy, and Italian poetry, in addition to treatises on medical questions, ranked as things that gave him "extraordinary satisfaction" (other pleasing items of note included pens, gems, metal bowls, and rare books).[54] Physicians read broadly, both out of personal interest and as part of their professional identities.[55]

In chapter 5 I will trace readership of prohibited medical books individually and collectively by analyzing requests for licenses like that of Mercuriale. Based on the requests for reading licenses in the early seventeenth century, it becomes clear that physicians felt that their credentials qualified them to read prohibited books related directly to medicine and surgery, and also texts ranging from histories to natural histories, philosophy to philology, and banned books about astrology and iatrochemistry.

Early modern physicians' voracious personal and professional appetites for reading and collecting books and their relentless drive to read widely and to accumulate knowledge would present a huge challenge to the system of Catholic book censorship in the wake of the Reformation.

As Cardano's reflection on his life reminds us, physicians were producers of knowledge in addition to collectors of it. Widespread interest in medical humanism meant that many of the most popular texts written by physicians were editions of, or commentaries on, the works of classical authors, such as Hippocrates and Galen.[56] These texts often featured acerbic criticisms of the Arabic commentaries upon which European physicians had relied for much of the medieval period.[57] Nicolò Leoniceno's nearly eighty years of teaching in Ferrara trained generations of physicians who were concerned with carefully editing and retranslating medical texts from ancient Greek.[58] The next generation of prominent medical humanists were predominantly Protestant physicians from Northern Europe, including Leonhart Fuchs, Conrad Gessner, and Janus Cornarius, among many others. Physicians sought their editions, translations, and commentaries throughout the sixteenth century.

In addition to editions of classical texts, Fuchs and Gessner also published lavishly illustrated botanies and wrote extensively on preparing medications. Gessner dabbled pseudonymously in publishing on distillation, which was useful for manufacturing medications, and in the genre of medical secrets, for which his student Levinus Lemnius was best known.[59] Physicians also wrote illustrated anatomies like those published by Jacopo Berengario and later Andreas Vesalius, who advocated that physicians not only write anatomy books but also conduct their own dissections.[60] Physicians wrote pedagogical materials in addition to hefty volumes, including lecture notes and commentaries on pathology and therapeutics.[61] Lest we think that students only purchased the books on their syllabi, some new medical texts proved to be wildly popular. The Polish physician Joseph Struthius's *Ars sphygmica* (*The Art of the Pulse*, 1555) is said to have sold eight hundred copies in a single day.[62] The other clearly medical genre in which physicians published prolifically in both manuscript and print in the sixteenth century was the short treatise on topics such as the plague written in both Latin and vernacular languages.[63]

Sixteenth-century physicians also wrote and published many books that were not primarily medical.[64] Thomas Erastus wrote extensively on the relationship between religion and the state.[65] Hadrianus Junius worked as a physician in the Netherlands and published lexicons, an octolingual dictionary, annotations on classical works, heraldic analysis, andteven religious poetry.[66] François Rabelais was trained as a physician but is best known for his satirical *La vie de Gargantua et de Pantagruel* (*The Life of Gargantua and of Pantagruel*). Though Rabelais's prologue to this work suggested that wrapping the book in warm cloth with a poultice of dung would be more effective than the remedies of physicians, the rest of the content is certainly not in the traditional genre of therapeutics.[67] Girolamo Rossi, the physician and censor from Ravenna who features in chapters 3 and 4, was most famous for having written a history of his native city from documents he consulted as a young adult in the Vatican Library while traveling as a humanist scribe.

Nancy Siraisi has extensively explored the generic and epistemological connections between historical and medical writing, highlighting the deep connections across genres of books in this period.[68] In addition to history there are a number of other mixed-genre books written by physicians in the sixteenth century. Girolamo Fracastoro's work on syphilis, for example, is famously written in the form of an epic poem.[69] Girolamo Cardano's *De vita propria liber* (*Book of My Life*) is at once biographical, medical, and bibliographical, interspersed with accounts of the historical and political events of his own life.[70] Cardano is also the supreme example for exploring the overlap between medical and astrological writing, as Nancy Siraisi and Anthony Grafton have shown, though he was far from the only physician publishing in astronomy and astrology.[71] We need only to think of Nicolaus Copernicus and his disciple, Georg Joachim Rheticus, both of whom were trained as physicians and who wrote on astronomy.[72] The genre of anthropologies, written by physicians but also by natural philosophers and theologians, examined how the body and soul were connected and how the body revealed religious and moral truths.[73] In addition to the many arts of which medicine was comprised, physicians' writings were as broad and varied as their reading.

How, then, did early modern physicians classify their work and the medical discipline at large? The early modern period was characterized by a proliferation of learned information, and scholars relied on management systems to make the search, storage, and retrieval of that information possible.[74] Further, early moderns were conscientious about classifications. Let us turn now to how the medical bibliographers of the sixteenth century described the contours of this field. These very bibliographies, the lists of titles and authors, would also become targets for censorship over the course of the sixteenth century.

The overlap between early modern medicine and information management is best personified in the life and work of Conrad Gessner. Gessner was the most prolific bibliographer of the early modern period. He was also a physician with wide-ranging interests that force us to think carefully about how one might classify the part of his studies and output that was, strictly speaking, medical.[75] We might also start by considering that Galen served as Gessner's model both for understanding how to diagnose and treat disease, and for considering the role that books and authorship played in medicine. Galen wrote his treatise *De libris propriis* (*On My Own Books*) to curb the circulation of books with his name listed as the author but which he had not actually written. His autobibliography provided inspiration for Gessner's universal bibliography (the *Bibliotheca universalis*) and certainly served as a model for Girolamo Cardano's account of his own books.[76]

Gessner's *Bibliotheca universalis* was more ambitious than Galen's personal bibliographical account. Over 1,200 folio pages, it included the authors and titles of all known books in Latin, Greek, and Hebrew. As a consummate humanist and philologist, Gessner knew his Galen. His entry on Galen's books in the *Bibliotheca universalis* emphasized that the works of the great physician of Pergamon were necessary not only for medicine but for other disciplines as well, a fact that Gessner noted in the margin of his own copy of the book.[77] Gessner summarized Galen's broad definitions of medicine by remarking merely, as Galen had in the title of one of his own works, that "the best physician is also a philosopher." The universal approach to knowledge in the *Bibliotheca universalis* reflected a long tradition of physicians' broad understanding of their own discipline.

Gessner followed the publication of the *Bibliotheca universalis* with companion volumes called the *Pandectae* (1548–49), in which he sorted the original alphabetically ordered entries into subjects and schemas. The volumes, organized by subject, were to serve as tools for scholars so that they would be more familiar with what had already been written on a given subject. According to Gessner, this awareness would, in turn, combat the "silliness of useless writings in our time" and "forestall the production of further useless books."[78] However, when Gessner, reader and bibliographer of Galen and physician-scholar extraordinaire, finally sat down to define the expansive field of early modern medicine, he came up short. When the *Pandectae* was published, it contained blank folio numbers for the sections on medicine and theology which were to appear separately. In 1549, the volume on theology appeared, but the bibliography of medicine never followed. We might imagine that Gessner, who was especially aware of the essential interdisciplinary nature of medicine and the rapid changes taking place in the field, found himself too overwhelmed to definitively catalog his own discipline as he had so many others.

Nevertheless, the *Pandectae* offers a few insights on the subjects that Gessner might not have classified as medicine, since he placed certain books under other headings. Despite Mercuriale's reading recommendations, according to Gessner, grammar, literature, and poetry were their own fields, and despite the historical writings of many physicians, history, too, was its own distinct discipline. Gessner maintained a distinction between philosophy and medicine, and he divided works of astrology and astronomy both from each other and from medicine, although contemporary medical practice included substantial overlap with the astrological arts at courts and universities.[79] Gessner's *Pandectae* also created a miscellaneous field of "Different arts, mechanics, and other things useful to human life," which included mechanical arts and engineering.[80] By describing these kinds of applied knowledge as useful arts, Gessner casts light on another discourse of utility: the art, or applied nature of the work. Medicine was a distinct theoretical discipline with access to philosophical truth, and the physician's work was also a practical applied art that was a source of utilitarian knowledge.[81]

In addition to Gessner's classifications, his *Pandectae* also provides us with another source for examining early modern book categorizations. Gessner was a superlative giver of thanks, and he strategically included many dedications in his published volumes.[82] In the *Pandectae*, Gessner dedicated every section to a different printer-publisher, the enterprising men who facilitated public conversation and castigation in the republic of letters. In addition to the dedication, Gessner reproduced a recent book list for each dedicatee, providing free advertising for the many volumes each bookman was selling. The booksellers' catalogs in the *Pandectae* represent another, though much narrower, view of the production and consumption of physicians. These lists featured primarily texts by and commentaries on classical and medieval medical authorities (Hippocrates, Galen, Celsus, Avicenna, Rhazes) in addition to manuals on plague, plants, anatomy, and diet.[83] This array of subjects represents the books that medical students were likely to purchase, including printed materials that were not books, such as tables of the veins and arteries.[84] The medical titles on Johannes Frellon's list are all printed in small formats (listed as octavos and sextodecimos), and Sebastian Gryphius's catalog includes only a few listings in quarto, none in folio, and the vast majority in octavo or smaller.[85] These medical books were not collectors' editions but were aimed at a broad cross section of practicing physicians. They were textbooks and the kinds of medical books that the physician could carry in a pocket to the university or to a patient's bedside. Booksellers defined the realm of medicine based on a low risk assessment of what they thought would sell at a reasonable price to the masses of trained scholars and professionals. Their classifications never strove to document medicine's complexity with the theoretical sophistication that Gessner's *Pandectae* might have if he had ever finished it.

While Gessner never succeeded in definitively categorizing medical books, in the early 1590s two medical bibliographies appeared on the book market, drawing their information from Gessner's *Bibliotheca universalis*. Le Coq's 1590 *Bibliotheca medica* divided medical books by subject within the discipline. The book opened with a list of the 1,224 authors writing in Latin that he cited in the book; it also included authors who did not write in Latin but whose books had been translated into Latin. Next followed short biographical sketches of authors "who have illuminated the art of medicine with their writings." This generous list included editions of these authors' works and where they were printed, lavishing praise on the likes of Leonhart Fuchs, Janus Cornarius, and especially Gessner, whose work Le Coq admired greatly. Le Coq's tribute not only noted the books that Gessner wrote but also claimed that the concepts relevant to medicine (*argumenta*) came from "his admirable and incredible works."[86] This kind of praise of a Protestant author like Gessner would necessitate a careful expurgation of copies kept in Italian, Spanish, and Portuguese libraries.

Le Coq followed his bibliographical list with separate appendices of contemporary French, German, and Italian medical authors. The book then changed tack, turning to various medical subjects. The first subjects were ancient and medieval authors, each followed by a list of authors who had published editions or commentaries on their works. The appendices continued with traditional categories of medicine: surgery, anatomy, medical herbals, and pharmacopoeias. The final four appendices to the bibliography included sections on the practice of medicating, on medical consilia, and lists of authors who had written on plague and venereal disease. Several of Le Coq's lists were borrowed or compiled from lists Gessner had published during his life. The eighty pages on medical herbals and pharmacopoeias were actually an essay lifted directly from a preface Gessner had written for the 1552 edition of Hieronymus Bock's book on plants.[87]

Israel Spach's 1591 *Nomenclator scriptorum medicorum* (*Names of Medical Writers*) was published in Frankfurt a year after Le Coq's *Biblio-*

theca medica, and it took a completely different approach to classifying medicine.[88] Doing away with complete bio-bibliographies, Spach approached the problem of sorting medical books instead by grouping authors under lists of headings he considered important for medicine. Of course, many authors appeared under multiple headings. Spach's list of medical subjects represented a view of medicine that better reflected the state of the discipline by the end of the sixteenth century, whereas Le Coq's classification had more closely followed the approach of mid-sixteenth-century humanists like Gessner. Spach's classification included traditional headings such as the general "medicine" category and sections on therapeutics, anatomy, and surgery, but he also thought more specifically about the practices and tools of physicians and more broadly about the body. There are substantial sections of the bibliography about astrological and even chemical medicine.[89] Readers could find resources in Spach's volume for the study of the body by following headings on the humors, temperament, sleep, age, and dietetics (which included two separate sections on food and drink). A specific heading listed works about the physician ("Medicus") and another about signs for prognosis. The book concluded with an index of author names and an index of subjects that would direct readers, for example, to the subheading "urine" in the larger section on signs.

From Spach and Le Coq, we see that by the 1590s there were two mainstream systems for classifying medical books and defining the discipline. The first (represented by Le Coq's bibliography) focused on classical authors and their commentators, with additional sections for pharmaceutical and surgical/anatomical materials. This breakdown corresponded roughly to the interests of three groups of medical practitioners: physicians, apothecaries, and surgeons. The other classification, exemplified by Spach, defined the field of medicine as including books about the body and things that affected the body, such as chemical medicine and possibly astrology. These two distinct approaches testify to the nonsimultaneity of the spread of information and ideas in the print world. Scholarly networks facilitated the sharing of information, expertise, and books, but these two descriptions of medicine at the end of the sixteenth century also demonstrate that contrasting visions of the field of medicine existed contemporaneously at the turn of the seventeenth century. These bibliographies underscore the evolving field of medicine and the potential for scholarly resistance to change as well as excitement about innovation. At the same time that ecclesiastical authorities were intervening in physicians' reading, writing, and scholarly networks, physicians across Europe were grappling with immense changes internal to their field of study. The fixity of interpretation and information that censorship sought to impose was fundamentally at odds with a field of scholarship in a radical state of flux.[90]

The works of classical and contemporary literature that physicians both wrote and read were not, strictly speaking, medical under either Le Coq's or Spach's classification of the field. Nor were the still more difficult to classify natural histories of birds, animals, and fish, which were the lifelong projects of many physicians, including Gessner. However, both classical works and contemporary projects were fundamental to physicians' libraries and the ways in which they spent their time and engaged with the broader learned community. These aspects of social presentation and scholarly sociability were essential parts of what it meant to be a learned physician in the early modern world.[91] Not every physician was a Gessner or a Cardano in scope of thought or breadth of scholarly connections, but many learned physicians participated in communities of learning that facilitated their interaction with the world of printed books in their own libraries. Books that were essential to the social world of physicians were at some level also medical books.

Throughout this book we must bear in mind these changing genres and the broad interests of physicians. I am inclined to be as humanistic in my approach to medical learning as the physicians of this period. I have adopted a broad definition of *medical books* to include all those texts that early modern physicians considered to be relevant to their work as doctors and to the role of physicians in society. This broad definition makes space for both Le Coq's and Spach's models of medical bibliography and also gives us the opportunity to take seriously Mercuriale's prescription that doctors should read poetry, Galen's belief that doctors should be philosophers, and Fuchs's assertion that it would be impossible to do his job without his library. Medical books were books that physicians used for their work as doctors and to consolidate that professional position in society. As Janus Cornarius explained, conceding the universality of medicine's goal, "Medicine truly seeks the particular nature, the disposition from boyhood, the doctrine of language, literature, philosophy, mathematics, and all knowledge."[92] If early modern medicine was a discipline that sought to master all knowledge, Gessner's *Bibliotheca universalis* (*Universal Library*) might indeed have been the only definition of medicine that encompassed this vast realm. Yet, both Gessner's works and his ecumenical approach to knowledge were at odds with Catholic censorship. Regardless of medicine's universal goal, the reality of practicing learned medicine in the Catholic world is better represented by Francesco Redi's self-admonishment and the expurgated copy of Le Coq's bibliography. The desire for forbidden knowledge may have been timeless, but the practices, strategies, and evasions to which physicians resorted to read prohibited medical books reveal the particular challenge that the universal conception of medical knowledge presented in Counter-Reformation Italy.

OVERVIEW

My research draws on archival research conducted in libraries and archives primarily in Italy, the Vatican, and the United States. While documents in the Archive of the Congregation for the Doctrine of the Faith in Vatican City form the backbone of this project, the reception history that so interested me could not be told from only the administrative papers in the Vatican. I have followed leads from Vatican archives to scholars' private papers and books in public libraries across Italy, in towns from Lecce to Milan, with long sojourns in Rome, Venice, Padua, and Bologna. While the majority of the relevant libraries and archives are in Italy, most of the prohibited texts whose reception I am tracing were actually written and published in Northern Europe. Catholic censorship has made the libraries and archives of the Italian peninsula a particularly visible context for understanding a broader European culture of learning.

This book begins by examining the community of physicians in the sixteenth-century medical republic of letters and how this community was targeted and affected by the 1559 Pauline Index of Prohibited Books. Drawing on papal edicts, the correspondence of early modern scholars, and inquisition trial documents, chapter 1 reveals the personal networks to which Italian physicians turned to obtain editions of newly prohibited texts and maintain scholarly ties across religious divides in Europe.

Chapter 2 focuses on the period between 1596 and 1607, when the Catholic Church called on theologians and lay professionals throughout Italy to work together to develop official expurgations of useful prohibited books. Following the formation and subsequent unraveling of the local Congregation of the Index in Padua, the greatest center of medical learning in sixteenth-century Europe, we see how Padua's university professors evaded, undermined, and manipulated Rome's order that they aid in expurgating works of philosophy and medicine.

Although the lay censors at Padua subverted Catholic expurgatory efforts, a physician in Ravenna, Girolamo Rossi, diligently wrote, copied, and dispatched to Rome expurgations of over a dozen popular and useful prohibited books. Chapter 3 considers how Rossi saw his own participation in the expurgation of medical books as an opportunity to participate actively in Catholic reform as a lay professional. In addition to his ex-purgations of prohibited books, Rossi's papers testify to his acts of self-censorship in which he took up his pen to purge his own writings of ref-erences to heretics. The expurgatory moment thus not only reconfigured texts, it also changed the culture of reading and interpretation in Italy, turning every lay reader into a possible censor and repurposing the tools of humanist study to the ends of the Catholic Church.

Conflicting bureaucracies and individual interests ultimately pre-vented the production of an official index of expurgations until the Master of the Sacred Palace, Giovanni Maria Guanzelli, spearheaded the effort on his own. Chapter 4 analyzes Rossi's expurgations and those of other Italian censors alongside the expurgations that were officially adopted in Guanzelli's 1607 *Index Expurgatorius*. The content of these expurgations focused primarily on astrology, demonology, and indications of confes-sional difference, although the different expurgations also reflect the pri-orities of individual censors.

While the pope, the Master of the Sacred Palace, the Holy Office of the Inquisition, and the Congregation of the Index all worked to detect and disrupt the circulation of prohibited books, they also simultaneously issued licenses to approved readers permitting them to "keep and read" books that were otherwise banned. Chapter 5 examines nearly six thou-sand requests for reading licenses, approximately 10 percent of which were granted to physicians. Using these licenses, we follow the impact and re-ception in Italy of important books of medicine, botany, astrology, and chemistry. Examining these licenses individually and collectively reveals the personal impetuses for physicians to read prohibited books and the collective trends in subjects and particular authors of professional inter-est. These licenses show that reading prohibited books was a widespread part of Catholic professional behavior in the sixteenth and seventeenth centuries.

The process of selective censorship and licensing resulted in a vast, dispersed archive of expurgated objects that have been "corrected" with pens, knives, glue, and paper. Copies of expurgated medical books are the primary source base for chapter 6, which explores how Catholic authori-ties understood the printed book as an intellectual threat and also a physi-cal object that could be manipulated and regulated. Combining historical and bibliographical approaches, we can reconstruct the ways that readers encountered texts and negotiated the unstable relationships between read-ing, writing, and orthodoxy in the sixteenth and seventeenth centuries. The names removed from censored books in this chapter reflect a practice of expurgating authors' names from texts as a form of *damnatio memoriae* (damnation of memory) that ritually remembered the desecrated memories of Protestant physicians.

Chapter 7, the final chapter, locates expurgated texts on the bookshelves and in the library catalogs of the Vatican Library, the Biblioteca Ambrosiana in Milan, and the Biblioteca Marciana in Venice, where they found homes in the seventeenth century. By the middle of the seventeenth century, Catholic authorities widely accepted that it was both possible and useful to rely on prohibited books to further Catholic learning.

In the epilogue, I turn from medicine to follow the themes of utility and professional expertise in the Catholic Church's response to Copernicanism in 1616 and in Galileo's reply to Copernicus's censor in his *Dialogue* of 1632. The decision to expurgate Copernicus's *De revolutionibus* centered on the work's perceived utility. Galileo was acutely aware of these contemporary medical and philosophical disputes concerning expurgation, expertise, and the professional utility of knowledge, and his responses to Copernicus's censors deployed these discourses. Ultimately, this discourse of the utility of scientific knowledge emerged from fraught encounters with ecclesiastical censorship and was employed as a justification for scientific works long before the Enlightenment and far outside the Protestant context of Baconian empiricism.

This study of censors and scholars, books and libraries, and above all the contested status of medical knowledge reveals the complex interplay between intellectual control and the demand for prohibited knowledge in Counter-Reformation Italy. Within this context, the utility of knowledge became an essential feature of discussions about the new and controversial developments in scientific thought. From the illustrated herbal of Leonhart Fuchs to the reconfigured revolutions of Nicolaus Copernicus, utility became the justification for keeping prohibited books circulating in Catholic society. Knowledge in an age of censorship was a product of ongoing negotiation between ecclesiastical authorities and learned scientific practitioners. By accommodating professional needs and recognizing the value of lay expertise, the Catholic Church developed a process of intellectual control which highlighted the ambiguities, contradictions, and paradoxes of censorship in a world enthralled by the possibilities of new knowledge. The study of censorship as a learned dialogue in Counter-Reformation Italy has much to teach us about medicine, about science more broadly, and above all about the utility of knowledge in the world of early modern learning.

CHAPTER ONE

The Medical Republic of Letters and the Roman Indexes of Prohibited Books

In a letter dated October 15, 1558, Ippolito Salviani, the personal physician to Pope Paul IV, remarked to the young naturalist Ulisse Aldrovandi that Conrad Gessner's book *De piscibus* (*On Fish*) would soon be available in Italy, "even though it cannot be read without a license from the inquisitors since all of his works are reprobate."[1] Aldrovandi and Salviani had exchanged letters for several years on the topic of fish, about which the two physicians were preparing books. They sent each other information and specimens from their respective fish markets in Bologna and Rome, and they kept abreast of recent publications by other naturalists across Europe interested in aquatic species. Salviani's 1558 letter indicates that even in the months before the papacy of Paul IV issued the first Roman Index of Prohibited Books, learned physicians had begun to discuss and prepare for the banning of works written by important Protestant colleagues. As a follower of the reformer Huldrych Zwingli, Gessner believed the virtues of actions taken during his life did not influence his salvation; likewise, the excellence of his publications did not prevent his forthcoming damnation in the Pauline Index of 1559. Nevertheless, from his position of influence in Rome, the papal physician, Salviani, was certain that he would be granted a license to continue reading the Swiss Protestant's works, remarking to Aldrovandi, "I know that they will give me a license."[2]

The Pauline Index was an aggressive effort by Catholic authorities in Rome to assert papal authority over the circulation of knowledge in Europe. It prohibited the works of nearly six hundred authors, including many of the most prominent physicians of Northern Europe. Yet, the Index neither fully prevented medical scholars from reading the prohibited works of their heretical colleagues nor kept them from discussing these works, sometimes even with the prohibited authors themselves. The 1559 Pauline

Index of Prohibited Books placed explicit and punishable boundaries on intellectual relationships within the European medical community, in addition to the practical obstacles that existed due to strong personalities, distance, and war. Despite these challenges, the medical republic of letters in which Aldrovandi, Salviani, and Gessner took part would continue to facilitate the exchange of books, ideas, and letters in the face of Catholic prohibitions.

This chapter describes the learned medical community in early modern Europe and the effects of Catholic censorship on the Italian members of the medical republic of letters. The Indexes of Prohibited Books formally proscribed the work of many scholars who had trained as physicians or published on medical topics. Over the course of the sixteenth century, subsequent Indexes of Prohibited Books continued to prohibit the works of important physicians from circulating in Italy, and Italian physicians continued to protest and find creative ways to evade the prohibitions, often by way of contacts maintained through the exchange of letters. Beginning with a description of how the medical republic of letters functioned before the Pauline Index, the chapter then turns to the Index itself and examines members of the medical community who were banned and the reasons for their prohibition. While the Indexes affected the ways that Italian scholars accessed the learned community across the Alps, the medical republic of letters remained a community that provided Italian physicians with continued access to newly prohibited works. Ultimately, Catholic physicians would seek and obtain authorization to correct and then read prohibited volumes despite the Indexes, leading to systems of expurgation and licensing that would become the ongoing means by which physicians accessed prohibited medical books in Italy.

PRINT AND THE MEDICAL REPUBLIC OF LETTERS

In the sixteenth century, letters crisscrossed Europe, carried in sacks by ordinaries and servants, forwarded by colleagues and family members, and more often than not arriving at their desired destinations thanks to increasingly regularized and centralized mail systems.[3] Letters were a form of scholarly treatise in addition to being "conversations between absent friends" (as famously described by Erasmus), reuniting scholars who were personally as well as intellectually connected.

Physicians formed personal connections with colleagues from many countries during their studies in the elite medical universities of Europe located in Bologna, Padua, Paris, Montpelier, and Leiden. This *peregrina-*

tio medica was the fundamental foundation of the medical and natural historical republic of letters.[4] The Swiss physician and bibliographer Conrad Gessner, for example, spent most of his career in Zurich, but he also studied in Bourges, Paris, Lausanne, and Basel, and he traveled in France and Italy. In 1543, he spent a month as the guest of the Spanish ambassador to Venice, using the lagoon city as a jumping-off point for side trips to Bologna, Verona, and Ferrara.[5] Correspondence and travel served, in Brian Ogilvie's words, as "the warp and woof" that wove an international assortment of physicians and naturalists into an international community.[6] Letters conveyed ideas and helped create networks of ongoing intellectual exchange.[7]

Letters were an essential form of scientific as well as personal exchange at a distance. Giovanni Mainardi, a physician from Ferrara, published the first collection of printed medical letters in 1521, drawing on both the medical tradition of *consilia* and humanist models of published correspondence.[8] Mainardi's newly revived genre found an audience, and medical letters became a popular, new form in the age of print.[9] There were various motivations behind the publication of medical letters. The German physician Johann Lange's *Epistolae medicinales* (*Medical Letters,* first published in 1554) were compiled to publicize the depth of his erudition, while for other physicians, such as the Protestant Johannes Crato, whose seven-volume *Epistolarium Cratonianum* took twenty years to complete, the publication's goal was to serve as a compilation, bringing together different views about the same subject.[10] The *Epistolae medicinales diversorum authorum* (*Medical Letters of Various Authors,* 1556) included not only the Lutheran Lange and the Catholic Mainardi but also the Venetian physician Nicolò Massa, the Brescian doctor Luigi Mundella, and Giovanni Battista Theodosi, who was born in Parma and spent much of his career teaching in Bologna.[11]

Preexisting scholarly networks primed the market for these books of medical correspondence.[12] *Epistolae medicinales* were both the material of learned books and an assurance for publishers that there was an audience willing to buy the texts. Printing was an expensive and risky business and, as Giovanni Odorico Melchiori wrote to Ulisse Aldrovandi in 1554, "Rare are those who print a new work if they do not first have a pattern of success in some other thing."[13] The medical community created the market and was in turn reinforced as a community by printed books.[14]

While printed volumes of medical correspondence were primarily a project of the first half of the sixteenth century, manuscript letters continued to create community and drive interest in scholarly publications into the seventeenth century and beyond. Even when scholars did not publish letters in their lifetime, they often archived them as resources for scholarly debates, and their students proudly collected and sometimes even published them posthumously. Conrad Gessner's letters were published in print by his heirs after the famous doctor's death in 1562.[15] The botanist Luca Ghini's Neapolitan student Bartolomeo Maranta claimed that the weekly letters he had received from his teacher (supposedly more than four hundred in total) were proof that he, among the great botanist's many students, was the worthy inheritor of the rest of Ghini's unpublished papers.[16] Ulisse Aldrovandi, another student of Ghini, preserved his own correspondence in manuscript—sometimes in original, sometimes recopied by his scribes into small notebooks. Aldrovandi then curated the correspondence to give prominence to the letters he had received from "illustrious men" and had them bound together in volumes.[17]

Humanist scholars in the republic of letters used this ephemeral epistolary community to consult colleagues and other manuscripts, and to establish the meanings of difficult or missing passages in texts.[18] Physicians and natural historians similarly relied on both printed works and letters to benefit from the expertise of their colleagues. Girolamo Rossi, a future censor for the Roman Inquisition but at the time a young physician and historian from Ravenna, first introduced himself to Aldrovandi by letter in 1581. Rossi declared that despite the fact that they had never met in person, Aldrovandi's name was "so bright and celebrated that it makes you known and seen everywhere and by everyone."[19] The young physician went on to describe in great detail a fish that had been caught in Ravenna two days earlier that he had been unable to identify in the works of the French physician and ichthyologist Guillaume Rondelet. The first step for Rossi had been to compare what he found in nature to published reference books. From there he entered into the manuscript conversations of the medical republic of letters. Manuscript correspondence and print were mutually reinforcing. Scholarly books published in Latin helped physicians establish international reputations that in turn invited correspondence from specialists across the continent.

Letters were a vehicle for exchanging not only expertise but also news, books, and specimens—the empirical data of early modern scientific commerce. Correspondence connected colleagues with the material culture of medical practice. Letters traversed Europe replete with early modern attachments such as plant samples, eyewitness observations, and curiosities. In 1556, Leonhart Fuchs, the famous Lutheran botanist and medical professor at Tübingen, wrote to the Catholic Rondelet at Montpellier with a list of sixty-two plants that he was hoping the Montpellier professor would be able to send him. "You can send them to me by way of Basel or by way of Strasbourg," Fuchs suggested. "In Basel, the printer [Michael] Isingrin or [Johannes] Oporinus will forward them to me; in Strasbourg, the doctor of medicine, Sebald Havenreuter will do the same."[20]

The medical republic of letters relied on soon-to-be-prohibited printers and medical practitioners in multiple capacities. The printers like Isingrin and Oporinus who produced books also facilitated the scholarly exchanges that led to publications. So, too, less famous physicians such as the Lutheran Havenreuter assisted in the scholarly practices that led to publications without publishing themselves.[21] Author-scholars and enterprising printer-publishers worked together to reach the book-obsessed learned medical community. Through the interaction of these individuals we can better appreciate how medical learning was yoked to commercial interests and patronage networks.[22]

There was much to be gained, in terms of both specimens and social status, by participating in the medical republic of letters; however, these potential gains also entailed great social and professional risk. The international epistolary community was a venue for rabid polemic as well as fulsome praise.[23] Leonhart Fuchs burst onto the publishing scene in 1530 at the age of twenty-nine with his provocative first parry in the medical republic of letters: *Errata recentiorum medicorum* (*Errors of Recent Physicians*).[24] In 1531, Fuchs published his *Compendiaria ac succincta admodum in medendi artem [eisagoge] seu introductio*, which earned him the immediate and lifelong enmity of Janus Cornarius, the German humanist based in Basel who saw himself alone as the translator who brought Greek medicine to Germany.[25] Cornarius's initial reaction to Fuchs was due to his belief that Fuchs had plagiarized his own 1529 publication, which bore the strikingly similar title *Universae rei medicae [epigraphe] seu enumeratio compendio tractata*. Cornarius did not let this slight pass unmentioned. By 1538, Cornarius had taken to referring to Fuchs in print as *Vulpecula* ("little vixen," playing on the German translation of *Fuchs* as "fox"), and by the end of his life, three of his publications exclusively savaged Fuchs.[26] In a similarly aggressive episode, the reaction within the medical republic of letters to Giovanni Argenterio's *Varia opera de re medica* (*Various Works on Medicine*, 1550) was so hostile that his student Reiner Solenander reported that "all of France, the whole of Germany, and Italy . . . began to rise up in order to suppress [Argenterio's] labors and industry." Aggression toward his anti-Galenism eventually contributed to his decision to leave his post in Pisa and take up a position in Naples.[27]

Relationships between medical scholars in Italy were also fraught with antagonism and antipathy. Pietro Andrea Mattioli, the physician, botanist, and tireless editor of Dioscorides, was a notorious sower of discord in the medical republic of letters. Mattioli's vision of his community was Catholic, Italian, and focused on the moral failings of his detractors and competitors.[28] In his correspondence, Mattioli referred to Amatus Lusitanus, the Portuguese physician and likely crypto-Jew who was placed on the Index in 1590, almost exclusively as "Marrano," a term of abuse that could carry dangerous consequences.[29] He accused Melchior Wieland, the director of the botanical garden in Padua, of luring his colleague, the anatomist and physician Gabriele Falloppio, into sodomy.[30] Wieland responded both publicly and privately to this polemic. In a copy of Mattioli's printed letters, Wieland slipped from his normal, formal Latin marginalia into Italian to inveigh against his colleague, "You are still an asshole, Mattioli."[31] (See figure 1.1.)

Falloppio, himself the unhappy recipient of epistolary and published abuse within the medical republic of letters, bemoaned this aggressive congress among scholars. In a letter to Aldrovandi on January 13, 1561, he wrote that the field of natural history was "sown with traitors, who continuously provoke each other, the one writing against the other. Not Rondelet, not Salviani. Consider Mattioli, Amatus, and Melchior. Consider Fuchs. Consider Gessner. There is only hate where there should be

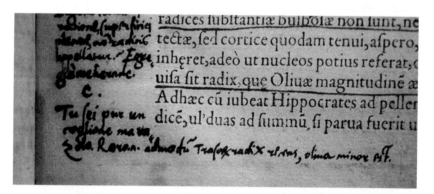

Fig. 1.1. Melchior Wieland's marginalia in a copy of Pietro Andrea Mattioli's published letters documenting the vitriol that was at least as common as praise in the medical republic of letters: "You're still an asshole, Mattioli" (*Tu sei pur un coglione mattiolo*). Pietro Andrea Mattioli, *Epistolarum medicinalium libri quinque* (Prague, 1561), 163, in Biblioteca Nazionale Marciana, Venice, 18.D.20. With permission from the Ministero per i Beni e le Attività Culturali—Biblioteca Nazionale Marciana. Reproduction prohibited.

love."[32] Indeed, Gessner's best-selling *Historiae animalium* (*Histories of Animals*), the first volume of which was published in Zurich in 1551, proudly displayed long lists of books, ancient and modern, that he had consulted to complete the work. In so doing, he acknowledged both the "recent books edited in reasonable or distinguished Latin style whose authors lived or are living in our recent memory or are recently departed from life" and also the "catalog of learned men who adorned this, our work, and the republic of letters."[33] Gessner's list of men of letters included a mix of Catholics, notably Luigi Mundella and Luca Ghini, and a number of famous Protestant authors, among them Achilles Pirmin Gasser, Guglielmo Grataroli, Sebastian Münster, and the printer Johannes Oporinus. Though he praised many colleagues liberally, this ecumenical list conspicuously left out Leonhart Fuchs, Gessner's greatest rival.

The European medical community interacted in person, in print, and in manuscript and created a venue for scholarly collaboration and debate that bridged confessional divides in Counter-Reformation Europe. Printed books brought together the expertise of many individuals and documented what Paula Findlen has described as the "shifting parameters of an emerging scholarly community."[34] Famous physicians were central to this community, but they also mobilized printers, publishers, and lesser-known colleagues in discussions and debates related to their work on health, medicine, the classical medical authors of antiquity, and the allied disciplines of natural history and botany.[35] The community they forged through conflict and collaboration, in print and manuscript, was the transnational and multiconfessional medical republic of letters.

ROME'S FIRST INDEX

The Catholic Indexes of Prohibited Books directly affected the European medical community. The first Index of Prohibited Books was published in Paris in 1544 and was followed shortly thereafter with the Indexes of Louvain, Venice, Spain, and Portugal. The Venetian Index of 1554 first banned Conrad Gessner's complete works, which was likely why Salviani and Aldrovandi suspected an imminent prohibition by Roman authorities. The papacy was a relative latecomer in 1559 when it published the Index of Paul IV, usually referred to as the Pauline Index. Even before the Pauline Index, however, papal authorities in Rome had engaged in censorship through the Roman Inquisition, which Paul III founded in 1542 and charged with controlling printed materials and the circulation of writings.[36]

Before 1559, there were efforts to compose a Roman Index of Prohibited Books based on the Parisian model. In 1547 and 1550 authorities in Rome provided regional inquisitions with lists of prohibited authors. Rome had avoided turning these early efforts into an official Index in the hope that the Catholic and Protestant Churches could be reunited at the second convocation of the Council of Trent in the early 1550s.[37] However, in September 1555, the Peace of Augsburg represented a turning point, after which the Catholic Church acknowledged its permanent separation from Lutheranism. The newly elected Paul IV Carafa turned his attention almost immediately to creating a Roman Index of Prohibited Books, which was finished in November 1557 and printed that December.[38] The printed Index survives in only one example, which suggests that it was never intended to be published widely and may even have been suppressed by Paul IV, though the documentation remains scarce.

During 1558 the commission charged with creating the list of prohibited books amended, augmented, and reprinted the Index. Finally, on December 30, 1558, the Roman Index of Prohibited Books was published and affixed to the doors of Saint Peter's Basilica and at the palace of the Roman Inquisition in Campo dei Fiori.[39] In January 1559, the Pauline Index was distributed across Italy condemning all works written, edited, printed, or commented on by heretics, works published recently and anonymously, and the vernacular Bible. It also required booksellers to present lists of their merchandise to the local inquisitor and priests to deny absolution to people who owned prohibited books. The Pauline Index was printed in five editions in Rome, and also reprinted in Coimbra, Bologna, Rimini, Novara, Naples, Venice, and possibly also Palermo.[40]

The Pauline Index divided authors into three classes. The first class included a list of authors whose written works were banned in their entirety. Authors listed in the second class had only certain of their books banned, while others were permitted. The third class consisted of anonymous books or books of uncertain authorship. The Pauline Index placed all writings by heretics in the first class, which included all Protestant authors even if the content of their works was not religious, thereby tying the authors' religious identities inextricably to all of their writings.[41] The second class of prohibitions included books that were threatening to the Catholic faith but written by authors who were not heretics. As soon as the Index was published, physicians including Aldrovandi and Salviani petitioned to keep certain prohibited medical texts, and Catholic authorities in Rome started to devise alternatives to the complete prohibitions placed on these books. In February 1559, less than two months after the Pauline Index was published, the Vatican released the *Instructio circa indicem* (*Instructions Regarding the Index*). This order attenuated the prohibition on the botanical works of Leonhart Fuchs, stating that the texts were prohibited but could be allowed if the author's name were removed.[42] From the outset, the prohibition of medical books represented an exceptional case that demanded a more nuanced approach to these texts than book burnings.

PHYSICIANS IN THE PAULINE INDEX

The Pauline Index placed prohibitions on forty-seven authors who were trained or practicing as physicians (see the appendix at the back of this book for a complete list of their names). All except four medieval authors—Arnald of Villanova, Pietro d'Abano (also known as Petrus Aponensis), Marsiglio di Padova (Marsilius of Padua), and Raimundo Sabande (Raymond of Sabunde)—were contemporary authors who lived into the sixteenth century. The reasons for their inclusion in the Index varied, but nearly all were banned for their participation in heterodox religious groups rather than the medical content of their writings. Many of these men were Europe's leading physicians and members of the medical republic of letters.

Once again, Leonhart Fuchs is a telling example. The physician and polymath Girolamo Cardano famously described his prohibited colleague Fuchs as the exemplar of the modern physician. He "commands fairly polished Latin, is skilled in Greek, writes briefly and with minimal confusion, and is clear and well organized, knows the works of Galen well, knows a great deal about simples, has written a lot, freely teaches what he knows, and is certainly a hardworking and erudite man."[43] In 1544, Cosimo de' Medici offered the Lutheran Fuchs a teaching position associated with the newly established botanical garden at the University of Pisa, which Fuchs declined, deciding to stay at the University of Tübingen, where he was professor from 1535 until his death in 1566. Although Cardano did not define Fuchs based on his reformed piety, it was clearly central to how others understood him. At his death, Fuchs's funeral oration paid nearly equal attention to the "vexations" Fuchs suffered at the hands of monks and other Catholic detractors as it did to extolling Fuchs's classical commentaries and medical writings.[44] The intellectual authority of physicians like Fuchs was established within a multiconfessional European context, but by the middle of the sixteenth century, the confessional lines across Europe were becoming more firmly established. While Fuchs's medical expertise remained irrefutable, his confessional identity became an increasingly important part of his professional identity over the course of the sixteenth century.

Fuchs's professional identity was as a physician above all else, but a number of men trained as doctors in the first half of the sixteenth century were known primarily for their theological positions, and the Pauline Index prohibited them because of their religious agendas. Michael Servetus is credited with the earliest theory of the pulmonary circulation of the blood, but he was included on the Index because of his antitrinitarian writings. Catholics and Protestants alike across Europe scorned Servetus, and, fleeing Italy, he was famously tried for heresy, convicted, and burned at the stake with his books in Calvin's Geneva.[45] Justus Velsius, like Servetus, was a practicing physician and humanist, but by 1559 he was best known as a self-proclaimed prophet and was ridiculed across Europe for his religious beliefs.[46] The Sienese preacher Bernardino Ochino studied medicine in Perugia before becoming an exiled evangelical preacher, and his colleague, the physician and humanist Francesco Stancaro, traveled with him in exile before making his own way to Poland.[47] While some physician-reformers espoused radical theologies, Wolfgang Fabricius Capito was a mainstream theologian who first studied medicine before going on to work closely with Martin Bucer, and Leo Jud began a degree in medicine before meeting Bucer and switching to study theology.[48] The intellectual trajectories of these physicians have led some historians to question whether medical training and thinking may have inclined people toward religious radicalism.[49] However, given the even larger numbers of physicians who did not become religious dissidents, the greater lesson is that we must acknowledge the close relationships among learned intellectuals in early modern Europe and take seriously the range of their interests beyond their scientific work.

In the field of botany, several prohibited authors were theologians turned medical authors, rather than physicians who became religious reformers. Otto Brunfels was an apostate Carthusian monk and reformed theologian. His name appeared in a 1550 list of the most important heretics, and he took up the study of medicine and plants only later in his life, obtaining his degree in medicine in Basel in 1532.[50] Brunfels made explicit what he saw as the connection between the religious Reformation and the upheaval in medical practice with the title of his 1536 German publication, *The Reformation of Pharmacy*. Brunfels believed botany to be a form of divine revelation, providing people with the tools to heal the body, just as Scripture provided the means to save the soul.[51] Georg Aemilius (Oemler) combined the study of plants and theology in his poetry. Though he had studied in his university days with Valerius Cordus, he was primarily occupied as a Lutheran schoolteacher and religious poet. He drew on his botanical training in some of his poetic compositions, sending poems about the plants in his own garden to Conrad Gessner.[52]

Gessner was one of the most prolific physicians listed on the Pauline Index and is emblematic of a class of physicians who were intimately connected with the leading religious reformers. Gessner was a Swiss physician, naturalist, and bibliographer; he was also a godson of Zwingli. Gessner participated in the intellectual life of the medical republic of letters and openly acknowledged the contributions of Catholic Italians in his published works.[53] Caspar Peucer, a physician, mathematician, and astronomer, also had close personal relationships with leading reformers. Peucer had lived in Phillip Melanchthon's house as a student at Wittenberg and then married his daughter in 1550.[54] Melanchthon himself had taken some medical courses during his extensive university training, and his circle contained a number of important physicians including his friends Jodocus Willich, Sebald Havenreuter, Viet Winsheim (Ortelius), and his student Bruno Seidel, all of whom were prohibited on the Index of 1559.[55]

Many of Northern Europe's great medical humanists found their works prohibited on the Pauline Index, initially including even their editions of classical authors. Janus Cornarius, Fuchs's adversary, was a licensed physician, a skilled linguist and translator, a friend of Erasmus of Rotterdam, and listed in the 1559 Index as "a sincere theologian of the Germans."[56] In 1555, the Venetian guild of bookmen petitioned to remove the prohibitions on his work, arguing that they were "all of great help to medicine and held in great esteem by good physicians."[57] Robert Constantin and Johann Günther von Andernach (Johannes Guinterius; Winther von Andernach) were also well-known and suddenly prohibited humanists who often turned their editorial and philological attention to medical works. Constantin was a student of Julius Caesar Scaliger and worked with Gessner and the prohibited humanist printer Robert Estienne. Johann Günther von Andernach published numerous editions of medical volumes but remains best known for his role mentoring two students: the famous Catholic Andreas Vesalius and the infamous heretic Michael Servetus.[58]

Several men prohibited on the 1559 Index had trained in medicine before pursuing careers in other fields, serving as yet another reminder of the humanist backbone of medical training and the flexible, varied careers of physicians. Achilles Pirmin Gasser studied at universities across Europe before taking over his father's medical practice. Gasser was banned for his Protestantism and in particular his praise of Luther and close associations with Melanchthon.[59] The most notorious prohibition to involve him was the 1616 ban on Copernicus's *On the Revolutions of the Heavenly Spheres*, for which Gasser wrote a prefatory letter to the second edition.[60] Luca Gaurico earned his medical degree in 1503 or 1504 but is best known for his rivalry with Cardano and in general for his astrological work, which landed him on the Pauline Index.[61] The itinerant physician-theologian-soldier Heinrich Cornelius Agrippa von Nettesheim's occult and theological interests led Catholic authorities to ban him as an author whose books and writings were all prohibited in their entirety.[62] Hieronymus Schurff followed a more unusual career trajectory, studying first medicine and then the law; his works were banned for his Protestantism and close relationship with Martin Luther, whom he served as a legal adviser.[63]

A few physicians who considered themselves to be pious Catholics must have been surprised to find themselves proscribed in 1559. The Catholic physician and mineralogist Georg Agricola (Bauer) was prohibited on the Pauline Index, with Roman authorities noting his praise of a Protestant patron in the dedicatory epistle of his *De re metallica* as the reason for the ban.[64] However, authorities must have realized their error; his name was "tacitly" removed from future Indexes without record of deliberations.[65] Hadrianus Junius, the Dutch physician and humanist, took an active role in amending his position on the Index. He was listed in the first class in 1559 for having dedicated books to the Protestant King Edward VI, though in fact Junius's own religious beliefs might better be described as that of a tolerant Catholic. Concerned that his appearance on the Index had tarnished his reputation, Junius wrote to the Committee of Cardinals in 1569 to defend his Catholicism and petition for removal from the Index.[66] Junius's petition was successful, and his name was removed from the Antwerp Index of 1571.[67] Pompeo della Barba was the papal physician to Pope Pius IV, and the compilers of the Index were aware of della Barba's Catholic faith and thus prohibited only his *First Two Dialogues* (listed as *De secretis naturae*) in the second class of the 1559 Index. While the majority of the physicians prohibited in the Pauline Index were banned in the first class because of their author's faith, physicians such as della Barba were listed in the second class with particular named works that were banned because of their content.

As these brief biographical sketches indicate, many of the physicians prohibited on the Pauline Index were banned for reasons that had little to do with the content of their work. These men were part of overlapping religious and intellectual communities, and many were interlocutors in the learned republic of letters. Some were central nodes in the medical republic of letters, which spanned Europe's institutions of learning, rising up and then unraveling around moments of intellectual controversy. The Pauline Index of 1559 was one such moment that elicited a flurry of activity and a change in how the members of the medical republic of letters understood their community.

REACTIONS TO THE PAULINE INDEX

Italian physicians responded immediately to the limits imposed on their multiconfessional medical republic of letters by the Pauline Index. This community was the most bookish part of late Renaissance medical practice, and physicians resented the limitations that the Index imposed on the books and libraries so necessary for their profession. Andrea Pasquale, the physician to the Duke of Tuscany, wrote of the Pauline Index in 1559, "From this is now born a huge inconvenience, that all doctors who for thirty or forty years have studied and thirsted for their books with their vigils and studies, are now deprived of them." He further complained, "If there had been physicians or philosophers present, it would not have happened like this," but because the Index was created by monks, matters of medicine were of little importance.[68] Physicians felt that their scholarly work was being undermined by ecclesiastical authorities. The medical community needed its books; it relied on relationships established through travel and letters and thrived on learned debate, new editions of texts, and large reference volumes, all published in print. In the years immediately following the Pauline prohibitions, the medical community responded by finding new ways to maintain the scholarly relationships in person and in print that had been so vital to developments in medical knowledge in the early sixteenth century.

The Pauline Index was published in the city of Bologna in late January 1559. Located across the Apennines to the north of Florence, Bologna was the second most important city in the Papal States (following Rome) and was the seat of the oldest university in Europe. Under the porticos of the bustling city, the University of Bologna was wrapping up a decade of its highest ever number of degree conferrals in arts and medicine and boasted the largest faculty and likely the highest student enrollment of any Italian university.[69] The city Senate, responsible for the oversight of the university, reported to its ambassador in Rome on January 25, 1559, a few days after the publication of the Index, to communicate that the Index had "generated much displeasure." Not only were professors distraught about the possible evisceration of their libraries, but they also feared that their student population would abandon their studies in Bologna and turn instead to a university that was more lenient about how the rules of the Counter-Reformation applied to those of "different groups." Further, the Senate wrote that it was "certain that the students will not stay here, and already some have left for this very reason."[70] The Senate was concerned for the sake of the city's economy, which relied on the many foreigners, both Italian students from other cities and students from Northern Europe, who had long produced substantial revenue for the town.

The implication of the Senate's letter was that students would leave Bologna and decide instead to study at Padua, the university of the Venetian state which was rumored to be something of a haven for non-Catholic foreigners in Italy.[71] The universities at Padua and Bologna were in constant competition for students and faculty, and the new Index of Prohibited Books threatened Bolognese intellectual supremacy. The Senate suggested that the professors be allowed to read books "from the humanities and all the other sciences that do not speak in any way about religion or faith" and urged the ambassador, Giovanni Aldrovandi, to hurry to procure an edict to this effect since some professors had already run to the vice legate, the bishop, and the inquisitor about this same issue.[72]

Rather than presenting their books to be burned as the Index required, professors in Bologna instead presented lists of books for which they hoped to receive exemptions from the prohibitions. Throughout February and early March, the leaders of the city and university of Bologna repeatedly voiced the concerns of the learned men of the city to Ambassador Giovanni Aldrovandi in Rome.[73] Professors of medicine, law, and the arts proposed long lists of "most useful and necessary authors" that did not deal with matters of religion and requested that the volumes be excused from the far-reaching Pauline prohibitions (see figure 1.2).[74]

Aldrovandi, a prudent ambassador, confessed wistfully in a letter dated February 4, 1559, "To be completely honest in my description of their lists [of books] that do not cause harm, it is that wanting too much, they will not achieve anything."[75] As predicted by the ambassador, the Master of the Sacred Palace confirmed that it would not be possible, based on the list provided, to obtain a general license for the whole university. Instead, doctors would have to apply individually for licenses to read the books that were "most necessary for their profession."[76]

With the advent of the Pauline Index, university professors joined a dissenting chorus of professionals whose livelihoods were at risk due to

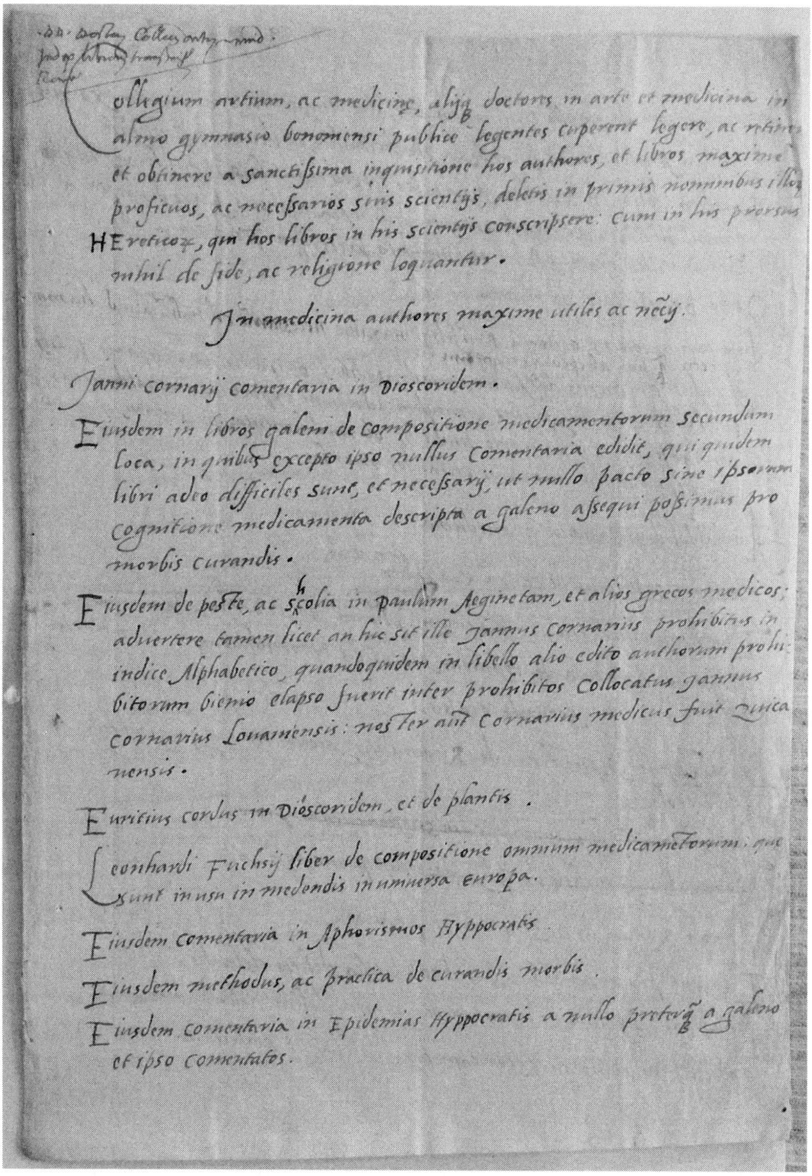

ollegium artium, ac medicine, aliiq doctores in arte et medicina in
almo gymnasio bononiensi publice legentes cuperent legere, ac retinere
et obtinere a sanctissima inquisitione hos authores, et libros maxime
proficuos, ac necessarios suis scientiis, deletis in primis nominibus illo[rum]
HEreticos, qui hos libros in his scientiis conscripsere. Cum in his praesertim
nihil de fide, ac religione loquantur.

In medicina authores maxime utiles ac neces.

Jani cornarii comentaria in Dioscoridem.

Eiusdem in libros galeni de compositione medicamentorum secundum
loca, in quibus excepto ipso nullus Comentaria edidit, qui quidem
libri adeo difficiles sunt, et necessarii ut nullo pacto sine ipso[rum]
cognitione medicamenta descripta a galeno assequi possimus pro
morbis curandis.

Eiusdem de peste, ac scholia in paulum Aeginetam, et alios grecos medicos;
advertere tamen licet an hic sit ille Jannus Cornarius prohibitus in
indice Alphabetico, quandoquidem in libello alio edito authorum prohibitorum biennio elapso fuerit inter prohibitos Collocatus Jannus
Cornarius Louaniensis: noster aut Cornarius medicus fuit Zuica-
mensis.

Euritius cordus in Dioscoridem, et de plantis.

Leonhardi Fuchsii liber de compositione omnium medicametorum que
sunt in usu in medendis in universa europa.

Eiusdem comentaria in Aphorismos Hyppocratis.

Eiusdem methodus, ac practica de curandis morbis.

Eiusdem comentaria in Epidemias Hyppocratis a nullo preterq[ue] a galeno
et ipso comentatos.

Fig. 1.2. Excerpt from a list of the "most useful and necessary authors in medicine" whose works had been included in the Pauline Index of Prohibited Books of 1559, as presented by the professors of arts and medicine at the University of Bologna in the hope of seeking an exemption from the papal order to burn those texts. Works by Janus Cornarius and Leonhart Fuchs topped the list. Archivio di Stato di Bologna, Studio, busta 353a, f. 1r. With permission from the Archivio di Stato di Bologna and the Italian Ministry of Cultural Heritage and Tourism.

ecclesiastical censorship. In early 1559, the guild of bookmen in Venice banded together to pressure the papacy to moderate the Pauline Index and insisted that its members toe the party line. The Venetian government supported the bookmen in 1559, declaring that if the Inquisition wanted to burn books, it would first need to buy them.[77] In Florence, Duke Cosimo I de' Medici reacted similarly, suggesting that the burning of books be a "fire for show" so as not to threaten the "poor booksellers."[78] In Rome, Cardinal Ghislieri was unsympathetic to these state efforts to protect booksellers. He protested that in times of plague, people willingly burned the goods in infected houses to protect the city despite financial losses.[79] Ecclesiastical language comparing heresy and plague highlighted the risk that the Church contended forbidden books could pose for spiritual health, but for printers, publishers, booksellers, and scholars, printed books offered intellectual and fiscal opportunities that outweighed the risk.

The scholarly books that physicians read and wrote were particularly expensive to produce. These reference books regularly bankrupted printers and authors alike since they required extra fonts and greater labor costs to employ correctors, indexers, and even authors.[80] Printers and publishers actively sought out internationally recognized medical authorities and commissioned works from them. For example, Conrad Gessner was paid fifteen florins to write a preface to Galen's works.[81] The expertise of physicians drove an economy that printed and sold books within Latin-speaking Europe.

The personal libraries of reference books that professors assembled to help them undertake their professional work represented massive investments of capital. Large folio reference books were essential tools for Renaissance scholars, especially those practicing law and medicine.[82] Legal compendia allowed jurists to compare current cases to historical precedent. Medical books published in folio included beautiful, deluxe editions of works by classical authors with translations and commentaries by the great humanist physicians of the early sixteenth century (including Fuchs, Gessner, and Cornarius). The highly illustrated botany, anatomy, and natural history books that were so popular in the mid-sixteenth century were both status symbols and essential points of reference for physicians who increasingly considered themselves to be the experts in the medical fields of surgery and pharmacology. These texts were also exorbitantly expensive. Gessner's first volume of the *Historiae animalium* with its eighty-two figures cost a hefty four florins for a colored copy, although it could be bought for two florins if uncolored.[83] The economic implications of censorship motivated the professionals involved in the book trade to protest the restrictions of ecclesiastical censorship. Increasingly these professionals included professors and physicians who were involved in both the production and consumption of printed books.

The Bolognese physician Ulisse Aldrovandi followed his correspondent Salviani's example by applying for licenses to keep and read the prohibited books in his library. Based on a 1558 list of his books, Aldrovandi already owned works written by authors such as Gessner who were about to be banned on the Pauline Index.[84] In January or February of 1559 Aldrovandi applied, with unknown result, for a license from his Jesuit confessor Francesco Palmio. The request was forwarded to the General of the Order, Jaime Laynez, and listed as a warrantor Aldrovandi's brother Teseo, who was in Rome as the procurator of his order, the Canons Regular of San Salvatore.[85] We know for certain that Aldrovandi was granted one reading license in 1566 and another in 1595–96, and that on August 4, 1603, the inquisitor of Bologna wrote a special license into Aldrovandi's copy of Zwinger's *Theatrum vitae humanae* granting him permission to read it at his Villa Saint Antonio.[86] Physicians took pains to protect their libraries in the wake of Catholic prohibitions.

In 1565, Cardinal Gabriele Paleotti of Bologna famously remarked that in his city, "the Index of Prohibited Books has done little." In the case of Aldrovandi, archival evidence suggests otherwise.[87] In a letter dated April 1, 1559, approximately two months after the Index was published in Bologna, Aldrovandi's friend Alfonso Cattanio, professor of medicine and natural philosophy at the University of Ferrara, consoled Aldrovandi on the recent incineration of part of his library. Cattanio expressed his "great sadness" upon hearing of "the burning of your books." He continued, "I would offer you some comfort if I did not know you to rise again in the things of this world, as the palm does even when it is oppressed, after you heard of the loss of many of your labors, studies, and vigils, and through this the damage of the many, very rare works that you sought with so much yearning in order to furnish your study."[88]

While the letter does not explicitly state that these book burnings were a direct result of the Pauline Index, it seems likely that Aldrovandi's past encounters with the Roman Inquisition and new status as professor in the university could have forced him to comply with rules that other scholars largely ignored.[89] It also seems likely that not all of Aldrovandi's books written by heretical authors and banned on the Pauline Index were actually burned. The last page of his expurgated copy of Gessner's *On Rare and Wonderful Herbs* is signed in Aldrovandi's own chicken-scratch hand "Totum perlegi" ("I read it all") and dated July 23, 1556.[90] It seems likely that most of Aldrovandi's medical books were spared but that he was forced to destroy religious works written by authors on the Index. While Paleotti was right that the Index of Prohibited Books had not accomplished its full goal of ridding Bologna of prohibited works, it brought certain individuals, and especially their libraries, under close scrutiny.

The implementation of the Pauline Index also disrupted book distribution networks in Italy. In the 1550s Aldrovandi exchanged several letters with Baccio Puccini, a physician in Pistoia, about procuring books, including texts that were prohibited on the Pauline Index. Puccini reported to Aldrovandi on March 6, 1559, that the books he had asked him to buy were bound "in a way such that they will not be ruined" and were being sent to Bologna.[91] By mid-April, Aldrovandi still had not received the books and wrote to Puccini to inquire as to their whereabouts. Puccini responded on April 25, 1559, that he had discovered why they had been waylaid: "It is in part because they were missing certain pages and in part because they couldn't be had because of the indisposition of the age. And then it happened that because of the press they needed to be sent to the inquisitor along with many of your books." Puccini's description suggests that the books, written or edited by prohibited authors, may have shipped without their title pages. Decrees from Congregation of the Index required local customs agents to be alert for this trick and not allow books to enter Italian cities with missing or fake pages that might disguise the fact that the printer, editor, or author was prohibited. Puccini assured Aldrovandi that he would send for the missing pages from Venice as soon as possible. As for the books in the hands of the inquisitor, since those editions were "translated or printed by those [heretical] men," he had heard from the duke that the books were not to be returned unless readers wrote to Rome. However, Puccini continued, he had just heard this morning that they would be able to keep the books, but he was uncertain when they would be sent back to him.[92] We might imagine that the missing pages to which Puccini alluded were title pages that were shipped separately from the body of the text to disguise the fact that the books were prohibited when they arrived at customs. Aldrovandi and his agents were taking new precautions and facing new uncertainties during the tumultuous spring of 1559.

As previously established methods for obtaining prohibited books became less reliable, readers exploited personal relationships for obtaining the range of reading materials to which they were accustomed. Readers like Aldrovandi used personal connections in the medical republic of letters to procure the prohibited texts they wanted to read. Giovanni Mario

Guidoli, one of Aldrovandi's students, wrote to Aldrovandi in 1560 informing him that his brother was friends with Conrad Gessner and had visited him at his house in Zurich.[93] He reported that his brother would bring a copy of Gessner's "works on Birds and Quadrupeds" to Aldrovandi. Guidoli continued, "At this time, my brother said he awaits the work on fish." In exchange for this favor, he asked Aldrovandi, on his brother's behalf, to send some medical simples with his brother on his return trip to see Gessner again in four months. He added that as part of this exchange, Aldrovandi was to write to Gessner to inform him that Guidoli was Italian and his student.[94] Books, plants, personal connections, and letters of introduction were the currency of the medical republic of letters, which the Guidoli brothers, Aldrovandi, and Gessner negotiated despite the new bans imposed by ecclesiastical censorship.

Though Aldrovandi and his correspondents safely navigated within and around censorship laws, importing prohibited books could be a dangerous business. Girolamo Donzellini was a Brescian physician and Protestant who spent most of his life in Venice.[95] He exchanged letters with Theodor Zwinger, Joachim Camerarius the Elder, and Pietro Perna, the Italian Protestant printer in Basel who not only published most of Paracelsus's works but also appears to have supplied Donzellini with prohibited books.[96] Donzellini was tried for heresy four times.[97] In a written defense during his inquisition trial in 1560, Donzellini reflected on how his medical career had prepared him to read prohibited books with a critical eye:

> Just as in philosophy and medicine I have read Averroes and Avicenna, which are full of errors and that derive from the shared religion which these authors read; so too I was drawn to read [evil] books by the fury of our corrupted age. And just as I have read Averroes through the lens of Saint Thomas [Aquinas] and Avicenna through the lens of Galen, so too I read these [prohibited] books seeing them always through the holy and orthodox doctrine of the Catholic Church.[98]

Donzellini posited that as a physician he was uniquely trained to separate the mistaken contexts of works from their useful content.

During Donzellini's third trial in 1574, the inquisitor of Verona reported to the Venetian Inquisition that upon their orders he had gone to Donzellini's study and had found some medical books. He listed works by Joachim Camerarius, books of medical syllogisms, and "three volumes by Conrad Gessner on the history of animals with the name of the author covered with paper."[99] The inquisitor's comment about the physical state of the book reveals that Gessner's volumes were expurgated according to the rules of the Index to remove the author's name.[100] The inquisitor additionally noted that there were medical works translated by authors in the first class; we might assume that these were editions of classical Greek authors translated by a talented Protestant humanist like Fuchs or Cornarius. The inquisitor concluded his report, "I did not find any book dealing with religion by a damned author." In fact, the inquisitor had noted a book by Christoph Corner, a Lutheran philosopher and theologian who had, a year earlier, become a professor of theology at Frankfurt. His printed works included Latin commentaries on Aristotle and more dangerous commentaries on the Psalms of David and the theology of the evangels.[101] Given the inquisitor's indication that the books were not religious, this must have been Corner's Aristotelian text instead. Though he reported finding a half sack of unnamed writings intended for the market, all of the books that the inquisitor examined (save that of Corner) dealt with medicine. Failure to turn up prohibited religious books did not quell the inquisitor's suspicion of Donzellini. He concluded his letter by suggesting that Donzellini's wife had been warned before his arrival and indeed even before Donzellini's incarceration and that she had been "awaiting his visit."[102]

It seems that Donzellini and his family were indeed well prepared for the inquisitor's search. In questioning on December 4, Donzellini testified that he had been granted a license to read medical books by a previous inquisitor of Verona, Fra Angelo Quogadro, and that the license was "written down and attached to those same books."[103] No wonder, then, that the inquisitor of Verona was less concerned about the works of Gessner in Donzellini's house. Donzellini's possession of a license to read these medical books suggests that local religious officials condoned some of Donzellini's reading of prohibited books. These medical books were related to Donzellini's well-respected professional life. Although prohibited books were central to his inquisition trial, his possession of medical books authored by Protestants was of less importance than other religiously based accusations such as Donzellini's close ties to Protestants and his involvement in the flight of two nuns from their convents.

The Venetian Inquisition recognized Donzellini's exceptional professional expertise as a physician when, in 1576, he was released from prison to work in the city combating the recent outbreak of plague. However, after his release, Donzellini continued to import and distribute prohibited books until 1587, when he was tried again, convicted as a relapsed heretic, and drowned by the Inquisition in the Venetian lagoon.[104] The same fate awaited Pietro Longo, who shuttled books and manuscripts between

Frankfurt, Basel, and Venice, including a copy of Gessner's banned *Biblio-theca universalis* destined for the collection of the famous Catholic phy-sician Girolamo Mercuriale.[105] Ludwig Iselin, a professor of law in Basel, wrote of Longo's death that he had been "thrown into prison at the com-mand of the Venetian magistracy and drowned at night twenty days ago for the same reason that the physician Girolamo Donzellini was executed last year."[106] The executions of Longo and Donzellini for their book smug-gling were severe and unusual outcomes for inquisition trials dealing with prohibited books. Indeed, reading prohibited medical books was fairly common among physicians, and doing so became safer over the course of the sixteenth and seventeenth centuries as the system for obtaining read-ing licenses became increasingly standardized.

CONTINUED CONVERSATIONS AND CONTINUED PROHIBITIONS

In the decades following the Pauline Index, prohibited medical books con-tinued to be a subject of open discussion for many of Ulisse Aldrovandi's correspondents. Gian Vincenzo Pinelli, the great collector and polymath from Padua, wrote to Aldrovandi in April 1572 to inquire about his cor-rections to sections of Conrad Gessner's volumes. Pinelli had received Al-drovandi's list of notes on the books about quadrupeds and birds, but the notes about the volume on fish were missing and it seemed to Pinelli "a difficult thing that in the said volume of similar material he [Gessner] might have been more moderate." Pinelli went on to mention a book by an English herbalist (perhaps William Turner, himself a close friend of Gessner) which included many new plants not yet depicted by others but complained that it was not "a book done by the hand of a master."[107] In June 1567, Alfonso Cattanio rhapsodized about reading Julius Caesar Sca-liger: "This year I am almost fantasizing about reading Theophrastus's *De causis plantarum*. Having seen Scaliger on him, I want to know what else he has done."[108] Cattanio seems to have read Scaliger's commentary on Theophrastus, which was published posthumously in Geneva in 1566 and became a subject of ecclesiastical attention between 1567 and 1573.[109] It was finally prohibited in Rome in the Sistine Index of 1590. Cattanio was regularly more explicit than most of Aldrovandi's correspondents about his interactions with prohibited books. He wrote to the Bolognese natural-ist again on July 10, 1567, to remind Aldrovandi that "I would still like to learn if one can have the things [works] of Fuchs, that is, all that he wrote himself or commented on Galen, and how much it would cost because I was asked by a student friend of mine."[110] Prohibited medical books were on the minds and in the letters of Italian physicians at the same time that they were being removed from bookshelves and consigned to inquisitors across Italy.

Aldrovandi and his colleagues not only continued to talk about prohibited physicians after 1559, they also continued to talk with them in exchanges of letters that regularly crossed the Alps. While Aldrovandi excised Gessner's name from the volumes in his library, he also maintained lists of plants that Gessner requested from him.[111] He similarly carried on a correspondence with the physician Joachim Camerarius the younger, the son of the great prohibited humanist by the same name, who studied medicine first in Padua and then in Bologna in the early 1560s. Aldrovandi also exchanged letters with Jakob Zwinger, the son of another famous humanist Theodor Zwinger, both of whom had studied medicine in Padua.[112] The Calvinist Theodor and then Jakob edited the *Theatrum vitae humanae* (first published in 1565), which was banned in the Antwerp Index of 1571 and again the Clementine Roman Index of 1596. Medical epistolary networks continued to link European scholars even as inquisitors seized medical books from libraries and customhouses and burned them in city squares. Physicians' personal connections existed above and beyond the form of the book.

Aldrovandi was hardly the only Italian physician to correspond with colleagues across the Alps. Mattioli and Gessner incisively disputed the accuracy of Mattioli's image of a flowering plant with medicinal properties called *aconitum primum* between 1555 and 1565 in private and public forms. The Swiss physician wrote to Crato von Krafftheim in 1562 to threaten that if Mattioli could not send him a specimen of the plant, he would "refrain completely from mentioning Mattioli's name, or delete it from the places where I have previously named him."[113] The language of censorship had begun to creep into language between colleagues in the medical republic of letters.

In the chapter of his autobiography titled "Testimony of Illustrious Men Concerning Me," Girolamo Cardano named a long list of authors, many of whom were heretical and prohibited, who had cited him in their books, including Conrad Gessner, Leonhart Fuchs, Philip Melanchthon, and Caspar Peucer.[114] Despite his prior praise of Fuchs's skills as a linguist, the Tübingen professor's criticisms of Cardano's grammar had soured the Italian's opinion of him. Cardano explained away the criticisms of Julius Caesar Scaliger and the banned astrologer Luca Gaurico on the basis that their attacks were solely "for the sake of making a reputation for themselves."[115] Cardano reserved special praise for Guglielmo Grataroli, a physician from Bergamo, who had warned Cardano against "lodging in a hostelry infested with the plague" on his visit to Basel and thus saved his life.[116] While Cardano's trip took place in 1552, several years before the Pauline Index, Grataroli had already been convicted of heresy and fled to Strasbourg and then Basel—he was certainly persona non grata in Catholic Italy, and his works were banned on the Roman Index in 1590.[117] The surviving evidence indicates that although the Pauline Index altered reading practices on the Italian peninsula, it did not entirely sever the ties between medical colleagues of different faiths.

Although Girolamo Cardano was Catholic, he too stood trial with the Inquisition in Bologna in 1570, and his works were constantly under suspicion and finally prohibited by edict in 1574.[118] Even before Cardano's works were banned, the controversial astrological content was obvious to readers. As early as July 8, 1560, Fabio De Amicis, physician and friend to the future Pope Paul V, requested and received a license to read Cardano's not-yet-banned works.[119] As one of the few medical authors prohibited primarily for the content of his works, rather than the confessional identity of the author, Cardano's case and its legacy loomed large among Italian intellectuals.[120]

One of the most famous transalpine relationships among learned physicians in the sixteenth century was between Girolamo Mercuriale, the famous professor of medicine at Padua and then at Bologna, and Theodor Zwinger. Between 1573 and 1588 the two physicians exchanged more than ninety letters. Mercuriale used this relationship to promote his work in Northern Europe, to keep abreast of publications in Basel, and as a conduit for his letters to other Northern European scholars such as Thomas Erastus, another "dear friend" of Mercuriale whose works were on the Index.[121] Erastus—a Swiss theologian, follower of Zwingli, and physician—had been prohibited in the Antwerp Index of 1569 on account of his theological writings, though he was not prohibited on the Roman Index until 1590.[122] Mercuriale successfully maintained these relationships by studiously avoiding discussions of religious topics.[123] Although Donzellini's case serves as a reminder that overly close relationships with the Protestant world could be fatal, the vast majority of cases were similar to those of Aldrovandi and Mercuriale: carefully orchestrated contact and continued scholarly engagement.

Through illicit reading and carefully worded letters, the scholarly community of the medical republic of letters persisted despite the disruption and risks caused by the Pauline Index of 1559 and the lasting politi-

cal and cultural fissures of the Reformation. It is tempting to attribute
the medical community's openness to people of diverse faiths and their
ideas as a universal embrace of all possible knowledge. In fact, it is probably more accurate to consider the medical republic of letters as engaged
in self-policing and resistant to outside control from nonspecialists. The
most obvious example from this era is that of Theophrastus von Hohen-
heim, better known as Paracelsus. Paracelsus was famously belligerent
(toward nearly everyone but toward ecclesiastics in particular), and his scientific projects were intimately connected with his theological agenda.[124]
The antiparacelsian backlash in the medical republic of letters followed
the posthumous publication of his works by the Perna press. This flurry
of publications about Paracelsus also brought his works to the attention of
the Congregations of the Inquisition and Index in 1574, more than thirty
years after his death.[125] His works were first officially prohibited in the
1580 Index of Parma and he was finally prohibited in the first class in the
Roman Index of 1596.[126] Paracelsus's late appearance on the Index indicates that the heterodoxy of an author's beliefs or views was not the only
factor in ecclesiastical prohibition.[127]

Members of the medical republic of letters not only threatened each
other with erasure from their books, as Gessner did in his polemic with
Mattioli, but they also appropriated the threats of ecclesiastical censorship as a means of retribution. Mattioli counseled Gabriele Falloppio in
1558 that books attacking him should never even have been published and
that further they should be taken from libraries and burned. "I know that
you are a man of such great authority," wrote Mattioli, "that you could
have prohibited such cowardice as an infamous libel and against every
law, decency, and civil order, and prohibited it not only from being published, but even have it removed from libraries and burned as a spiteful
thing full of every treachery when you could not prevent it from coming
to light."[128] While the boundaries set by ecclesiastical censorship were
shunned by the multiconfessional, European medical community, the
methods of censorship—prepublication prevention, postpublication erasure, and even book burnings—were part of the active discourse of scholars who also sought to control knowledge and their own reputations in the
republic of letters.

In the coming years, further Catholic prohibitions of physicians appeared on the Tridentine Index of 1564, on additional lists of banned authors issued by the Master of the Sacred Palace Paolo Costabile in 1574
and 1576, and then on the Roman Indexes of 1590, 1593, and 1596. The local Index of Parma of 1580 took a particularly tough stance on a number

of popular physicians, banning the works of Johann Lange and Johannes Jacob Wecker and the *Centuriae* of Antoine Mizauld.[129] Though Mizauld and Wecker were not officially prohibited in the sixteenth-century Roman Indexes that followed, Lange was listed in the first class in the Roman Index of 1590, while in 1596 Levinus Lemnius, the Dutch physician and student of both Gessner and Vesalius, found his *Occulta naturae miracula* (*Secret Miracles of Nature*) banned pending expurgation on the Roman Clementine Index of 1596.[130] Beginning with the Tridentine Index of 1564, the future Indexes of Prohibited Books allowed for books to be expurgated or corrected to remove heterodox content. Prohibitions of works by physicians continued well into the seventeenth century, but alongside these prohibitions systems of expurgation and licensing were established to "reform" prohibited texts and make them available to worthy readers.

CONCLUSION

The medical community in sixteenth-century Europe was bound together by networks of correspondence and print that crossed political and religious divides. Across Europe, Catholic communities were issuing Indexes of Prohibited Books to control reading in their jurisdictions. The papacy published its Pauline Index of Prohibited Books in 1559, which banned forty-seven physicians and opened conversations about what it meant to be a pious consumer of medical texts in Counter-Reformation Italy. The Pauline Index had profound but not insurmountable effects on the scholarly community in Italy. Physicians continued to correspond with their colleagues and students across the Alps, though there were dangers involved in keeping prohibited books, as the cases of Ulisse Aldrovandi's library and Girolamo Donzellini's trials demonstrate. One of the most interesting effects of the Pauline prohibitions was the immediate move to limit the effect of the prohibitions by permitting certain readers to be licensed and by selectively expurgating texts rather than burning them.

Successive Roman Indexes of Prohibited Books would ban still more authors, although there was a simultaneous effort under way to revise texts and make them available to Catholic readers. Following the prohibitions on the works of Girolamo Cardano between 1572 and 1574, the physicians in the town of Asti had become frustrated by the limits placed on their professional reading. They approached their local inquisitor, Girolamo Caratto, who in turn wrote to Cardinal Scipione Rebiba, the dean and vice-prefect of the Holy Office, in Rome. "The doctors," wrote Caratto, "are protesting that they are confused about how to medicate without

Fuchs's *De medendis morbis* and *Paradoxa medicinalia.*"[131] Caratto suggested a timely solution to the cardinal: "If it appears expedient to you, Sir, that I see, correct, and then consign them [to the physicians], I will do it. When I can't, they must be patient."[132]

The tides had been turning in the months and years following the initial Pauline Index of 1559. Despite increasing prohibitions on medical books, there was a growing sense among scholars and clerics that selectively expurgating texts was a solution that could make books available to the professionals who needed them. Medical books, after all, were not works of theology or religion, and many of them were written by the most well-respected physicians in Europe. The inquisitor Caratto's offer of expurgation reflected a broader impulse to take on the onerous work of transforming useful but heretical books into orthodox Catholic objects, a process that, as Caratto foresaw, would be difficult for Church officials to do on their own and for which there would ultimately be little patience from any of the parties involved. It is to this exhausting and consuming process of expurgating, or "correcting," medical books that we now turn in the coming chapters.

CHAPTER TWO

Locating Expertise, Soliciting Expurgations

In a letter dated January 16, 1597, the inquisitor of Vicenza, Girolamo Giovannini, wrote to Cardinal Agostino Valier, a member of the Congregation of the Index of Prohibited Books, to discuss the continued alliance between physicians and printers in the city of Vicenza. "These physicians," wrote Giovannini, "are soliciting the printers of this city. They want to print certain books of their medical profession, among them the works of Leonhart Fuchs, and they have spoken to me of it again."[1] At the end of the sixteenth century, almost forty years after the publication of the first papal Index of Prohibited Books, the Congregation of the Index had just rolled out Pope Clement VIII's Index of 1596. The debates about this Index had been fierce, and following its adoption the regulations surrounding banned medical authors were still in doubt. Giovannini wanted to clarify how to proceed. He understood that some titles listed on the Index of Prohibited Books could not circulate at all, that others needed corrections, and that some could be read if the reader possessed a license, but could books that were corrected then be reprinted and distributed? Was it possible to selectively censor highly useful, professional texts written by heretical authors and then reprint them in Catholic Italy? The Vicentine inquisitor's questions were not his alone but reflected a general sentiment among both physicians and ecclesiastics that the medical community needed access to certain prohibited works. The solution, and common ground between clerics and lay medical practitioners, was a system of selective censorship within particular texts, a process referred to at the time as expurgation.

The Counter-Reformation movement to expurgate books began with the Tridentine Index of 1564, which listed certain works as prohibited until corrected (*donec corrigantur*). However, despite the efforts of bishops such as Gabriele Paleotti in Bologna, efforts to expurgate books did not gain widespread traction in Italy until the 1580s.[2] As with the Pauline Index, Roman authorities had arrived late to centralized efforts to expurgate prohibited books. Expurgation as a method of censorship was utilized elsewhere in Catholic Europe with the first Indexes of expurgations published in Antwerp in 1571 and in Spain in 1584.[3] However, these efforts were not formalized by Roman authorities until the Congregation of the Index adopted an official policy to expurgate books in 1587.[4] The years between 1587 and 1594 were a complicated and chaotic period for the Index. The conservative agenda of Pope Sixtus V was followed by the rapid turnover of three popes between 1590 and 1591 and then by the liberal agenda of Clement VIII. This tumultuous period contributed to the Congregation of the Index's decision to delegate the work of composing official expurgations to local dioceses instead of undertaking the work exclusively in Rome.[5] Upheavals and changing papal agendas led to both instability and the inability of these congregations to reach binding legal decisions. After the promulgation of the Clementine Index in 1596, collaboration between lay and ecclesiastical authorities was made official, and the Congregation of the Index began to delegate the task of expurgating books to groups of scholars and ecclesiastics in cities across Italy.

Originally, the term *expurgate* referred to people who, through confession and penance, purged themselves of a sin. *Expurgate* was also the word of choice when discussing a textual purge, such as the removal of prohibited books from bookstores. On August 23, 1571, Alvise Valvassori, a Venetian bookseller who was brought before the Venetian Inquisition for possessing prohibited copies of Aretino's dialogues, explained that he did not know which books were in his shop because he was always moving and going to book fairs. He even claimed not to know "if I was in Venice at the time of the expurgation." When the Holy Office in Venice decided to "expurgate the shops and stands of the booksellers," Valvassori delegated the task to a young man working for him.[6] In the sixteenth century, purging followed a censure, and this was true of people, of bookstores, and of individual texts.[7]

As a form of censorship, expurgation was the process of removing parts of books that were problematic, a kind of redemptive editing as Catholic censors understood it. Unlike book burning, expurgation provided an opportunity for parts of prohibited books to be corrected and made available to readers. But who should make these difficult decisions about what could stay and what needed to be removed in order for a book to be reviewed, corrected, and then made available to physicians?[8]

This chapter focuses on the processes and people involved in expurgating medical books in the late sixteenth and early seventeenth centuries.[9] The setting is the Padua of Galileo Galilei and Cesare Cremonini, where Andreas Vesalius had written his *De humani corporis fabrica* (*On the Fabric of the Human Body*; 1543) and William Harvey was still merely another young, foreign medical student of Protestant persuasion crowding into the brand-new anatomical theater to hear the lectures of Girolamo Fabrizio d'Acquapendente. The Roman Congregation of the Index turned to experts in Padua to assist with the expurgation of medical texts precisely because of the university's long-standing tradition of excellence in the fields of medicine and philosophy. In doing so, the Congregation of the Index was paradoxically soliciting the services of professors whose philosophical positions bordered on heretical and who had close ties, both economic and intellectual, with Protestant Europe.

The Paduan congregation of censors drew on ecclesiastical and lay experts, reflecting the recognition by Catholic authorities that they could not manage on their own the enormous project of reforming knowledge in the mirror of faith.[10] However, the expurgation of books in Padua is not primarily a story about cooperation; it is a tale of resistance from the Paduan intellectual community, which was international, heterodox, and answered to the government of Venice rather than Rome. This case study brings to light the interpersonal and bureaucratic problems of expurgatory censorship as a negotiated program between ecclesiastical authority and lay expertise.[11] While the Catholic Church actively tried to leverage the professional expertise of lay physicians and philosophers to create expurgations, these efforts were largely unsuccessful. The independent character of university and intellectual life in Padua combined with ecclesiastics' logistical hurdles to undermine collaborative efforts and ultimately forced the Church to look elsewhere to complete the project of expurgation.

THE UTILITY OF EXPURGATION

In the eyes of early modern physicians, booksellers, and ecclesiastics alike, medicine was a subject particularly worthy of the immense effort that expurgation required. The needs of lay, professional readers reached Catholic Church authorities in Rome through direct interaction with local Catholic officials and indirect communication through written petitions for permission to read prohibited books. The letters and conversations between physicians and ecclesiastical officials added to a discourse about the utility of prohibited medical books. Both parties, ecclesiastical and lay, came to understand these books as essential to physicians practicing in Catholic society, and Catholic authorities in Rome openly acknowledged this lay input as they decided which texts were high priorities for the effort and expense of expurgation.

An anonymous list, written by a censor in Rome sometime before 1590, included a variety of "Damned authors that learned men of natural science and medicine might desire."[12] The German physician and botanist Leonhart Fuchs's name tops the list on the basis that his works "offer great truth to those who study nature," and "many learned men esteem the use of the commentary on the history of plants." Fuchs's book, the *Institutiones medicinae* (*Principles of Medicine*), which the doctors of Asti had emphasized as essential for their work, was marked by this censor as "very useful to have." The censor went on to point out that Fuchs was very learned in Latin and Greek and that his works had now surpassed those of his predecessor, the apostate friar Otto Brunfels.[13]

The utility of Brunfels's works, despite his religious beliefs, had been established several years earlier by the Bolognese bishop Gabriele Paleotti. While composing the Tridentine Index, Paleotti suggested on January 26, 1562, that the *Onomasticon* of Otto Brunfels, a lexicon of plant names and terms in Latin, Greek, and German, should not be prohibited because it was "an indespensible lexicon, for we have none but this one [to replace it]."[14] However, three decades later, the anonymous censor in Rome clarified that since there were now other comparable texts available, the utility of Brunfels to Catholic scholars lay only in those of his works that could not be replaced by the better, newer treatises by Leonhart Fuchs. Similarly, the works of the German Catholic Georg Agricola, whose books on metals had been mistakenly prohibited because of confusion about his patron and Germanic name, were hailed by the anonymous Roman censor as a treatment of a subject on which "no [author] ancient or recent has written."[15] The utility of a work increased when there were no comparable texts available, and the sense of uniqueness was a factor in determining whether a book was to be selectively expurgated or prohibited in its entirety.

The anonymous censor from circa 1590 also identified the importance and utility of the works of the physician, natural historian, and bibliographer Conrad Gessner, despite his status as a follower of the theologian Huldrych Zwingli. The censor described Gessner's catalog of plants as "in use by many different men." Gessner's *Historiae animalium* (*Histories of Animals*), which drew from the expertise of his transnational and multi-

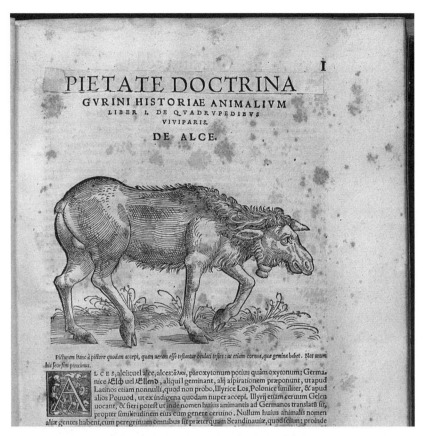

Fig. 2.1. Expurgated page from Conrad Gessner's book on animals in which Gessner's name has been censored by pasting the printed words PIETATE DOCTRINA (piety doctrine) over the author's name, a method of expurgation that preserved the useful image of the moose and perhaps elicited a laugh from future readers. Conrad Gessner, *Historiae animalium* (Zurich, 1551), a[1]r. Call number 55.8.l.1, vol. 1. Reproduced with permission from the Biblioteca Nazionale Centrale di Roma.

confessional correspondence networks, was, in the estimation of this censor, "strengthened by the authors from whom it is compiled." The censor praised the expensive woodcuts in the highly illustrated volume, asserting that "the pictures that represent animals are of great help to those who read it."[16] In figure 2.1, the owner of a copy of Gessner's *Historiae animalium* cleverly obscured Gessner's name from the top of the page by pasting over it with a printed slip of paper bearing the words *pietate doctrina* (piety doctrine) that were cut from another text. "Correcting" the text by obscuring Gessner's name prevented the reader from having to remove the whole page and destroy the image of the moose. His decision to reform the text by changing it from a Protestant author's name to a declaration of Catholic piety maintains the naturalistic image of the moose as the focal point of the page and mocks the process of expurgation by juxtaposing the words "piety doctrine" and the lumbering quadruped. The naturalistic images in Gessner were useful for readers who sought visual representations of animals and were combined with textual accounts of his conversations with correspondents about the animals and his careful reading in the textual traditions surrounding these creatures.[17]

Enthusiasm for the utility of Gessner's medical and natural historical works extended even to his especially controversial bibliography of all books, the *Biblioteca universalis* (1545). The anonymous Roman censor who produced the list of "damned authors" determined that Gessner's lists of books could be appropriate for scholars once they were purged to leave only "good authors."[18] In its unexpurgated form, Gessner's *Biblioteca universalis* was useful to the Congregation of the Index, which included the work prominently on a list of books to be purchased for the Secretary of the Congregation at the Frankfurt book fair.[19] In 1601, Cardinal Bellarmine explained that an expurgated version would be "of great future utility."[20] Expurgation could preserve the utility of books while also maintaining the general prohibitions on works by Protestant authors.

The views articulated by the anonymous censor around 1590 are paradigmatic of the views of many other Catholics who participated in the banning and subsequent correction of books. Vincenzo Bonardi was a consultor for the Congregation of the Index—a position appointed for life by the pope that placed ecclesiastics as advisers to the Congregations of the Index and Holy Office.[21] The Dominican Bonardi provided input similar to that of the anonymous censor in a treatise called *Discorso intorno all'Indice da farsi de libri proibiti* (*Discourse about the Index Regarding Prohibited Books*) in which he explained that the works of Gessner, Fuchs, and Cardano (among others) were "especially desired."[22] Members of the Congregation of the Index also acknowledged a link between what doctors wanted to read and what was useful. When Agostino Valier wrote to the inquisitor of Pisa in December 1599 to inquire about the progress of the city's censorship projects, Valier suggested that the inquisitor should first censor the medical and philosophical works listed in the Index as *donec expurgentur* and then move on to "the rest of the books of the same profession, that are desired by many and judged useful and that need censoring."[23] In the eyes of early modern ecclesiastics, in order for medical texts to be expurgated rather than banned outright, they had to be both requested by professionals and considered useful.

Finally, because medical texts were generally theologically unthreatening and therefore prohibited primarily because of their authors' religious beliefs, discussions of their utility were fairly straightforward compared with the more complicated distinctions between helpful and heretical knowledge in texts whose subject overlapped with theological concerns, such as prognostication and astrology. One anonymous adviser to the Index wrote a tract sometime after 1596 in which he attempted to resolve the complications about texts that engaged in prognostication. A book dealing with the topic of physiognomy, the author suggested, should be permitted "when it serves to judge the phlegmatic complexion, or that of the stomach or the blood." However, when physiognomy is "abused to tell the future or for palmistry, it is prohibited."[24] This fine line had a long history in the Catholic Church's relationship to the astrological arts, which distinguished between "natural" astrology (used for medical purposes, navigation, and generally understanding nature) and "judicial" astrology (which predicted the future and thus attempted to usurp God's power).[25]

Readers and ecclesiastical officials were well aware of this distinction but perpetually unclear about how to enforce it. Gaspare Mosca, the canon of the Salerno Cathedral, wrote to the Congregation of the Index between 1596 and 1597 to ask whether astrology books that dealt with predicting the future "can be permitted under the pretext that people want to make use of them for agriculture, navigation, and medicine."[26] The bishop of L'Aquila, Giuseppe Rossi, wrote to Rome almost three years later with the same query: "Among the books of astrology I do not know if anyone can make use of the ones that deal with judiciary astrology in the realm of medicine, agriculture, and navigation, since these subjects are connected and these treatises cannot be separated from the others which are in the same books about the deeds of men."[27] Expurgation was a solution to the problem of useful knowledge in prohibited books. However, it was a process that Rossi noted would require hard work and skill, and in the end, it might still prove impossible to separate problematic material from safe content.

Unlike the bishop of L'Aquila, Agostino Valier believed in the process of expurgation and the possibility of extracting from texts to create new, licit versions that were safe for Catholic readers. When Valier wrote to the inquisitor of Cremona, Alberto Chelli, in December 1599 to praise his timely response and help in creating an expurgatory Index, Valier suggested optimistically that the inquisitor could also "censor some books of

medicine or philosophy, or maybe from books of astrology you can choose
from them what can be useful for navigation, agriculture, and medicine,
trimming off all that is superfluous and pernicious and putting together
what good there is from one and from the other."[28] The prolific censor Alfonso Chacón also weighed in on the complicated nature of the mixed astrological disciplines. Chacón recognized that astrological classics were
also often essential texts in astronomy and that prohibiting astrology
wholesale would damage and discredit authors of legitimate works.[29] To
Valier and Chacón expurgation was a process through which the Church
could maintain the utility of books, and in particular of medical books,
while minimizing the risk of readers encountering heretical ideas. The
purpose of expurgation was to preserve knowledge and create books that
took advantage of the learning of the previous generations without com-
promising faith in an omnipotent Catholic God.

The project of expurgation took place in communities across the Italian peninsula. However, the Catholic Church's effort to correct Protestant medical books in the 1590s was delegated first and most importantly
to Padua, the university town of the Most Serene Venetian Republic and
home to arguably the most famous medical faculty in Europe (the University of Bologna was always in competition for this status). The Congregation of the Index in Rome outsourced the censorship of books of medicine
and natural philosophy to the ecclesiastical and lay authorities in Padua
in the final years of the sixteenth century. The outsourcing of censorship projects was a climactic moment in the story of negotiation between
church and lay authorities to produce orthodox, useful books. Church authorities relied on the energy and expertise of bishops, inquisitors, and laypeople outside Rome in order to recommend changes to the texts they had
decided should be revised.

Assigning censorship projects to lay experts in Padua was initially an
opportunity for cooperation, but it ultimately precipitated the dissolution
of the process of productive, expurgatory censorship.[30] Politically Venetian, religiously diverse, and overflowing with books, the characteristics
that made Padua a great European university city and center of medical
learning also stood in the way of local attempts at Catholic censorship.
The learned and cosmopolitan community of philosophers and physicians
was almost exclusively uninterested in catering to the intellectual ambitions of Counter-Reformation Rome, and despite occasional symbolic ges-
tures of obedience, the doctors of Padua saw to it that efforts at expurgatory censorship in their city effectively failed.

THE EXPURGATORY CONGREGATION IN PADUA

The city of Padua was at its zenith in 1596, when the Clementine Index was published with its long list of works prohibited *donec corrigantur*. Catholic, Protestant, and Jewish students from across Europe descended upon Padua to enroll in classes with the renowned faculty, especially in the fields of medicine and law.[31] Among the jurists, approximately 19 percent of the student body was German, and between 1546 and 1630 more than ten thousand German students studied at Padua.[32] The careers of German students are particularly well documented, but the cosmopolitan collection of students in Padua also hailed from England, Scotland, Hungary, and Poland—and even included French Huguenots.[33] As Thomas Coryate observed in 1608, "More students of forraine and remote nations doe live in Padua, then in any one University of Christendome. For hither come in, many from France, high Germany, the Netherlands, England, &c. who with great desire flocke together to Padua for good letters sake."[34] During the middle of the sixteenth century the architect Andrea Moroni had overseen extensive renovations at the heart of the university, the Palazzo del Bo (figure 2.2), and the celebrated anatomy theater was inaugurated by the professor of medicine Girolamo Fabrici d'Acquapendente in 1595.[35] This was the Padua of Gian Vincenzo Pinelli, who had accumulated one of the richest and most important private libraries ever seen in Italy.[36] Pinelli opened his book and manuscript collection to the likes of Ulisse Aldrovandi, Nicolas-Claude Fabri de Peiresc, and Galileo Galilei, who were eager to consult his books and benefit from his patronage. Padua in 1596 was also home to two of the next century's great natural philosophers, Galileo and Cremonini, who over the course of their careers would both run afoul of the Italian Inquisitions due to the theological implications of their philosophical and scientific theories.

In 1596, Padua was not only a thriving university city full of books and the intellectuals who read and debated them; it was also a vibrant site of Tridentine reform under the new leadership of the dedicated bishop Marco Cornaro, who in 1596 was just beginning his nearly thirty-year episcopacy.[37] The branch of the Venetian Inquisition in Padua had contributed to prosecuting several well-known intellectuals, including the mathematician and magician Francesco Barozzi and, a few years later, the philosopher Cesare Cremonini.[38] Ecclesiastical reform was also reaching the Veneto through the Society of Jesus. The professors at the University of Padua were at arms about what they perceived as illegal encroachment

Fig. 2.2. The new façade of Palazzo del Bo, the main building at the University of
Padua (*Gymnasium Patavinum*). Padua was widely recognized as one of the centers of
medical learning in Europe at the end of the sixteenth century when professors there
were called to help censor books of medicine and philosophy. "Gymnasium patavinum:
The university." From *Gymnasium patavinum Giacomo Filippo Tomasini* (1654).
Courtesy of Wellcome Library, London. http://wellcomeimages.org/works/sqzzaxx8.

on teaching by the Jesuit College that had been steadily gaining students
since its founding in 1542.[39] In 1591, led by Cremonini, Alessandro Piccolomini, and a group of naked students with guns, the professors and students alleged that the Jesuits had established an *antistudio*, or rival university, and eventually succeeded in closing the Jesuit College later that
year.[40] Still over a decade prior to the climax of the Venetian Interdict, the
uneasy tension between the ecclesiastical powers of Rome and the economic and intellectual interests of Venetian subjects were already being
played out in the university town.

It was into this climate that on March 8, 1597, the Congregation of the
Index issued a decree regarding the expurgation of medical books: "Written instructions have been sent to the Bishop of Padua to conduct the ex-

purgation of philosophical and medical books from that celebrated University [of Padua], employing the Consultors, and to make use of the services of the Inquisitor of Vicenza."[41] As the decree indicated, earlier that same day, Cardinal Agostino Valier had written from Rome to both the bishop of Padua and the inquisitor of Vicenza to inform them of their new duties. This decree came as a response to the inquisitor of Vicenza's query earlier that winter about the prohibitions on medical books and as part of a widespread effort to solicit expertise and labor in drafting official expurgations of books. The Congregation of the Index sought help from ecclesiastical and lay experts with the *honorata impresa*, the "honorable enterprise" of censoring books.[42]

For the bishop and inquisitor in Padua, the prospect of undertaking the Herculean task of expurgating the collective library of the city of Padua must have been as appealing as the proverbial task of cleaning the Augean stables. The renown of Padua ensured that scholars across Italy would take note of this theoretically cooperative endeavor to transform Protestant knowledge into books that could be safely read by Catholics. Following in the wake of the 1593 publication of Antonio Possevino's ideal Catholic library, the *Bibliotheca selecta*, the honorable enterprise of community expurgation was an experiment in renewing community-wide participation and endorsement of a censorship program that had been under way for nearly forty years.

The Congregation of the Index chose the city of Padua and its officials to censor books of medicine and philosophy because of the university's reputation for excellence in medicine. Other university cities in Italy had also been selected to censor books in other disciplines that were in great demand. The expurgation of astrology books was assigned to Venice; historical texts to Milan; books on dueling to Parma, Piacenza, and Cremona; canon law to Bologna; civil law to Perugia; and Italian literary works to Florence.[43] Rome's intent was to call on cities with specific areas of expertise to take part in the *honorata impresa*. Clerics and theologians had their own areas of expertise, but as one censor of legal books wrote, as nonexperts they would "decide on these matters like a deaf man on music or a blind man on colours."[44]

When the decree of the Congregation of the Index arrived in the hands of Girolamo Giovannini, the recently appointed inquisitor of Vicenza, he was likely unsurprised. As we saw in the opening to this chapter, he had written to the cardinal in January asking for advice about precisely this issue. We might imagine that he did not, however, expect Valier's response to come in the form of an official command. Giovannini, in coordination with the bishop of Padua and with the help of the educated men of Padua and Venice, was instructed to expurgate the "serious and useful" books of medicine that they wanted to read and send the proposed corrections to the Congregation of the Index in Rome for approval.[45]

Valier's letter granted Giovannini the authority he needed to devote himself to the censorship of the texts that his flock requested. However, the Congregation of the Index's missive ended with a of warning:

> [Your work on these expurgations] will not, however, be universally embraced, if it is not first approved by this Sacred Congregation. You can communicate about this project with Monsignor the Bishop of Padua, since we have given the responsibility to that university of making an Expurgatory Index of the books of philosophy and medicine. But neither your expurgations nor those of the University of Padua will be universally embraced if you do not first send them to Rome and they are approved by our Congregation, etc.[46]

Even as the Congregation of the Index reached out to peripheral dioceses to take on the project of censoring texts, its members reiterated that the jurisdiction granted was still subject to Rome's approval.[47] On March 8, 1597, Cardinal Valier also sent a letter to the bishop of Padua, Marco Cornaro. Unlike Giovannini, Bishop Marco Cornaro was no Dominican theologian seeking advice about how to selectively censor prohibited books—he was a member of a famous patrician family of Padua, a devoted Tridentine reformer, and a patron of art and music.[48] Cornaro was also in a difficult position as bishop because while he labored to improve parochial schools and train more attentive priests for the provinces, he also presided over a period of great religious tolerance for foreign students at the University of Padua.[49] Cornaro may have seen Valier's call to participate in the *honorata impresa* as an opportunity to further integrate his goals of reform, education, and tolerance.

Nearly three weeks later, however, Cornaro wrote to Valier to explain that although he would have liked to, he had not been able to work on his assigned censorship project. He had "encountered several impediments." Cornaro reported that the father inquisitor of Padua, Felice Pranzini, had said that he would do something about expurgating the books and would report to Rome, though Cornaro did not know exactly what steps he would take.[50] The Congregation of the Index, this time in the person of Cardinal Marcantonio Colonna, responded promptly to Cornaro in a letter dated April 10, 1597, in which he stated that "we assure ourselves that you, Sir, with your prudence and authority will be able to overcome every difficulty."[51] The Congregation's message was clear: Valier, Colonna, and the Congregation of the Index had delegated the duty of creating an expurgatory Index of medical and philosophical books to the bishop, inquisitor, and learned men of Padua, and they were expected to accomplish their task.

Marco Cornaro responded to Colonna and Valier on April 25, 1597, explaining that he had already begun his task and would do "everything that he knew, and everything that he could, so that a project so useful and important to Christianity will germinate that good aim, that you desire with great piety."[52] Cornaro's position at the head of this project was further emphasized in a letter from the Congregation of the Index to Felice Pranzini, the inquisitor of Padua, on April 26, 1597.[53] The letter began with praise for Pranzini, describing how members of the Index spoke of the zeal that Pranzini showed when he had burned a great quantity of prohibited books.[54] However, the majority of the missive bore warnings about who was in charge of the censorship project. The Congregation reminded the inquisitor Pranzini, that "the work of the Index is no less important than that of the Holy Office [of the Inquisition], from which the Index derives."[55] We see here the tensions over ecclesiastical jurisdictions not only between bishops and the Congregation of the Inquisition, but also between the Congregation of the Index and the Paduan Inquisition. The Congregation of the Index was also emphasizing the importance and urgency of making Catholic knowledge available in addition to curbing heretical behavior. The Congregation of the Index further warned that on this project it was necessary that Pranzini "work united in the service of God with Monsignor the Bishop, . . . [showing] the very same zeal and diligence that you exhibit in the work of the Holy Office. We await the expurgatory Index of medical and philosophical texts."[56] The Congregation of the Index had transferred the censorship project onto the shoulders of Cornaro and Pranzini, who together were to devote their attention to producing the materials Rome requested.

As competent as Pranzini and Cornaro must have been as theologians and ecclesiastical leaders, there is nothing in the letters to suggest that this task—correcting medical and philosophical texts—was assigned to Padua because of these two men in particular.[57] On the contrary, Valier's initial letters to Cornaro and Giovannini made explicit reference to the "much celebrated and illustrious university, which has always been filled with learned men" and the "learned men that can help you in Padua and Venice."[58] Padua was selected to participate in creating expurgations because of its men of learning, rather than its theological expertise. The learned, lay professionals who were enlisted in the expurgation of books were an eminent cross section of the University of Padua's renowned faculty.

The first evidence of official involvement of lay experts in Padua was a letter from Ercole Sassonia to the Congregation of the Index, dated July 11, 1597, which survives in an autograph copy.[59] Sassonia was the Dean of the College of Philosophers and Physicians in Padua and was widely respected as a professor and medical practitioner. He had served as the personal physician to Holy Roman Emperor Maximilian II of Austria. In his letter, the doctor reported that at the university they "had not wasted time in responding to the letters." Instead, together with Bishop Cornaro and Inquisitor Pranzini, the college had "elected twelve doctors for the correction of books, six in medicine . . . and six in philosophy, who have all, with great promptness, embraced this duty."[60]

The individuals elected to work on the censorship project were an impressive representation of Padua's robust intellectual community. The six doctors—Orazio Augenio, Ercole Sassonia, Girolamo Fabrici d'Acquapendente, Alessandro Vigonza, Annibale Bimbiolo, and Niccolò Trevisan—were a distinguished group. All of them taught medicine at the University of Padua in subjects ranging from theoretical medicine to practical medicine to anatomy. As a group, they had treated or would go on to treat such famous patients as Maximilian II, Carlo de'Medici, the Duke of Urbino, Galileo Galilei, and Paolo Sarpi.[61] Only two years earlier, Fabrici d'Acquapendente had established the first permanent anatomical theater in Padua, which had further elevated the elite reputation of medical learning in the city. The philosophers elected were no less impressive and included Francesco Piccolomini, Faustino Sommi, Girolamo Zacco, Benedetto Dottori, Schinella Conti, and Michele Brazolo.[62] The philosophers belonged to a number of academies and published philosophical and literary works. Of these twelve professors, only three had formal training in theology (one doctor and two philosophers). These "learned men" of Padua were selected to bring their expertise in fields outside of theology to bear on the task of expurgation.[63]

The reputation of the University of Padua should not be underestimated when considering Rome's decision to select Padua for this censorship initiative. Paul Grendler has suggested that between the arrival of the philosopher Pomponazzi in 1488 and Galileo's departure in 1610, the University of Padua was enjoying "the most illustrious period of any university in the Renaissance or in modern Europe."[64] It is also important to keep in mind that Padua was not particularly well known for its theology faculty. In Italy, unlike France, universities placed a greater emphasis on law and medicine than theology, the student body was slightly older (age eighteen to twenty-five), and the faculty consisted mostly of married laymen (as opposed to northern Europe, where the faculty members were mostly members of the clergy).[65] While it would be wrong to describe the University of Padua as secular, it was independent from the administration of the Catholic Church. Furthermore, although three of the twelve professors selected to participate in expurgating books of medicine and philosophy had received training in theology, none of them were members of the clergy. The Congregation of the Index turned to Padua because of the lay expertise at the university. Additionally, the outsourcing of censorship projects by Rome indicates an acknowledgment on the part of the Catholic Church that there was what Sabina Brevaglieri has described as an "institutional and cultural polycentrism" emerging in Italy.[66] Although the papacy did not recognize authorities higher than itself and the Congregation of the Index made the same claim of primacy over peripheral centers of expurgation, outsourcing censorship projects was an example of Rome's recognition of distinct and dispersed sources of expertise.

Between May 1597 and March 1598, the inquisitor Felice Pranzini submitted the first results of Padua's censorship project to the Congregation of the Index in Rome: an *Index selectus*.[67] The list was the starting point for the expurgation efforts taking place in the city. It did not yet include the actual expurgations or notes about content to be removed but instead resembled a to-do list of the books that needed to be censored. Pranzini divided the list into three categories: medical books, philosophy books, and mathematics books. The category of medical books included fifty-nine books by eighteen authors including Fuchs, Gessner, Paracelsus, Lusitanus, Grataroli, and Arnald of Villanova. Interestingly, Cardano's medical works do not appear since they were technically permitted, though twelve of his texts (including his treatises on proportion and arithmetic) appear among the fifty-seven works on the list of philosophy books. The twenty-five mathematics texts to be corrected included Georg Joachim Rheticus's *Narratio prima*, which was published both on its own and alongside Copernicus's *De revolutionibus* in 1566. The inclusion of Copernicus on this list is due to Rheticus's Lutheranism rather than heliocentrism, of which most sixteenth-century censors appeared largely unaware.[68] In the coming years, Pranzini would go on to submit actual expurgations for only a few of these texts. The project of Paduan expurgation was ambitious, and the project would ultimately fail to accomplish its lofty goals.

Pranzini's initial efficiency was remarkable, however. It seems that he had worked mostly alone compiling this list and not in collaboration with the selected team of censors. Already within a short time, Pranzini's work made clear that there was great potential for the Index to reap the benefits of having delegated the task of censorship to other centers. Perhaps it would be possible to assign censorship tasks to locations and authorities outside Rome and achieve timely results. In fact, these early accomplishments by Pranzini set a precedent of efficiency that in the future would be difficult to meet. Over the next year, questions of jurisdictional authority again resurfaced as the Paduan censorship project moved forward. On July 16, 1599, Simone Tagliavia, cardinal of Terranova and a member of the Index, wrote a letter responding to now-missing questions from Padua, in which he warned Pranzini to use every diligence with the bishop when seeking advice from the College of Doctors. He further emphasized that the primary responsibility for this project lay with the bishop and with the inquisitor and not with the college.[69]

The Congregation of the Index in Rome had been careful to emphasize from the outset that it held the ultimate authority over the expurgation of books. But who was responsible for the operation of the local, delegated censorship project? Gigliola Fragnito has described the relations of power in peripheral censorship bodies as fairly rigid. Local tribunals were always headed by the bishop, with the inquisitor (if there was one) in a subordinate capacity, supported by consultors able to contribute useful expertise.[70] Fragnito explains that there were political as well as practical motivations for this hierarchy in the period following the Council of Trent. The Congregation of the Index, comprised of powerful cardinals hoped to enhance episcopal powers, while the Roman Holy Office wanted to "monopolize the post-Tridentine project of the acculturation and moralization of the faithful, and in so doing bring the episcopate under its sway."[71] Tagliavia's letter chastising the bishop and inquisitor suggests that another risk of outsourcing censorship projects was the possible confusion about ecclesiastical and lay power dynamics and expertise in the post-Tridentine era.

Some of the difficulties Cornaro and Pranzini were encountering in Padua were also purely logistical. On March 14, 1598, Pranzini had written to Valier to apologize that he had not at first been able to devote himself fully to the work of the Index. It was because, as Pranzini explained, Bishop Cornaro had called the first meeting on a Wednesday, but "two days of the week, Monday and Wednesday, were dedicated to gathering together to attend to the work of the Inquisition, [and] on those days I could not adequately attend to anything else."[72] In addition to scheduling conflicts, the censors of Padua had a liquidity problem.[73] Cornaro and Pranzini realized that it seemed that no one owned copies of the books they were supposed to be censoring, and acquiring the texts would be costly. Cornaro addressed a letter to Tagliavia on July 30, 1599, explaining, "The theologians claim to not have the books, neither do the physicians they say; since these are books of some importance, perhaps it is necessary to send them to get [the books] outside of Italy at some expense."[74] The lack of books and the cost of procuring them were significant impediments to the censors' progress. Pranzini echoed Cornaro's concerns in a letter to Tagliavia two weeks later, observing that "there remains the need to procure the rest of the books to censor, which will cost a good sum of money; I am a poor friar, and because of my poverty I cannot buy this type of book, nor do I know where to direct myself to someone who would take on this expense."[75]

Tagliavia's response from Rome to these practical concerns was distinctly unsympathetic:

A variety of censures of medical and philosophy texts have been presented to our Congregation sent from places where there is not an abundance of learned men and where there is also a scarcity of books. Thus it appears difficult to believe that in Padua, where there are so many renowned men and so many famous bookstores, it would be necessary to spend so much and send for books in distant countries.[76]

After airing his contempt and disbelief, Tagliavia turned threatening, suggesting that he might reconsider the reading licenses that had been granted to Cornaro and Pranzini. In addition, Tagliavia continued, "There are many books of medicine and philosophy to censor that are sufficient to occupy many Consultors. Give priority to those necessary tasks or else, not seeing the desired outcome, all of your licenses will be revoked."[77] Tagliavia's comment about there being many books to censor was true on both general and practical levels. There certainly existed many books written by Protestants that were good candidates for correction. There were also many copies of these books in the libraries of Padua. It was not necessary to go abroad to find copies of prohibited medical books because the libraries of professors, individual students, student nations, and learned citizens such as Gian Vincenzo Pinelli were rich repositories of sixteenth-century printed books in this center of medical learning. Tagliavia's response is also revealing because, in a state of frustration with the ineffectiveness of peripheral expurgation projects, he seems to have abandoned the ini-

tial impetus behind correcting these medical texts. Tagliavia was no longer concerned with making specific, important, useful texts available.
He simply wanted progress. Any book would do, and there were plenty of
books in Padua to keep the censors busy if they wanted to be.

Despite the abundance of prohibited books in Padua, Cornaro's excuse
of "no books to censor and no money to get them" is a plausible response
from the perspective of the bishop or inquisitor. Although clandestine copies of prohibited texts certainly existed in Padua, it is unlikely that people who had taken the energy to procure, hide, and preserve these books
would hand them over to Catholic authorities. It is also possible that the
supposed lack of books could have been an excuse censors leveraged to
evade the difficult and undesirable work of expurgation. While this excuse
may have been a pretext for the elected twelve lay doctors and philosophers, Pranzini's separate letter corroborating Cornaro's story indicates
that the bishop and inquisitor felt logistically constrained by the imposi-
tions of the project of expurgation.[78] While lay censors may have fabricated
the excuse of "no books to censor" to deliberately obstruct the work of
the Catholic Church, it is unlikely that Cornaro or Pranzini shared this
motivation. Cornaro, by all accounts, appears to have been dedicated to
Catholic reform, and Pranzini, as a Conventual Franciscan inquisitor from
Pistoia, surely saw the ecclesiastical stepping-stones laid before him—the
Index and the Inquisition could make one's career.[79] It is likely that Pranzini, in particular, stood to benefit from working hard to fulfill the orders
of both the Congregation of the Inquisition and the Congregation of the
Index. The rapidity with which he produced the initial list of expurgations
and the promptness with which he sent them to Rome further suggests
that he was dedicated to this work and that he and Bishop Cornaro were
indeed strapped for cash and lacked the tools they needed to carry out
their censorship projects. After all, even the Congregation of the Index in
Rome had to propose a list of prohibited books for purchase before it could
proceed to correct them.[80]

While Cornaro and Pranzini struggled to effectively expurgate texts,
the Congregation of the Index was maneuvering on several fronts to ensure
the success of its goals. As we already saw, the Congregation of the Index
had reached out to Giovannini at the same time it approached Cornaro,
and it continued to expand the base of people upon whom it relied to censor texts. On August 14, 1599, the Congregation of the Index decreed that
the task of expurgating medical and philosophical texts had been "handed
over" to the inquisitor and the vicar of the archbishop of Pisa.[81] Tagliavia
clarified his intent two days later in a letter to the inquisitor's vicar in

Pisa. The cardinal said that the Pisans should "institute a Congregation of various Consultors to this effect and go about censoring and correcting books of medicine and philosophy with the help of the inquisitor and of the many excellent men that are in this venerable university."[82] The letter ends with the reminder that the results of this work should be sent to Rome so that they could quickly publish an expurgatory Index. Following a period of neglect and closure at the beginning of the fifteenth century, by the middle of the sixteenth century, the University of Pisa had risen to prominence, especially in medicine, thanks to investments by Duke Cosimo I of Tuscany. In 1544, the duke had worked hard to attract Leonhart Fuchs to the college and made a similar attempt to bring the famed anatomist Andreas Vesalius permanently to the university, an offer which the Flemish scholar also refused. However, not all of the duke's efforts to bring famous physicians to Pisa were in vain. In 1543, the naturalist Luca Ghini had arrived in Pisa and founded the first university botanical garden, from which he taught courses in medical botany from 1544 to 1555.[83] Although Pisa never reached the level of international renown that Padua did, it was sensible for the Congregation of the Index to look to this university town, located within the more religiously compliant Grand Duchy of Tuscany, when the Paduan censors proved recalcitrant.[84]

Tagliavia also approached the vicar of Naples, delegating to his community the same task of censoring "various books of medicine, and philosophy."[85] Even if the Congregation of the Index considered projects in Naples and Pisa as secondary to that of the Paduans, Valier followed up on Tagliavia's requests in December 1599, reminding the vicars of Pisa and Naples that they awaited copies of their expurgations in order to print revised editions.[86] While Padua and Pisa were seats of medical learning, we might well question who would have been the lay experts to assist this expurgatory work in Naples. Naples had an active contingent of philosophers including Bernardino Telesio, Giambattista della Porta, Nicola Antonio Sigliola, Tommaso Campanella, and Giordano Bruno, the latter of whom was imprisoned by the Holy Office in Rome, though he had yet to meet his fiery end in Campo de' Fiori. The work of Naples's best-known scholars of philosophy and medicine ranged from irreverent to occult to downright heretical. The choice to turn to Naples instead drew on the proven capacity of ecclesiastical censors in Naples to produce expurgations. Led by the Augustinian Cherubino Rato (usually known as Cherubino Veronese), a team of censors in Naples had already compiled an extensive collection of expurgations that they would, in turn, send to Rome in the coming year.[87] The Congregation of the Index's appeal to medical censors in Naples reflected an approach to expurgation that valued production and reliability over subject area expertise.

The Congregation of the Index did not hide its efforts to recruit multiple cities to accomplish the same censorship tasks. In the letter of August 16, 1599, to Marco Cornaro, in which Cardinal Tagliavia firmly suggested to the Paduans that they accelerate the pace of their censorship projects, the cardinal leveraged the fact that multiple institutions were now working on the same projects in order to motivate Cornaro. However, this strategy was largely detrimental to making progress on expurgations in Padua. The Catholic Church had initially conferred what it saw as a special status and responsibility on the Paduan authorities by selecting them above all other cities in Italy to lead the expurgation of medical and philosophical texts. However, as multiple congregations began working simultaneously on the same expurgation projects, local congregations were indignant because the overlapping assignments undermined their authority. The bishop of Cremona reflected in a letter on October 13, 1603, that some of the members of his local congregation were "pained to have been assigned the correction of the same medical books that were already corrected and named in Padua and Milan."[88] The authority that came from expurgation had been qualified from the beginning, and the Congregation of the Index in Rome always had the final say, but now it seemed even less likely that the opinions and work of local congregations' censors would stand as authoritative readings of texts.[89]

The overlapping expurgation assignments led Italian censors to question whether they were wasting time and resources on projects that were regulated arbitrarily from Rome. The Roman authorities, eager to outsource the intellectual labors of censorship, had also foisted upon the Paduans the logistical costs of executing that task. When the process failed to produce efficient results, the authorities in Rome appeared to change tack from a strategy that drew upon the most respected intellectuals in particular communities to a diversified effort to maximize the number of expurgations produced. The problem of duplication of effort seemed less troublesome in Rome than the very real possibility that the elite learned communities of Italy would never follow through with the expurgations they had been assigned. As Gigliola Fragnito has argued, peripheral censors felt as though they were "pointlessly labouring on texts for which there was no longer a market and which would never be reprinted."[90]

On December 1, 1600, Felice Pranzini and Camillo Peltrari, a member of the inquisitorial tribunal in Padua, sent a series of censures of works by the Calabrian philosopher Bernardino Telesio, the thirteenth-century physician and religious reformer Arnald of Villanova, and the Lutheran botanist Leonhart Fuchs to the Congregation of the Index in Rome, while also promising to soon send expurgations of works by Girolamo Cardano, Italy's beloved but unorthodox astrologer, physician, and philosopher. Once again, the inquisitor Pranzini begged forgiveness of the Congregation in Rome, explaining that "if we had copies of the medical and philosophical books to expurgate we would do much more, but because of our poverty we are not able to get them."[91] In light of Pranzini's plea, we might suspect that some of the censors elected to the congregation of censors in Padua were deliberately and intentionally obstructing the ability of the inquisitor to carry out the expurgatory task. Prohibited books were, after all, so notoriously widespread in Venice and its environs that a certain Emmanuel Mara had plausibly explained in a 1568 inquisition trial that the prohibited manuscript in his possession was one that he had found "in Padua, at the Bo, in the place where I went to urinate."[92] The idea that someone of the stature of Ercole Sassonia, who shared close ties with the German Nation (the German students' organization) at Padua, would not have had access to works by Fuchs, seems especially unlikely. Foreign students who enrolled in the University of Padua took advantage of the emphasis on practical medicine taught there by physicians like Sassonia, and they came from Germany equipped with copies of essential books, especially those that they knew it might be difficult to purchase in Catholic Italy. Commenting on the particular resources available at the University of Padua, an annalist of the German Nation remarked, "We also have books at home which we can just as well read there as here. It is the study of practice that has led us to cross so many mountains, and at such great expense."[93] Indeed, the records of the German Nation library show, unsurprisingly, that the German students owned editions of these works and brought them to Padua during their studies there.[94]

Pranzini's continued, if slow, progress despite the lack of cooperation from his lay censors showed unusual dedication to the work of expurgation. Most of the university men nominated to take part in the "honorable enterprise" were either uninterested in participating or actively undermining the process through their participation. Other cities across Italy charged with expurgation of other fields of knowledge confronted similar problems. The inquisitor of Mantua asked if perhaps people could be allowed to read at least the big works of law by Ulrich Zasius since "I consider making such an expurgatory book to be very difficult and I am certain that no one here wants to take on this task."[95] In Bologna, the inquisitor Stefano da Cento lamented that he had convened the College of

Theologians and exhorted them to expurgate books, but "most of them said that they could not attend to it . . . because at the time they were impeded by preaching, confession, and lectures." Later he divided the books between the scholars and found that "among all of them, none had finished his work, and few had begun."[96] The inquisitor of Genoa, Giovan Battista Lanci, shared similar frustrations, stating simply that the learned men of the town were instead "attending to their own things" and they had been "very little help" with the expurgations.[97] Vicenzo Castrucci, the inquisitor of Perugia, knew that lawyers in Bologna had also been commissioned to correct books (though he evidently was unaware that it was not proceeding well), and he hoped to use this knowledge as leverage to kindle the enthusiasm of his own College of Jurists who "say that they are without doubt very busy with all of their lessons, work, and other previous family matters." In a letter to Rome, he suggested that perhaps it is "necessary that you [the Congregation of the Index] write a letter to the College . . . and they will be excited and will emulate Bologna."[98] Across Italy, the attempt to involve lay professionals in expurgation was failing. While expurgation may have been the priority of the Congregation of the Index, the incentive for most lay scholars was limited, though as we will see in chapter 3, occasionally physicians did rise to the meet the Church's expectations.

The effort involved in expurgation was also wearing on the ecclesiastics charged with these corrections. Alfonso Soto, the Dominican professor of theology at the University of Padua and a member of the Paduan congregation for expurgation, was charged with correcting the philosophical works of Giacomo Zabarella, the famous Paduan Aristotelian philosopher who had died less than ten years earlier.[99] The task of correcting Zabarella's discourse on the soul was, Soto agreed, "a labor of importance," and he was grateful to the Congregation of the Index and to his colleagues for having chosen him for the task. However, the duty was also difficult, and he pleaded that his old age (sixty-seven) and the fact that he left issues of philosophy years ago in favor of his present profession should exempt him from the work. Additionally, he continued, they ought to find a person who was "free, and not obligated like I am" since it would require "turning the whole work upside down with different interpretations, making different links, and carefully putting together different discourses. If this job consisted of only noting the places that are repugnant to the truth, then it would be a thing that could be done by many."[100] Expurgatory censorship asked a lot of censors financially, logistically, and even intellectually. The censor's work was a challenge that required careful analysis of a text and painstaking unraveling of the many possible meanings of a given work. The *honorata impresa* was too much for even the devoted people who worked hardest to implement it.

CREMONINI AND THE PARADOXES OF CENSORSHIP

When Felice Pranzini did finally send more expurgations to the Congregation of the Index in December 1600, the list of necessary corrections was undersigned by a congregation of fifteen men, none of whom were part of the original twelve experts elected to participate in 1597.[101] Although two of the three sets of expurgations that Pranzini sent to Rome were for medical works, of the fifteen censors who undersigned the corrections, none were physicians. In fact, the vast majority of the censors (thirteen out of fifteen) were ecclesiastics or theologians.[102] In an added twist, the two philosophy professors who added their names to the expurgations were none other than Camillo Belloni and Cesare Cremonini.

We might pause here to appreciate the irony that a close study of the past can offer. Camillo Belloni was extraordinary professor of philosophy and had taken up his post in Padua in 1591, the same year that Cremonini moved from Ferrara to fill Zabarella's chair in philosophy. It was Belloni who, on April 16, 1604, would denounce his fellow censor and higher paid colleague, Cremonini, to the Paduan Inquisition with the charge that Cremonini argued for the mortality of the human soul.[103] However, the 1604 denunciation was only one of more than eighty Inquisition files opened against Cremonini beginning as early at 1598. As Edward Muir has pointed out, this makes Cremonini "one of the most, if not the most, thoroughly investigated thinkers in the early modern Catholic world."[104]

What should we make of the eternally censured turned censor?[105] How is it that the libertine Cremonini signed off on expurgations when so many more explicitly pious professors refused? Did Cremonini feel obligated to take part in order to show himself as aligned with the projects of the Catholic Church? In my view, Cremonini's decision to partake in Catholic efforts at expurgation was opportunistic and in accordance with his personal philosophy *Intus ut libet, foris ut moris est* (Think what you like, but say what is expected of you).[106] Cremonini must have concluded that cooperation in Padua's censorship efforts would help his image in the eyes of the Church, which was then slapping condemnations on his fellow philosophers Francesco Patrizi, Bernardino Telesio, Giordano Bruno, and Tommaso Campanella.[107] Cremonini was also well aware of the controversies surrounding Pietro Pomponazzi in the previous century and the strong resonances between Pomponazzi's ideas about the soul's mortality and his own Aristotelian teachings.[108] Facing investigations into his own piety, Cremonini joined forces with the inquisitor and bishop of Padua in the *honorata impresa* of book expurgation.

In 1598 the Roman Inquisition had ordered the inquisitor of Padua, Felice Pranzini, to investigate Cremonini for holding the heretical proposition that the soul was mortal. Confronted by these investigations and following an admonition from Pranzini in 1599 to keep his interpretations of Aristotle's *De anima* strictly within the decrees of church councils (to which he "reverently" agreed), Cremonini joined the inquisitor by publicly performing the work of a pious censor.[109] Cremonini became one of only two university professors in Padua to join Pranzini's congregation of censors and sign off on the *Index selectus*. Cremonini was making a good impression with local ecclesiastical authorities through his participation in the expurgatory efforts underway in Padua. In 1604, after Belloni denounced his colleague Cremonini to the Inquisition, the Holy Office in Rome ordered Padua's bishop, Marco Cornaro, to investigate Cremonini extrajudicially. The content of the letters that Cornaro sent in response is unknown, but they had the effect of halting the investigation of Cremonini for the next two years.[110]

Cremonini calculated that participating in the expurgation of medical and philosophical books was also an investment in his own future in the Veneto, where he would continue to live until his death in 1631. The University of Padua in turn invested heavily in Cremonini, both financially and politically. To give a sense of scale of the Venetian investment in Cremonini, we need only remember that after Galileo's pay nearly doubled in 1609 following his refinements to the telescope, Cremonini was still earning twice as much as his younger colleague.[111] Politically, the Venetians repaid Cremonini for his participation in civic life by protecting him against the Roman Inquisition during the more than three decades that he taught in Padua and was under investigation for his writings and beliefs. In 1604, following Belloni's accusations, Cornaro was forced to investigate Cremonini extrajudicially because Venetian authorities had refused to comply with the investigation, telling the new inquisitor of Padua that "information is not to be collected against Cesare Tremonini [*sic*]."[112] Even following the controversial philosopher's death, the Venetian doge tried to prevent Cremonini's papers from being shared with Roman authorities.[113]

It is possible that, like many of the university physicians who joined the efforts in 1597, Cremonini treated his participation in the task of censorship as merely a signature on paper, thereby participating in the process without doing the work. However, it seems more likely that for Cremonini the decision to participate was a strategic choice to ingratiate himself with the authorities investigating him. When the Congregation of the Index in Rome reached out to the learned men of Padua for help censoring books, they found help in Cremonini, who, in turn, manipulated the task to his own ends. In the coming years the Congregation of the Index would turn on this helpful, if insincere, censor, denouncing Cremonini's works repeatedly and vociferously. The ensuing inquisitorial investigations into this dilettante philosopher and diligent censor also became a springboard for the far more famous inquiry into the activities and beliefs of his colleague Galileo.[114] Cremonini's role in the expurgation of books in Padua reveals the ways that scholars dissimulated as they participated in the Italian culture of censorship. It also serves to remind us that the groundwork for the encounter between Catholicism and natural philosophy that would play out in the Galileo affair had deep roots in the censorship of medical and philosophical books.

CONCLUSION

The story of the congregation of censors in Padua ends in 1602. In February of that year, Marco Cornaro, the bishop of Padua, wrote a terse letter to Agostino Valier in Rome: "Many congregations for the expurgation of books were convened here in Padua, and many works were put into the hands of different men, who because of lectures and other business did not do anything other than begin. I advised the Congregation of the difficulties we had making progress without receiving a response."[115] Eight months later, in October 1602, Inquisitor Felice Pranzini was transferred to the post of inquisitor of Siena. He wrote to the Congregation of the Index in Rome on October 26 about the expurgations that he had written in Padua, promising officials in Rome that he would update the new inquisitor of Padua, Fra Zaccaria Orcioli da Ravenna, about the progress he had made.[116] In the following years, this new inquisitor of Padua wrote regularly to the Congregation of the Index in Rome, but he never once mentioned the project of correcting books.

It had finally become as clear to officials in Rome as it was to local censors that the project of expurgation in Padua had been an unsuccessful experiment. The Paduan congregation of censors had insisted that they could not find books in one of the most library-rich cities in Europe. They had taken advantage of the censor's right and duty to read broadly, and they may even have received funds to acquire prohibited books. Yet, these

uomini dotti, these learned medical men, had little to offer the Congregation of the Index. We must read this lack of production, this archival silence, in light of other censors' complaints and conclude that expurgating
medical texts was not a scholarly priority for these physicians and philosophers. Paradoxically, one of the few professors of medicine or philosophy
to fulfill his duties as a censor was the contentious philosopher Cesare
Cremonini, who, despite this show of allegiance to the Catholic Church,
would spend much of his life defending himself and his ideas against
charges of heresy. Ultimately, Cremonini's own works would be prohibited
by the Congregation of the Index, and parts of his oeuvre, forever stalled in
the bureaucracy of prepublication censorship, remained unpublished after
his death in 1631.

By enlisting committees of censors and local experts to aid in the *honorata impresa*, the "honorable enterprise," the Catholic Church tacitly
acknowledged lay professional expertise, though leveraging and institu-
tionalizing this expertise on behalf of the Church was largely unsuccessful. The fact that the Congregation of the Index selected communities and
people with reputations in specific fields to act as censors suggests that the
Catholic hierarchy acknowledged the intellectual authority of people independent of the Church. Physicians were not only scholarly men at universities; they were also practitioners and healers whose reputations relied on
their expertise both in and out of the classroom. Among the requirements
for a medical degree at Padua in 1496 was the expectation that the medical student work for a year with "a famous physician" to solidify both the
practice and the theory of his craft.[117] This practical training also provided
differentiation between the expertise of the physician and the knowledge
of church authorities. Official discussions within the Congregation of the
Index about the utility of medical knowledge reflect the willingness of
ecclesiastics to accept input from doctors and recognize the authority of
lay scholars and practitioners.[118] Rome's recognition of the need to engage
professional expertise indicates that we should consider censorship as another aspect of the Counter-Reformation that is best understood as a negotiated enterprise in intellectual and social control.[119]

The expectation in Rome that the Catholic intellectual community
would rise to assist the Church in its hour of need was ultimately disap-
pointed. It might seem strikingly naive that the Congregation of the Index
believed that it would receive assistance with expurgation from the very
scholars whose libraries had been gutted by the book burnings of previous decades. However, from the perspective of church officials in Rome,
expurgation was a form of compromise. The process allowed scholars to

obtain access to "corrected" books that were necessary for their professional activities. Rome expected its brethren to enthusiastically commit to the *honorata impresa*, and as we will see in chapter 3, some physicians in Italy did indeed take this work seriously. The University of Padua, however, despite and perhaps because of the many features that contributed to making it Europe's greatest seat of medical learning, never became the site of Catholic intellectual reform that the Congregation of the Index envisioned.

CHAPTER THREE

The Censor at Work

In chapter 2 we observed the learned men, the inquisitors, and the bishop of Padua working with and against one another to create expurgations of highly useful medical books. The professors of Padua were often obstructionist, but this was not the case with all lay censors in Italy. This chapter examines a particularly diligent and effective censor from Ravenna named Girolamo Rossi (1539–1607) for whom there exists unusually extensive documentation of his life and career as a historian, physician, and censor for the Roman Inquisition. Rossi's faith permeated his intellectual life. The constantly changing boundaries of late sixteenth-century Catholicism forced him to revisit, redefine, and even rewrite both his own scholarship and the work of his colleagues. The wealth of documentary materials concerning Rossi allows us to examine the work of a physician who was also a censor and to consider carefully the personal and intellectual nature of his work for the Roman Congregation of the Index.

Studies of censors and inquisitors are a popular and productive area of research for scholars of the inquisitions in Italy, Spain, and the Spanish Americas. The approach of these studies has often been to break down the Congregation of the Inquisition or Congregation of the Index of Prohibited Books into their constituent human actors. This methodology highlights the individual goals and motivations of censors and inquisitors and challenges the notion that these organizations had agendas and actions apart from the people who participated in them.[1] My study of Girolamo Rossi builds on this approach but takes as its subject the lay censors who were, for a short time, voluntary participants in the Catholic bureaucracy of censorship. Instead of focusing on professional ecclesiastics, I examine Rossi as a physician and lay professional who volunteered to work as a censor for the Catholic Church.

Formalized lay participation in censorship happened at a particular juncture in Counter-Reformation Italy between the publication of the Clementine Index in 1596 and the Roman *Index Expurgatorius* of 1607. This expurgatory moment, when the Catholic Church reached out to ecclesiastics and lay professionals across Italy, created a generation of learned readers who read and wrote with censorship in mind. The main character of this chapter, Girolamo Rossi, published books, composed expurgations, censored his own writing, and ultimately participated in the work of reforming medical and scientific knowledge in the long aftermath of the Reformation. Rossi's paper trail is uniquely rich and can serve as an example to help us understand other pious, Catholic readers for whom the record is less complete.

Despite Rossi's well-known importance as a historian of Ravenna, his career as a physician has never been the subject of sustained scholarly attention, and until Ugo Baldini and Leen Spruit published documents from the archives of the Roman Inquisition, no biographical sketches mentioned his role as a censor for the Congregation of the Index.[2] While learned physicians at Padua refused to participate in and even obstructed the censorship apparatus, Rossi volunteered with enthusiasm. In his story we see a different example of an individual weighing and balancing personal and intellectual motivations in response to the process of Catholic reform.[3] While for some intellectuals these choices led to conflict, for others, like Rossi, expurgatory censorship provided an opportunity for synthesis and accommodation between religious beliefs and medical practice.

Prior to his career as a practicing physician and as a censor of medical books, Girolamo Rossi was a historian, a published author, and a humanist secretary for a famous cleric. These themes of humanist learning, book publication, and devotion to the goals of the Counter-Reformation Church that we see in Rossi's early life remain relevant throughout his long career, which corresponded almost exactly to the period of the most dramatic changes in Roman censorship policy (1559–1607). Through Rossi's story we can also trace the continuity of scholarly tools used by physicians in the period between the Renaissance and the Counter-Reformation. The humanist practices of reading and note-taking that were essential for learned physicians were adopted and adapted for censorship and book expurgation. The *Index Expurgatorius* of 1607, to which Rossi contributed, should be understood not only as a tool for confessionalized reading but also as an anti-commonplace book. Physicians like Rossi drew upon their humanist educations to repurpose intellectual and didactic tools to serve the Counter-Reformation agenda.

GIROLAMO ROSSI AS HISTORIAN AND PHYSICIAN

Girolamo Rossi was born in Ravenna to Isabella Lodovicchia and Francesco Rossi and was baptized on July 15, 1539.[4] Little is known about Rossi's
childhood in Ravenna, though at age fifteen he was taken under the wing
of Archbishop Ranuccio Farnese after delivering a Latin oration in honor
of him. Farnese secured a place for the young Girolamo at the Collegio
Ancarano, but rather than beginning his studies there, he instead traveled
to Rome with his uncle Giovan Battista Rossi.[5] Rossi pursued studies at
La Sapienza in Rome under the supervision of Francesco Sempronio and a
certain "Bishop Giacomello." He returned to Ravenna in 1560, at the age of
twenty-one. The next year he obtained a degree in arts and medicine from
the University of Padua. Rossi's eighteenth-century biographer Pietro Paolo
Ginanni suggests that in 1561, the young graduate returned to Ravenna to
begin assembling documents for a book on the history of Ravenna. At this
point, Rossi's career again took an unexpected turn. In 1562, the uncle he
had once followed to Rome, Giovan Battista Rossi, was elected vicar general of the Carmelite Order. Girolamo did not become a Carmelite himself,
but he did follow the vicar general around the Veneto while he visited monasteries, helping his uncle with his public disputations and writing his letters, presumably in the capacity of a secretary. In 1564, Giovan Battista was
promoted again, this time to general of the Carmelite Order, and Girolamo
again joined him in Rome.[6]

It was during this second trip to Rome that Girolamo Rossi began
working in earnest on the history of Ravenna for which he would become
famous.[7] Rossi began his research for this work in Rome, making use of the
rich Roman libraries on his visits to his uncle's monastery. Rossi's early
archival research has made his *Historiarum Ravennatum libri decem* (*Ten
Books of Histories of Ravenna*) among the best known and most widely
referenced sources on the history of Ravenna. The Senate of Ravenna paid
for the printing of the first edition of the book at the press of Aldus Manutius the Younger in 1572 and then paid again for the book to be reprinted
in 1589 in Venice at the Guerra press, decrying that "one can no longer get
a copy much to the disgust of this city."[8] Rossi's scholarly work as a historian remained fresh in his memory years later. In a 1595 letter, Rossi reminisced about consulting a manuscript at the Vatican Library of a work by
Riccobaldo of Ferrara, "It's been about thirty years since I saw the work by
Riccobaldo, which was written in a Latin hand on sheepskin parchment
with pages about four fingers high. It was located in the Vatican library, on
the sixth bench if I remember correctly."[9] Rossi's description of the expe-

rience consulting manuscripts in the Vatican Library is spatial, specific, and embodied. Even though it had been thirty years since he consulted the work by Riccobaldo of Ferrara (who wrote about geography, including the history of Ravenna), Rossi recalled with some precision where it was located in the library. His description of the manuscript as "four fingers high" refers to the small format of the book, which he recalls in his hands as well as on the desk.[10] The expertise that Rossi developed as a historian continued to feature in his intellectual life and his self-perception, even as his career shifted toward the study of medicine.[11]

Whereas recent studies of Rossi's work have identified him first and foremost as a physician, little of his career as a doctor has been studied by historians.[12] Yet Rossi was a well-respected physician; over the course of his professional life he repeatedly turned down offers of university positions in Ferrara, Bologna, and Rome that would have forced him to leave Ravenna, though the honor of papal physician was eventually too great to refuse.[13] Rossi was also a member of the medical republic of letters, and he corresponded with the famous physicians Ulisse Aldrovandi, Girolamo Mercuriale, Fulvio Angelini, Gasparo Tagliacozzi, Marco degli Oddi, and Arcangelo Piccolomini. He also exchanged letters and treatises with two members of the Paduan congregation of censors: Ercole Sassonia and Girolamo Fabrizi d'Acquapendente.[14]

Rossi and Aldrovandi in particular shared more than exchanges of correspondence. They were both physicians and polymaths with many intellectual projects. They had close relationships with family members and patrons who were well connected in ecclesiastical circles. For Aldrovandi these relationships included his brother Teseo and his patron Paleotti, while Rossi's influences included his uncle Giovan Battista and his patron Cardinal Anton Maria Salviati.[15] By the early seventeenth century, Rossi's expertise was so widely acknowledged that Aldrovandi excerpted parts of Rossi's letters into his notebooks, copying extensively from a discourse that Rossi had written for the Bolognese vice-legate about a sixty-foot whale that had beached in Cesenatico.[16] Rossi's reputation as both a physician and a historian lent his description of this event particular value in the republic of letters.[17]

Although there is no evidence that Rossi, like Mercuriale or Aldrovandi, corresponded with Protestant medical colleagues living across the Alps, he was both flattered and worried when his book *De destillatione liber* (*On Distillation*) was reprinted in Basel in 1585 by Sebastian Henricus Petrus, the son of Henricus Petrus. Rossi had first published the book in Ravenna three years earlier with Francesco Tebaldini. The book is a sparsely illustrated, nearly three hundred-page manual initially dedicated to Francesco I de' Medici, the Grand Duke of Tuscany and a great supporter of alchemical research. The work explained the distillation of liquids and medicines and discussed chemical experiments.[18] The Basel press that pirated Rossi's text, the Officina Henricpetrina, was famous for having printed a number of books that had been or would be banned in Italy, including editions of Sebastian Münster's *Cosmographia* and the second edition of Nicolaus Copernicus's *De revolutionibus* in 1566. The Officina Henricpetrina must have seen an opportunity in this Latin volume to profit from the increasing interest in chemical medicine.

Rossi first became aware of the piracy long after it took place, and mention of it appeared in Rossi's correspondence only in 1595. However, shortly thereafter, in 1596, Rossi began to approach printers about having his work printed yet again.[19] Because the Basel reprint had occurred without his knowledge or participation, Rossi was intent on having his book republished with corrections to errors that appeared in the pirated edition. In May 1596, Rossi wrote to Fabio Paolini, the public reader charged with the task of prepublication censorship in Venice:

> I am sending my book *On Distillation*, which was already printed in this city in quarto and reprinted in Basel in octavo, and since the reprint occurred without my knowledge, I was not able to add certain additions that I send now. Since I do not want to spend more than is my credit, it could be printed either in octavo or in sextodecimo or in some other form. In either case, I greatly desire that it be corrected and that you do not find it unworthy of being reprinted.[20]

Rossi understood the costs and implications of reprinting his book. He indicated that he was willing to help finance the publication and that he knew that the format might need to be reduced yet again. Finally, Rossi implored Paolini that his great desire was that the work be "well corrected," specifying in a postscript that Paolini choose a type that "isn't too big nor too small, but that suits the page."[21] The pirating of his book by Protestant printers in Basel gave Rossi the opportunity to update his own work, which was republished in 1599 by Domenico and Giovan Battista Guerra. The work was reprinted yet again in Venice in 1604 by Giovanni Battista Ciotti, the printer and bookseller so highly favored by Giordano Bruno.[22] As Rossi advocated after the piracy, these later editions included additional material at the end of chapters and minor reordering of parts of the text.[23]

It is clear from the publishing and republishing of his book *On Distillation* that Rossi was also aware of the paradoxes of maintaining confessional divides in Europe. The pirated edition of his *On Distillation* simultaneously extended his readership, removed his sense of control over the text, and provided the impulse and opportunity to issue new editions of his own. His experience revising a text originating from a Protestant press, in this case his own book, was in some respects a complementary process to the expurgations that he was writing as a censor for the Catholic Church.[24]

GIROLAMO ROSSI AS CENSOR

Following the publication of the Clementine Index in 1596, bishops formed congregations of censors in cities across the Italian peninsula. While some cities, like Padua, were assigned particular censorship tasks based on the expertise of laypeople in the city, other congregations of pious individuals volunteered their service to Rome. In Faenza, a city in the Po River valley and the seat of the Roman Inquisition for the region of Emilia Romagna, the bishop Gian Antonio Grassi convened one such congregation of volunteers. Bishop Grassi received his copy of the Clementine Index and published it in his diocese in July 1596. On December 21, 1596, he wrote to the Congregation of the Index in Rome, stating that he had "the list of all the books from which to remove the bad and to correct what needs correcting. To that end, I have deputized learned men in all the sciences—that is, theology and philosophy, and canon and civil law, and the humanities."[25] The Dominican inquisitor of Romagna, Alberto Chelli, wrote repeatedly from Faenza to the congregation in Rome to update the Congregation of the Index on the progress of his censors.[26] This correspondence is an especially important example of how censorship efforts connected Rome with intellectual sites around Italy. Congregations of censors in this region also served to solidify Rome's political interests in a region that was in contention between Venice and the Papal States. The late 1590s were a particularly key moment for this shifting political frontier since Ferrara would become part of the Papal States in 1598.

Within two weeks of convening the congregation, the inquisitor of Faenza wrote to report that, "thank God," no one was lacking in diligence in attending to the expurgation of books. Some of the books they corrected were on the Index and others were not and, according to Chelli, "had not been observed in the past."[27] Along with the letter, Chelli sent a list of medical books that had been checked and corrected by Giovan Battista

Codronchi, a physician from Imola and a *qualificator*, or occasional consultant, for the Index.[28] Codronchi, who would take holy orders after the death of his wife in 1618, was at that time a lay physician, trained at the University of Bologna and living and working in Imola.[29] Codronchi was deeply involved in dictating the proper behavior of a Christian physician and composed and printed an entire book on the subject.[30] His expurgations are highly unusual because they are organized as a subject group: medical books with incantations and superstitious remedies.[31] Rather than choose a prohibited book and then expurgate it, Codronchi was appointed to the congregation of censors and turned to his reading notes or commonplace book to compile the passages from various medical texts that he had noted as superstitious. This is an unusual use of subject expertise in expurgation that started with one reader's deep understanding of a field of study and then turned to expurgation, rather than beginning with the list of authors on the Index and tackling their works one by one. Codronchi's expurgations were undersigned by three other officials of the local Inquisition: a vicar of the Inquisition in Imola, Michele da Lugo; the vicar general of the Inquisition in Imola; and Alberto Chelli, the inquisitor of Romagna (the regional head to whom Imola reported).[32] The Congregation of the Index responded enthusiastically to Codronchi's work, praising Chelli's efforts and the "convenience of the learned men in the district under his jurisdiction."[33]

By mid-February the inquisitor of Faenza had also received expurgations of medical books proposed by Girolamo Rossi from Ravenna, whom he described to the Congregation of the Index in Rome as "an intelligent physician, and one of those deputized in this city to expurgate medical books."[34] The expurgations included corrections to three important and regularly requested medical works: the Portuguese crypto-Jew Amatus Lusitanus's *Centuriae* and the ever-controversial Girolamo Cardano's supplement to the almanac and astrological commentaries on Ptolemy's *Quadripartitum* with its notorious birth horoscope of Jesus Christ.[35] In addition to supplying a list of words and sections to be removed from prohibited texts, Rossi had also explained why they needed to be expurgated so that no one could accuse him of having "removed them without reason."[36] Throughout the spring of 1597, Alberto Chelli continued to receive and forward expurgations written by Codronchi and Rossi, but by March 1598 Rossi was bypassing the intermediary of the local inquisitor and sending his expurgations of works by Guglielmo Grataroli and Merlin Cocai directly to Rome. Rossi's letter to the Congregation of the Index noted that he was still completing other expurgations, including the works of the physician and religious controversialist Thomas Erastus, which "I have already finished making note of but had to stop . . . because of my other occupations."[37]

Rossi's direct interaction with the congregation in Rome seems not to have been a problem for the inquisitor Chelli or his vicars, who also continued to forward Rossi's expurgations in the summer of 1599 when Rossi completed further censures of Erastus and Lusitanus.[38] However, in the fall of 1599 Chelli was reassigned to the post of the inquisitor of Cremona and was replaced as inquisitor by Pietro Martire Rinaldi. On December 3, 1599, Cardinal Agostino Valier wrote to Chelli in his new post to praise the speed with which he had responded to help create the *Index Expurgatorius*. Valier suggested that from this new position Chelli could continue his work producing expurgations. Describing the process of expurgation as "trimming off all that is superfluous and pernicious and putting together what good there is from one and the other," Valier hoped that Chelli would continue to create books that would be "useful."[39] Through careful expurgation of books by skilled censors, the Church could salvage the utility of books while curbing the circulation of harmful and heretical ideas. However, when Valier composed this letter to Chelli, he failed to acknowledge the expertise that had been so essential for Chelli's earlier productivity as a censor. The inquisitor was the spiritual authority undersigning the expurgations, but Rossi and Codronchi had been the medical experts who had trimmed from and recompiled the prohibited works that were necessary for Catholic doctors.

While Chelli and his team had been a productive congregation of censors, the new inquisitor Pietro Martire Rinaldi had different priorities for his office. He reported to Rome in August 1600 that the inquisition in Faenza was very busy and could not attend to the correction of books. A separate letter from the bishop Gian Antonio Grassi in November 1600 contradicted Rinaldi, confirming that a congregation for the correction of books would once again be instituted in Faenza.[40] As in Padua, the inquisitor and the bishop struggled to work cooperatively together. Adding further complication, much of the correspondence from the Congregation of the Index in this period was directed not to Faenza but to the archbishop's *vicarius* (vicar) in Ravenna.[41] Perhaps it is in light of these different ecclesiastical agendas that Girolamo Rossi began to send his expurgations directly to Rome in the early years of the seventeenth century, bypassing local intermediaries.

On July 31, 1602, Rossi responded directly to a letter from the Congregation of the Index in Rome, stating that with his letter he was sending the

expurgations that he had made "at different times of books of various professions and especially medicine." He continued, "I have collected them
into a volume [*corpo*] . . . and as soon as our Monseigneur Archbishop's
new *vicarius* arrives, which should be any day, he said he will review and
undersign the expurgations along with the three deputized theologians."[42]
Rossi considered this work important enough to have made and kept copies of these censures among his own treatises and correspondence, though
copies in Rossi's archives were messy drafts with cancellations and inter-
linear additions.[43] As it turned out, he was wise to have kept these dupli-
cates. In February 1602, the *vicarius*, Fabio Tempestivo, wrote anxiously
to Rome, expressing surprise that he had never heard back from the congregation about Rossi's expurgations of Cardano, which he had sent a year
earlier. "Please let me know," he urged, "because if it was not received I
will send another copy. And with this letter I send the correction of the
Scuola salernitana, done by the same Signor Girolamo Rossi."[44] The versions Rossi sent to the Congregation of the Index, and which Fabio Tempestivo feared were lost en route to Rome, were clean copies of the expurgations. Rossi might even have suspected in preparing these clean copies
that they were destined for the print shop. Cardinal Terranova had indicated that the goal of producing expurgations was, after all, to print them
for the public benefit.[45] However, before they could become official policy
and printed, these expurgations first had to arrive in Rome following the
proper congregational procedures.

Another change of inquisitorial personnel in 1602–3 heralded yet
again the reorganization of the Faenza congregation charged with correcting books. On October 24, 1602, the new inquisitor, Stefano de Vicariis,
complained that small cities like Faenza could not be expected to produce corrections of books. The "learned men" with whom he had spoken claimed that this task needed to take place in the "big [cities] and
the general universities."[46] Turning to the major centers of learning in
Italy had, of course, been the Roman Congregation of the Index's original
plan. However, that model had since been abandoned after having been
met with resistance in places like Padua and Bologna. Now communities
that were less intellectually prestigious but more reliably productive had
emerged as the main sites of correction, such as Rossi's congregation in
Faenza. The next summer, on August 6, 1603, a new inquisitor of Faenza
reported from Imola that in his territories there were not enough people
"capable of revising books" to make up a congregation. It is perhaps not
a coincidence that the very next day, August 7, Girolamo Rossi wrote to
Rome complaining of the difficulties of working with other people to pro-

duce the desired expurgations, "If the task of sending the censures were only up to me, according to my duty, they already would already be sent a long time ago . . . but since it depends on others, I don't know what I can do, except to complete my part of the said service perfectly, copying the censures, and to solicit, as I have already done, those things that the others deal with."[47] Cardinal Terranova responded from Rome with a letter to the archbishop of Ravenna, reminding him that though the Congregation of the Index was pleased with Girolamo Rossi's expurgations, they required that they be undersigned and approved by local theologians, the inquisitor, and the ordinary.[48] While Girolamo Rossi in Ravenna painstakingly expurgated books, his authority derived from his position as part of a congregation of censors, which was itself subservient to the authority of the Congregation of the Index in Rome. The congregational approach to censorship was fundamentally collaborative and frequently fraught and inefficient.

While Rossi's expurgatory efforts were directed primarily toward medical books, his expertise as a historian remained a central part of his reputation. In 1603, the Congregation of the Index in Rome asked Rossi, who was busy with his task of expurgating medical books, to also inform them if he saw anything that needed to be expurgated from foreign history books.[49] Rossi responded, affirming that he would take note of these books and assuring the congregation that if they "come to be in my hands, I will obey [and send expurgations] of those things that to me seem worthy of censure." However, Rossi continued, "It is very rare that new books arrive in this corner where I find myself."[50] Girolamo Rossi was a renowned historian, a respected physician, and a center of intellectual gravity in his own community, but he was the first to acknowledge that Ravenna was an intellectual backwater compared to cosmopolitan and erudite Padua. As Rossi the historian knew all too well, Ravenna was already a millennium past its golden age as the seat of the Roman Empire. Rossi's career as a historian, physician, and censor reminds us of the many kinds of expertise that learned physicians claimed. We also see Rossi as an example of the continued scholarly ambitions and aspirations of early modern people who lived outside traditional cultural centers.

From his home in Ravenna, Rossi persisted at his tasks in the following months, despite the apparent lack of coordination between Ravenna and the Holy Office in Faenza. In February 1604, Cardinal Terranova was so pleased with the final products of Rossi's expurgations that he queried the inquisitor in Faenza about what cities under his jurisdiction might be able to print Rossi's expurgations.[51] Later that year, Rossi traveled to

TABLE 3.1. Works Expurgated by Girolamo Rossi

Date Rossi's Expurgation Submitted	Author Name	Title of Work
1597	Girolamo Cardano	De supplemento almanach
1597	Girolamo Cardano	Quadripartitum Ptolomei
1597	Girolamo Cardano	De exemplis centum geniturarum
1597	Conrad Dasypodius	Quadripartitum Ptolomei
1597	Amatus Lusitanus	Centuriae
1598	Guglielmo Grataroli	Opuscula
1598	Merlin Cocai [aka Teofilo Folengo]	Macaronica
1598	[Teofilo Folengo]	Zanitonella sive innamoramentum Zaninae et Tonelli
1599	Thomas Erastus	Disputationes de medicina noua Philippi Paracelsi
1599	Thomas Erastus	De astrologia divinatrice epistolae
1599	Thomas Erastus	De putredine et de febribus
1601	Girolamo Cardano	De subtilitate
1602	Arnald of Villanova	De conservanda bona valetudine

Rome as an ambassador of his city, and Pope Clement VIII appointed him papal physician. Rossi's tenure as papal physician coincided with Clement's final months of life, and the death of the pope on March 5, 1605, brought about Rossi's return to his beloved home in Ravenna.[52]

In total, Rossi composed expurgations of thirteen books and sent them to the Congregation of the Index in Rome (see table 3.1). He provided the Congregation of the Index with in-depth descriptions of problematic passages and his reasons for removing them. Rossi's expurgations were primarily confessional rather than medical in nature.[53] He proposed the removal of passages that mocked the clergy, praised non-Catholic scholars, or included citations of Protestant authors. A few expurgations dealt with the practice of medicine, such as influence of God, demons, astrology, and magical words on the heath of the human body.[54] Reading carefully with confessional difference in mind was a skill cultivated through censorship and which the censor Girolamo Rossi had mastered.

In the end, the vast majority of Rossi's expurgatory efforts were never integrated into the expurgations that Rome eventually published in 1607. A line-by-line comparison of the censures that Rossi composed and the *Index Expurgatorius* of 1607 reveals that his corrections to the works of Guglielmo Grataroli, the Protestant physician from Bergamo, were formally adopted. Rossi's expurgations make up about three-quarters of Grataroli's entry in the 1607 *Index Expurgatorius*, but even though those passages appear verbatim in print, Rossi's manuscript suggestions included still more expurgations that were deemed unnecessary by censors in Rome.[55] Despite his prolific participation in censorship efforts, Rossi's contribution to the long-term integration of Protestant knowledge into Catholic Italy was extremely limited. Catholic authorities called on physicians like Rossi to lend their expertise to expurgation projects, but ultimately the authority for these decisions lay with ecclesiastics in Rome.

THE MIND OF THE CENSOR

Since Girolamo Rossi and his family had strong ties to the Catholic Church, it is no surprise that the vicar Fabio Tempestivo described Rossi as "a man no less full of zeal and Christian piety than of learning."[56] But was piety alone enough to induce a learned physician to become a censor? After all, most university professors in Italy had dragged their heels when asked to participate in the *honorata impresa*, that "honorable enterprise" of censoring books. The heterodox philosopher Cesare Cremonini, who had participated in censoring books in Padua, leveraged his position as a censor to promote his image with the local inquisitor, bishop, and government officials. Rossi was opportunistic like Cremonini, though his goals were entirely different. His biography and personal papers offer glimpses into the mind of this physician-censor and reveal Rossi to have been driven by a desire to influence the reformation of medical thought through censorship. The process of expurgation in turn shaped the way Rossi read and engaged with his own texts as a Counter-Reformation humanist reader who now read with pen in hand to revise and to censor.

Rossi's uncle Giovan Battista Rossi, for whom he served as a secretary as a young man, was an important and high-profile reformer in the post-Tridentine period.[57] Giovan Battista Rossi is remembered fondly in the *Foundations* of Saint Teresa of Avila as the head of the Carmelite order who allowed her to continue founding monasteries.[58] He was also one of three ecclesiastics assigned to correct the translation of the Vulgate Bible in 1568.[59] He was rigid in his applications of the rules of Trent and valued

strict obedience and hierarchy within the Church. A story about Giovan
Battista relates that he threatened to throw another Carmelite brother into
prison if the friar illicitly possessed a copy of the works of Saint Cyprian
with an introductory letter by Erasmus.[60] It is likely that some of the rigor
of Girolamo Rossi's Catholic piety was inculcated during his long stays as
a young man at his uncle's convent of San Martino ai Monti in Rome. We
know for certain that Girolamo Rossi respected and admired his uncle.
Although Giovan Battista died in 1578, he lived on for Girolamo Rossi as
an important character in his *Histories of Ravenna* and as the namesake
of his firstborn son.[61] Unlike his uncle or the physician from Imola, Giovan Battista Codronchi, Rossi never took up holy orders. Instead, he lived
his entire life as a lay member of society. He was the father of ten children
and was especially closely involved in the lives of his twin sons, Francesco
and Gerardo, who studied law in Padua and graduated in 1599.[62] His correspondence is also full of references and salutations to his wife, Laura, to
whom Rossi was married for thirty-seven years and whom he described
following her death in 1604 as a woman "of great genius in the administration of domestic matters, but of still greater devotion to God."[63]

    Girolamo Rossi is an exceptional figure in the history of medicine, sci-
ence, and censorship, since he is among the few lay censors who produced
a large number of expurgations in the period between 1596 and 1607. For
someone like Rossi, the Catholic Church's call for the expertise of lay practitioners gave a family man who never held an ecclesiastical position the
chance to participate actively in shaping the Counter-Reformation Church.
As a theologian and humanist, his uncle Giovan Battista volunteered his
expertise in philology and languages to correct the Vulgate Bible. Girolamo
did the same in the field of medicine, expurgating the works of prohibited
authors to make available and render orthodox the most useful medical
learning of a confessionalizing world.

    Rossi's efforts to correct medical books did not stop with those written
by Protestants that he expurgated as part of a local congregation of censors. Building on his previous work correcting the pirated edition of his
On Distillation, Rossi once again took up his pen to correct and censor
his own writing. In an undated draft of a document titled *Disputation of
Girolamo Rossi of Ravenna on the quantity of those qualities which are
attributed to elements*, Rossi removed his own references to prohibited authors.[64] In a passage describing the errors of Aristotle for which Girolamo
Cardano's works served as an important corrective, Rossi praised Cardano
as "a man most learned in every respect and of greatest ingenuity" (figure 3.1). Revising and rereading as a censor, Rossi must have felt that this

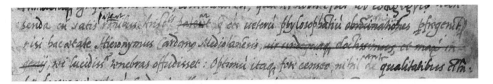

Fig. 3.1. Girolamo Rossi manuscript showing self-censorship where Rossi struck through his praise of Girolamo Cardano, "a man most learned in every respect and of greatest ingenuity." Composing expurgations transformed the humanist practice of reading with pen in hand into the Counter-Reformation project of reading with confessional difference in mind. BCRa, Mob. 3. 1 B, n. 4, f. 405r. Reproduced with permission from the Instituzione Biblioteca Classense, Ravenna.

praise was too unqualified to be applied to someone like Cardano, who had been tried several times by the Roman Inquisition and whose name appeared on the Index of Prohibited Books. While Cardano remained a light in the darkness, Rossi excised the praise of his intellect and learning with a thin line of ink.[65] Throughout the treatise Rossi edited himself, carefully removing his praise of Cardano's great learning (doctrina) and his eloquent writing.[66]

Rossi, acting simultaneously as censor, reader, and author, also removed some of his praise of Ortensio Lando, a student of medicine but best known as the author of the Paradoxes and translator into Italian of Martin Luther's works and Thomas More's Utopia.[67] Rossi had originally praised Lando as a man of "extraordinary power of speaking, of the greatest festivity and charm in addition." His censored version removed this high praise, leaving only the author's name and inserting the title of his book, the Paradoxes.[68]

Rossi also edited Giovanni Battista da Monte's name from his treatise, replacing it with the general and inoffensive "Hic."[69] Although Da Monte's works were never prohibited on the Index, he was renowned among Northern Europeans like Theodor Zwinger and his works were edited and published posthumously, many by Protestants including Johann Crato von Krafftheim, Girolamo Donzellini, and Valentinus Lublinus.[70] Rossi had read Da Monte's works closely. His own On Distillation mentioned Da Monte several times, describing him as "very learned" and specifying that he was from Verona. Rossi was too much of a bibliophile to suspect Da Monte and his texts of guilt by association. Instead, the acts of censoring and editing a text had become one and the same project for him.

The explicit evidence of self-censorship found in Girolamo Rossi's treatise is an example of a phenomenon that was likely widespread but

rarely left a paper trail.[71] Traces of self-censorship exist through intermediaries, such as printers seeking imprimaturs or the Italian translator of
Jean Bodin's *Démonomanie* (first translated into Italian in 1587), who censored his translation so that Bodin's work would be permitted.[72] Rossi's
expurgations-turned-edits provide an opportunity to reconstruct the kinds
of things that might have been said, even by pious, orthodox Catholics, in
an unrestricted intellectual climate.

We might also imagine the dissonance that Rossi felt as he expurgated
his own works alongside those written by esteemed, prohibited colleagues.
It seems significant that, with the exception of the works by Thomas Erastus and the German mathematician Conrad Dasypodius, all of the books
that Rossi censored were not only part of Italian culture but mostly written by Rossi's fellow Italians, and all of these authors spent time living
on the peninsula or studying in Italian universities. Several of the physicians whose works Rossi would later expurgate he praised glowingly in
the 1582 first edition of his own *On Distillation* though the index entries
for Conrad Gessner, Hieronymus Brunschwig, Girolamo Cardano, Jean de
Roquetaillade, Oribasius, Philipp Ulstad, and Ramon Lull described these
authors as "lapsus."[73] It is no wonder that taking up his pen to censor his
colleagues also led Rossi to look back critically at his own work. As Ugo
Rozzo has eloquently written of the changing status of book collections,
we have seen the "theological and cultural contradictions in which even
sincere Catholic intellectuals found themselves."[74] Living within this culture of censorship forced an unstable relationship between Catholic physicians and the works of not only their most celebrated colleagues but even
their own writings.

CONCLUSION

In the end, while authorities in Rome required censors to work in congregations, it did not take teams of censors to create effective expurgations. In
fact, the most efficient way of producing expurgations was by harnessing
the energy and expertise of enthusiastic individuals like Girolamo Rossi.
While many lay censors appear to have been unwilling to cooperate with
the congregations and did not produce expurgations, Rossi is an example
of how productive some individual censors were.[75] Rossi was a pious physician who embraced the task of expurgation and integrated the Tridentine goals of individual, social, and intellectual reform into his career as
a medical professional. In his global, comparative study of censorship,
Robert Darnton points out that coercion alone cannot sustain systems

of control and concludes, "All systems need true believers."[76] Girolamo Rossi was a true believer, working at the intersection of early modern Catholic bureaucracy, scholarship, professional expertise, and faith. There were also other prolific censors of scientific books, including Ambrogio Biturno and Alfonso Chacón, but both of these men were ecclesiastics. By contrast, Rossi presents an unusual case of an active, lay censor whose professional vocation was medicine. Through Rossi's efforts we can gain some intuition into how pious lay professionals responded to the expurgatory moment. The *honorata impresa* was an unwelcome burden to most scholars, yet Rossi seized on it as an opportunity to engage as a physician in defining the important, orthodox knowledge of Counter-Reformation Catholicism.

Christopher Black described the process of expurgatory censorship as "cumbersome, inefficient, and counter-productive" and suggested that the process ultimately worked against the goal of producing official, Catholic versions of texts.[77] If the process of expurgation was in fact ineffective, it was not for lack of expurgations but instead because the Church's apparatus of dispersed congregations and inflexible hierarchy prevented Catholic authorities from making the quick decisions that would allow for republication. However, Rossi's rich archival trail allows us to reframe the outcomes of Catholic expurgation in terms of readers rather than texts. During the last decade of the sixteenth and first decade of the seventeenth centuries, by reaching out to ecclesiastical and lay readers throughout Italy for help censoring books, the Catholic Church created a generation of scholars that was acutely aware of and even implicated in efforts to sanitize books by scholars of all religious backgrounds, with special attention to those written by Protestants. In Catholic Italy, these two decades created a culture of censorship in which readers—lay and ecclesiastical alike—were trained to read like censors.[78] Readers only rarely articulated this ethos that imbued their reading and writing, but we catch glimpses of it in the work of a censor like Rossi.[79]

In the realm of medicine, this expurgatory moment was thus constitutive both in the sense that it attempted to create newly Catholicized medical texts and in the sense that it taught physicians to read with an eye toward correction. Anthony Grafton has revealed how the humanist culture of correction entered into print shops through the role of correctors. This same humanist ethos to purify and correct texts also entered into censorship efforts through the process of expurgation, especially in the 1590s and early 1600s.[80] Girolamo Rossi's work as a censor changed the way that he read and worked as a physician; when revisiting his own manuscripts, he did so with a pen in hand to correct, revise, and ultimately expurgate.[81] Catholic physicians knew that when they read books by certain authors, they were obligated to circumscribe their interpretations of those texts. Similarly, when physicians like Rossi published books that they wanted to be read in Italy, they understood that they now needed to read and write more cautiously. While the Index of Prohibited Books was largely unsuccessful at producing medical texts that were purely Catholic, cleansed of their references to Protestant authors and theologies, the process of lay expurgation succeeded in training physicians and other professionals to actively censor themselves.

If the culture of censorship created readers who thought like expurgators, it was in part because of the long-standing humanist tradition of reading for commonplacing.[82] The humanistically trained Renaissance reader read with a pen in hand, not to censor, as Girolamo Rossi did, but to mark passages or copy them into commonplace books, like Seneca's industrious bees. It is a short step, then, to arrive at expurgation as the "dark side of commonplacing," marking passages ultimately for removal instead of preservation.[83] During the expurgatory moment, the Catholic Church turned traditional humanist techniques for extracting from books into a new, pious form of reading.

Expurgating was the dark side of commonplacing, and the *Index Expurgatorius* of 1607 became something of an anti-commonplace book, delineating what Catholics should not repeat. The volume lists fifty authors and the pages where readers could find the expurgations of their works. It included a recapitulation of the rules of the Clementine Index and the kinds of words, images, and ideas that needed to be removed from books in order to read them. Last followed the specifics, organized by author, title, and edition. From Amatus Lusitanus's *Curationum medicinalium*, Lyon edition of 1580, in *curatione* 9, page 99, readers were instructed to delete after the words "ante sua obitum multum vigilaverat" until "si qua vero noctes pars, etc."[84] The offensive parts of the expurgated passage are obscured through the instructions, and the only full passages that are reproduced in this dark commonplace book are those in which the text sets out how a passage should read after a section has been removed. As a commonplace book, this *Index Expurgatorius* was a compilation of readings to be, if not forgotten, then certainly not reused or repurposed by Catholic scholars.[85]

Girolamo Rossi died in Ravenna in 1607, the same year that the *Index Expurgatorius* was published in Rome. The work by physicians and other lay censors across Italy did not result in many of the expurgations published in the *Index Expurgatorius*, but this reference book of prohibited passages to be extracted from important texts laid the groundwork for how licensed readers throughout the seventeenth century would encounter prohibited books. Ultimately, participating in expurgation taught medical scholars to read like censors, and the eventual publication of the 1607 *Index Expurgatorius*, to which we now turn, served as a critical, official stepping-stone toward reintegrating prohibited and useful medical books into Italian libraries and collections.

CHAPTER FOUR

Censoring Medicine in Rome's
Index Expurgatorius of 1607

Ever since the Tridentine Index of 1564 marked certain books as "prohibited until expurgated," the Catholic Church in Rome had been trying, largely unsuccessfully, to compose precise and comprehensive expurgations of prohibited books. In 1571 and 1584 expurgatory Indexes had been printed in Antwerp and Spain, which included "corrections" to medical authors. While authorities in Rome were comparably slow to produce expurgations, there was a need and a market for expurgations in Italy. In Piedmont, the Quinctiano press in Alessandria issued a short "correction of Fuchs" as a broadside in 1580 and then as an appendix to an *Annotatio librorum prohibitorum* (*Note on Prohibited Books*) in 1585.[1] This set of expurgations to the popular and widely requested Fuchs was similar to the expurgations issued in Antwerp or Spain, though not identical in content or arrangement. In 1588, the Congregation of the Index in Rome solicited the printer Domenico Basa to print the expurgatory Indexes of Antwerp and Spain together in one volume to "more easily compose similar expurgations and bring them to perfection."[2] However, as we have seen, Rome's congregational approach was different from that of the theologians in Louvain and Spain, who composed the Antwerp and Spanish expurgations. In 1596, after the publication of the Clementine Index, the Congregation of the Index was largely unable to enlist lay support for the expurgation project in Padua but succeeded in recruiting the physicians Giovan Battista Codronchi and Girolamo Rossi as censors in a congregation organized by the inquisitor of Romagna.

We turn now from the process of soliciting and producing expurgations to the task of completing and promulgating a Roman Index of expurgations. The first successfully completed Index of expurgations in Italy was compiled in Naples in 1594, though it was ultimately never promulgated or printed. This collection of expurgations, with a profound focus on legal texts, came to the attention of censors in Rome in 1598 and was curiously abandoned around 1600. In 1604, the Master of the Sacred Palace, Giovanni Maria Guanzelli da Brisighella, stepped into the void left by the Congregation of the Index's failure to produce an Index and pushed through his own *Index Expurgatorius* in 1607.[3] This first volume promised a second volume to follow shortly thereafter, which never materialized.[4]

This chapter examines the concerns raised by censors about the content of prohibited medical books and the answers that Roman authorities provided through the *Index Expurgatorius* of 1607. Italian censors sought to settle medieval disputes about magic and astrology, transform works by unorthodox Italians into viable Catholic texts, and strip Lutheranism from the works of popular humanist physicians. Where possible I have identified the authors who composed the expurgations and examined their consistency as censors and their reading methods. I pay particular attention to Girolamo Rossi and the extensive and circumspect expurgations he submitted to the Congregation of the Index in the final years of the sixteenth century. Taken as a whole, the *Index Expurgatorius* of 1607 was an amalgamation of types and styles of expurgations, with emphases on particular content depending on the priorities of the censor who composed a given expurgation. Analysis of individual expurgations also reveals that the Master of the Sacred Palace and his associates ignored the input of many censors whom the Congregation of the Index had solicited for expurgations. The Catholic Church believed that heterodoxy was clear cut, but the many differing expurgations submitted to the Congregation of the Index in the years leading up to 1607 indicate that there was much room for interpretation and disagreement by, and among, readers.[5]

THE NEAPOLITAN EXPURGATORY INDEX

The first completed Italian Index of expurgations was undertaken by Cherubino Veronese, the coordinator of the congregation of censors in Naples, and was completed at least as early as July 1594.[6] Although little is known about Veronese's early life, he was an Augustinian friar who, in 1593, was charged with reporting on prohibited and suspect books that were confiscated in Naples. As part of a congregation of censors in Naples, Veronese's expurgations were undersigned by Vincenzo Bonincontro, Girolamo Zancaglione, Martino Alfonso Vivaldi, and Baldassare Crispo. Additionally, the heading on the final document ("Index of expurgated books by Cherubino Veronese . . . with the approbation granted to him by the Most

Illustrious and Reverend Cardinal Gesualdo, the Archbishop of Naples")
pointed explicitly to the authority granted by the local archbishop. Veronese, Bonincontro, Zancaglione, and Crispo were all ecclesiastics and
theologians, and Vivaldi was a secular priest and professor of theology in
Savona and Siena.[7]

The Neapolitan congregation succeeded in doing what Padua never
managed to accomplish; it compiled and submitted to Rome a large volume of expurgations.[8] The manuscript volume held in the Vatican is a
complete and authentic copy of the censures composed in Naples, which
were requested by Cardinal Agostino Valier in 1598, although the volume
itself was not sent to Rome until January 1600.[9] The Index contains expurgations of works by 145 authors ranging from Ovid and Erasmus to authors
of medical works such as Paracelsus and Girolamo Cardano. The list,
though nearly three times longer than the later Roman *Index Expurgatorius* of 1607, does not include censures of the frequently requested works
by Leonhart Fuchs, Conrad Gessner, Amatus Lusitanus, or Arnald of Villanova. Medical texts would feature much more prominently in Guanzelli's *Index Expurgatorius* of 1607. Instead, the Neapolitan congregation
had been tasked in particular with the correction of other works, especially the legal texts of Charles Dumoulin.[10] Neapolitan censors were considered to have particular subject expertise in law, as opposed to Padua's
expertise in medicine, and the Neapolitan Index reflected this focus.[11]

The predominance of legal texts included on the Neapolitan Index
raises an important comparison between legal and medical professional
communities in Counter-Reformation Italy.[12] The intersecting histories of
the prohibitions of legal books and medical books are a recurrent theme
throughout this book because lawyers were a professional group with a
similar degree of coherence as physicians. Like that of physicians, lawyers' work depended on a Pan-European corpus of texts that were then prohibited, in whole or in part, during the middle of the sixteenth century
and expurgated into the seventeenth century. Additionally, the works of
contemporary Italian and Spanish legal scholars were constantly subject
to review, prohibition, and correction.[13] While lawyers and doctors relied
on similar processes and justifications to read texts, there was almost no
overlap in the content required or sought by these two groups of professionals.[14] The mechanisms of censorship were the same, but the content
was completely different. The texts included on the Neapolitan Index,
which focused especially on legal texts, brings the relative importance of
medical texts to the foreground in the Roman *Index Expurgatorius.*

Although the Expurgatory Index of Naples was completed before the turn of the seventeenth century, it was never published, and it never became an official policy of the Catholic Church. When Agostino Valier and the Congregation of the Index in Rome received the expurgations in January 1600, the minutes from their meeting indicate that they wanted the expurgations to be printed and published. The congregation in Rome wrote to the vicar in Naples and to Cherubino Veronese urging them that "it will be greatly desired to see it in print for the benefit of the public."[15] The Congregation of the Index's archive in the Vatican also includes an eighteen-point set of instructions regarding required revisions to the Neapolitan Index prior to printing, including an admonition that the numerous grammatical errors present in the copy submitted to Rome ought to be corrected.[16] Over the next two years, Valier repeatedly sent letters to Naples encouraging the Neapolitan congregation to print the expurgations, but it never happened. According to one of Cherubino Veronese's expurgators, Donato Favale, "The door to hell is getting larger, since it expects that due to human curiosity people will not aspire to [follow] the prohibition, and many souls will fall down into its mouth. This could be easily remedied by publishing an expurgatory index."[17] Despite having assembled a team of highly productive ecclesiastical censors and having gained the approval and encouragement of officials in Rome, the expurgations from Naples never became a tool for revising copies of prohibited books. The doors to hell, which Favale thought could be closed so easily, remained open. The efforts of the efficient, productive team in Naples fell by the wayside.

MEDICINE AND GUANZELLI'S *INDEX EXPURGATORIUS* OF 1607

When Giovanni Maria Guanzelli da Brisighella, the Master of Sacred Palace, composed the 1607 *Index Expurgatorius* he was in the midst of a series of disputes with the Roman Inquisition and the Congregation of the Index regarding his jurisdiction. By 1604, Guanzelli had broken from the Congregation of the Index's congregational approach. He removed six volumes of expurgations that had previously been submitted to the Congregation of the Index from where they were being stored in the Vatican Library, and he cobbled together his own *Index Expurgatorius*, which was published in 1607 (see figure 4.1).[18] Guanzelli published the Index without the official approval of the Congregation of the Index and, as a result, it was technically not legally binding.[19] Nevertheless, reading licenses issued by

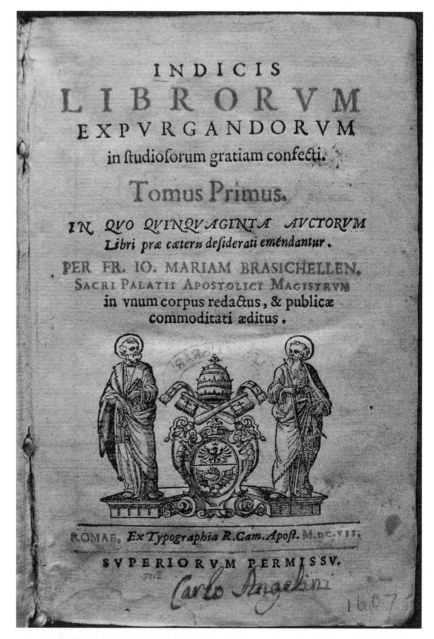

INDICIS
LIBRORVM
EXPVRGANDORVM
in ſtudioſorum gratiam confecti.

Tomus Primus.

IN QVO QVINQVAGINTA AVCTORVM
Libri præ cæteris deſiderati emēndantur.

PER FR. IO. MARIAM BRASICHELLEN,
SACRI PALATII APOSTOLICI MAGISTRVM
in vnum corpus redactus, & publicæ
commoditati æditus.

ROMAE, Ex Typographia R.Cam.Apoſt. M.DC.VII.

SVPERIORVM PERMISSV.

Fig. 4.1. Title page of Giovanni Maria Guanzelli da Brisighella's *Index Expurgatorius* of 1607, formally titled *Index librorum prohibitorum: Indicis librorvm expvrgandorvm in studiosorum gratiam confecti Tomus primus* (Rome, 1607). Call number *KB 1607 Index librorum prohibitorum. Rare Book Collection. The New York Public Library. Astor, Lenox, Tilden Foundations.

ecclesiastical authorities throughout the seventeenth century instructed readers to expurgate the works they were licensed to read according to the 1607 Index.[20]

Notwithstanding that he had access to the expurgations composed by censors across Italy, the Master of the Sacred Palace drew primarily (though not exclusively) from Indexes of expurgations already compiled and published in Spain and Antwerp. The medical authors corrected in the Antwerp edition included Conrad Gessner, Helius Eobanus Hessus, Euricius [Valerius] Cordus, Janus Cornarius, Johannes Lonicer, Julius Scaliger, and Leonhart Fuchs. Within the Antwerp Index of expurgations, the entry for each author listed the author's works, followed by indications about what needed to be removed from the text. Many entries did not require any correction, and the Index therefore had the effect of promoting that author's works by circulating his bibliography.[21] The Antwerp Index of expurgations is organized by subject. As such, authors appeared in multiple disciplines. Corrections for works by Janus Cornarius can be found under the headings of theology, medicine, and humanistic disciplines. By contrast, the Spanish Index of 1584 and Guanzelli's Index of 1607 were organized alphabetically by the first name of the author. The Spanish Index included expurgations of the medical authors Amatus Lusitanus, Arnald of Villanova, Conrad Gessner, Hadrianus Junius, Girolamo Cardano, Janus Cornarius, Johann Lange, Leonhart Fuchs, Levinus Lemnius, Theodor Zwinger, Theophrastus Paracelsus, and the classical author Theophrastus of Eressos, edited by Julius Caesar Scaliger.[22]

Guanzelli's Roman *Index Expurgatorius* of 1607 included official corrections for forty-two authors and seven texts without authors. Nine of the expurgated authors were physicians who had written medical books that were prohibited: Amatus Lusitanus, Arnald of Villanova, Guglielmo Grataroli, Girolamo Cardano, Janus Cornarius, Leonhart Fuchs, Levinus Lemnius, Francisco Vallés, and an entry under the heading "Theophrastus" that somewhat confusedly included both the text *De causis plantarum* by the classical author and edited by Julius Caesar Scaliger and also the *De chirurgia minore* by the sixteenth-century Theophrastus Paracelsus.[23] The expurgations of medical texts extend over 61 (noncontinuous) pages of the 743-page octavo volume. Each begins with the author's name, but beyond that, each entry reflects the format and degree of explanation preferred by the individual censor who composed the expurgation. Indeed, one of the most remarkable features of the expurgation of early modern medical books is the lack of consistency across Indexes from different Catholic jurisdictions, and even within a single Index. As scholars and censors grappled with the difficulty of imposing a single interpretation to a given text, the problem was further compounded by competing official edicts in various Catholic jurisdictions. Among the expurgatory Indexes of Spain, Antwerp, and Rome, the only medical authors found on all three lists were Janus Cornarius and Leonhart Fuchs, the feuding Lutheran humanist physicians. Given these differences, we must understand the particular Indexes and expurgations as reflecting concerns about dangerous theology and useful knowledge that were local and profoundly tied to the interpretations of individual censors.

MEDIEVAL QUESTIONS REVISITED

Only four of the forty-seven physicians prohibited by the 1559 Pauline Index lived during the medieval period: Arnald of Villanova, Pietro d'Abano, Marsilius of Padua, and Raymond of Sabunde. Of these four medieval medical authors, the 1607 Roman *Index Expurgatorius* provided expurgations only for Arnald of Villanova. This is not to say that medieval authors and the issues their works raised did not come under scrutiny in this period but rather that the Italian Indexes focused above all on contemporary authors and their networks. Prohibitions on medieval physicians focused on heterodoxy during the authors' lifetimes. At the Council of Trent, the Catholic Church confirmed that authors prohibited before 1515 would continue to be banned. When Catholic censors reexamined these authors' works at the end of the sixteenth century with an eye toward expurgation, they were particularly struck by the superstitious and magical nature of many of these works and by the recurrent discussions of astrology that could not be justified by use in medicine, navigation, or agriculture. I will briefly examine how these issues appeared to censors at the end of the sixteenth century and will then examine the process of correcting Arnald's work and the expurgations that the 1607 Index ultimately adopted.

From the earliest days of Christianity, theologians had been forced to address the relationship between Christianity and astrology, which in some forms could be incompatible with Christian belief. Augustine was opposed to astrology because he believed it could lead to a denial of free will and because to him it smacked of idolatry through ancient practices of the worship of planetary deities.[24] Despite this hostility, Augustine also allowed for astral influences on the body: the planets could incline bodies without determining action or depriving people of free will. This compromise—inclination, not determination—allowed for the continued study of astrology in Christendom. An influx of Greek, Arabic, and

Hebrew astrological texts were translated into Latin in the twelfth and thirteenth centuries and became widely used sources for astrology and astronomy in Latin Europe. Universities created an institutional Christian space for the study and practice of astrology in addition to its role in medicine and use in the royal, ducal, and papal courts. With the Renaissance recovery of ancient Greek and Roman texts, many astrological texts began circulating anew. Reevaluating the place of astrology in Christian society was a multiconfessional sixteenth-century project that was as much about integrating new texts into an established philosophical corpus as about the new confessional fault lines drawn in Christian Europe.[25]

While magic and learned astrology could often be distinguished intellectually and socially, these categories were officially collapsed in the Rules on Prohibited Books from the Council of Trent. Rule IX established widespread prohibitions on all magical writings and on astrological texts that sought to determine destiny through the stars:

> All books and writings dealing with geomancy, hydromancy, aeromancy, pyromancy, oneiromancy, chiromancy, necromancy, or with sortilege, mixing of poisons, augury, auspices, sorcery, magic arts, are absolutely repudiated. The bishops shall diligently see to it that books, treatises, catalogues determining destiny by astrology, which in the matter of future events, consequences, or fortuitous occurrences, or of actions that depend on the human will, attempt to affirm something as certain to take place, are not read or possessed. Permitted, on the other hand, are the opinions and natural observations which have been written in the interest of navigation, agriculture, or the medical art.[26]

Canon and inquisitorial law alike reduced astrology and magic to their divinatory principles, emphasizing similarity rather than difference between the disciplines. At the same time, the Rules of Trent dealt with astrology alongside other magical arts, simultaneously dividing astrology to carve out space for "natural astrology" that was distinct and separate from "judicial astrology." Rule IX formally decreed a distinction between astrological influences that inclined and those that determined or necessitated certain outcomes. In so doing, the Council of Trent acknowledged the institutional position of astrology while formally maintaining the Church's position on free will.[27] Even so, in 1571 the archives of the Congregation of the Index mention petitions from physicians to attenuate Rule IX, which was ultimately upheld.[28] On January 5, 1586, Pope Sixtus V issued a bull (*Coeli et terrae creator Deus* [*God Creator of Heaven and Earth*]) widely condemning divination and specifically judiciary astrology. In the next few years before his death in 1590, Sixtus attempted to put more stringent prohibitions on astrological texts, which subsequent popes revoked in the years following his papacy. In the final years of the sixteenth century, jurists and members of the Congregation of the Index debated how best to clarify censorship rules about astrology in light of the possible conflicting interpretations of the Council of Trent's Rule IX and Sixtus's papal bull.[29] Ultimately, the Congregation of the Index settled on a compromise by which Rule IX (and the additional gloss on it in the Clementine Index of 1596) defined the degree of certainty that astrology could attain, while Sixtus's bull laid out the penalties for violating the Rule. Although this compromise settled the theological dispute between Rule IX and the papal bull, the logistics of implementing these rules remained extremely unclear, and the Congregation of the Index fielded numerous questions from bishops and inquisitors about how to proceed.[30]

The astrological compromise also raised, yet again, the question of what to do about medieval astrologers. Since the definition of heresy required that a person have been instructed in the true faith (Catholicism), classical authors, Muslims, and Jews were not technically heretics, though their ideas could still be dangerous. Thus, while medieval Arabic astrologers were not prohibited as heretics, the line between natural and judiciary astrology was often inseparable in their works, providing grounds for prohibition. Ultimately, the Congregation of the Index did not establish a formal, consistent method for solving this problem. Baldini and Spruit have concluded that Catholic censors tended to censure Arabic astrologers primarily when they appeared in editions with commentary by Christian authors.[31] In 1598, the congregation in Naples compiled expurgations of Ptolemy with commentary by Haly Abenragel (Abū l-Ḥasan al-Shaybānī), of Alchabitius (Al-Qabisi) edited by John of Saxony, of the *Flores* by Albumasar (Abu Maʿshar), and of ephemerides and astrological tracts by a number of Renaissance authors.[32] These expurgations primarily focused on attenuating the degree of certainty possible in astrological predictions, and none were adopted in the 1607 *Index Expurgatorius*.

The one medieval author who did receive formal expurgations in the 1607 Index was Arnald of Villanova (1235–1311). Arnald was a physician, apothecary, and alchemist from Valencia whose works and beliefs were scrutinized by inquisitors throughout his life and in the four centuries that followed. During his life Arnald's unorthodox religious beliefs about central tenants of Catholicism (including the Mass, Christ, the Antichrist, and the end of the world) came into conflict with inquisitors in Spain and

France. He was forced to travel throughout Europe because he was repeatedly banished.[33] In 1316 the archbishop of Tarragona condemned thirteen of his treatises, and in 1559 the Pauline Index condemned him as a heretic and prohibited all of his works. In this sense, his prohibition fit squarely within the rule that previously banned authors should continue to be prohibited.

While Arnald was scrutinized by medieval censors due to his views on faith and astrology, early modern censors of his medical works grouped him alongside Fuchs, Lusitanus, and Paracelsus as having written "books which seem rather useful and necessary" and which needed to be corrected rather than completely banned.[34] Alfonso Chacón concurred that same year, including Arnald's *Regimen* on a list of books "that could be read with much utility once expurgated." The *Regimen* was a medical rather than theological text, and Chacón pointed especially to the fact that it was written in the form of a poem, which could be easily remembered, concluding that "it is indeed a book useful for medicine."[35] A few years later, Chacón went a step further, declaring that Arnald's *Regimen* "should not be prohibited because it has nothing bad in it."[36] We see yet again how the Pauline Index linked an author's religious beliefs and nonreligious works in ways that were incompatible with medical scholarship. Roman censorship at once conflated the author and the work and at the same time worked through expurgation to separate the two, identifying the utility in texts through reference to medicine.

In 1593 the Congregation of the Index took the unusual step of requesting an expurgation of Arnald's works from scholars in Salamanca.[37] Including Spanish scholars in the expurgation of Arnald was strategic politically as well as intellectually. The Spanish crown and Spanish scholars were particularly involved in the rehabilitation of Spanish intellectuals. The most complicated and notorious Spanish censorship case dealt with the medieval Spanish scientist and theologian Ramon Lull (1232–1315). Lull was a philosopher and theologian turned hermit and later missionary who worked to disprove Averroism and convert Muslims to Catholicism. He was put to death by stoning in Tunis in 1315. Lull's works were condemned by the medieval Spanish inquisitor Nicholas Eymerich, who promulgated an apocryphal bull (*Conservationi puritatis* [*For the Conservation of Purity*]) dated January 25, 1376, which identified two hundred heresies in his writings, condemned twenty works, and ordered an examination of the rest of his writings. In the following centuries, Eymerich's apocryphal bull continued to have ramifications for the circulation of Lull's works. In 1559 Roman censors banned Lull citing the apocryphal bull, but at the Council of Trent Spanish delegates successfully revoked the prohibition, showing that *Conservationi puritatis* was in fact fraudulent and never sanctioned by the pope. However, a new edition of Eymerich's *Directorium inquisitorum [Inquisitor's Guide]* was published in Rome in 1578, again drawing misinformed attention to Lull's works, which led once again to their condemnation by Roman authorities. Delegates of the Spanish crown negotiated directly with the Congregation of the Index to revoke the prohibitions. These disputes carried on throughout the following centuries.[38] Lull's case illustrates clearly the potential political implications that could follow from the condemnation of a famous author. The heroes of the Middle Ages were political touchstones in early modern debates that were further complicated by the spread of misinformation in printed books.

The censorship of Arnald was less contentious and more easily resolved than that of Lull. The Congregation of the Index had requested line-by-line expurgations of Arnald's works from Spanish theologians and from Italian scholars and theologians. According to Italian censors, one of the main problems in Arnald's medical works involved passages that censors tended to describe as superstitious. In 1596, Pietro Ridolfi di Tossignano, the bishop of Senigallia, wrote to Agostino Valier about expurgations of Arnald, describing the author as someone who had disseminated heresies in his time, but that these heresies appeared only sparsely in his texts. The notable exception for Ridolfi was Arnald's *Expositiones visionum quae fiunt in somniis (Explanations of the Visions which Occur in Dreams)*, which "speaks a thousand impertinent things."[39] Girolamo Pallantieri, a theologian in Padua, also took umbrage with this treatise. While Arnald claimed that a patient's dreams could be useful for physicians, Pallantieri contended, "But doctors consider dreams only for recognizing unwholesome humors, not for telling the events of the future with certainty."[40] Further, Pallantieri complained, Arnald's *Expositiones visionum* "is not therefore for supporting the art of medicine, as it promises on the cover, but for divination."[41] According to Pallantieri, Arnald's approach to dreams had crossed the line from acceptable medical practice into divination.

Girolamo Pallantieri's expurgations, undersigned by the team of censors in Padua, were the most thorough reading of Arnald by Italian censors. Pallantieri objected to Arnald's superstition, paying particular attention to when medical remedies drew on religious formulas or objects.[42] Similarly, Pallantieri was concerned when Arnald wrote about physicians using amulets, "since it contains incantations, and wondrous things, and superstitious amulets."[43] Other passages in Arnald, such as a long excerpt on the properties of stones, verged explicitly into the realm of magic.[44]

Pallantieri denounced this tendency unequivocally, writing that Arnald's works "manifestly inclined toward divination, augury, sorcery, incantations, oaths, magical amulets, uncovering the way of superstition, throughout evincing the inevitability of fate and the domain of judicial astrology."[45] These parts of Arnald's texts contradicted Catholic faith and were dangerous to the public.

Despite his many criticisms of Arnald's works, Pallantieri believed that Arnald's texts should not be destroyed in their entirety. Instead, the threatening content could, and should, be expurgated, and the texts could then be allowed to be seen by readers. Pallantieri especially believed that his expurgations should remove heterodox content while preserving the medical utility of the text for physicians. The parts of the texts that Pallantieri sought to remove were things that "do not assist the art of the doctor (as Arnald [of Villanova] claims) but dishonor it and stain it."[46] This approach to expurgation was fundamental to Pallantieri's comments on Arnald's approach to astrology where he wrote, "In the same place there follows a treatise on the judgements of astronomy which is utterly irrelevant to the art of medicine."[47] Pallantieri justified his approach at the end of his text before his signature: "This censure was carried out according to Rule IX of the new Index."[48] Girolamo Pallantieri's project of expurgation focused on medical content that verged on superstition and attacked astrology that did not contribute to useful medical practice, while justifying his approach with Rule IX of the Clementine Index.

In chapter 3 we watched as the process of expurgating books became part of how the humanist doctor from Ravenna, Girolamo Rossi, approached his own texts with his pen in hand. Here we have the opportunity to carefully read Rossi's expurgations of Arnald of Villanova. In early 1602, Girolamo Rossi submitted an expurgation of Arnald's book on regimen, *De conservanda bona valetudine* (*On Conserving Good Health*). Rossi provided full bibliographical details for the work at the heading of his entry; he was using the Johannes Curio edition printed in duodecimo in Venice by Giovanni Maria Leni in 1573. Rossi proclaimed Arnald's popular text to be one "in which many things are seen to be worthy of correction."[49] Rossi's expurgations mark a number of passages that named or quoted from prohibited authors including Ludwig Helmbold, Joachim Camerarius, Leonhart Fuchs, Janus Cornarius, and William Turner. Interestingly, Rossi held forth extensively against Desiderius Erasmus, who was quoted on folio 56 verso, describing him as "an author of the first class whose jokes about sacred things cleared a path for Luther to disseminate his heresies. As they say in German, Erasmus hinted, Luther rushed in:

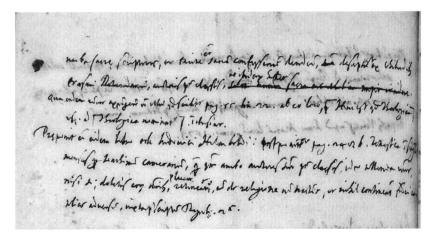

Fig. 4.2. Excerpt from draft of Girolamo Rossi's expurgations of Arnald of
Villanova's *Schola Salernitana*. The passage identifies Erasmus of Rotterdam as
a prohibited author and mentions Martin Luther, but without supplying the full
quotation that Rossi included in the final copy of the expurgations that he sent
to ecclesiastical officials in Rome. BCRa, Mob. 3.1 B, n. 3, f. 319v. Reproduced
with permission from the Instituzione Biblioteca Classense, Ravenna.

Erasmus laid the eggs, Luther hatched the chicks."[50] This last quotation
is not included in full in the draft copy in Rossi's personal papers (see figure 4.2). Instead the draft includes a short note indicating a connection
to Luther.[51] Rossi then filled in the rest of the text in the expurgations he
sent to Rome. The line about Erasmus laying the egg that Luther hatched
was a "popular quip," and although Rossi's transcription of these sayings
differs from the text in Antonio Possevino's *Bibliotheca selecta*, it is still
his most likely source.[52] If Rossi were reading Possevino as part of grounding his own work as a censor, he would have noticed Possevino's text slid-
ing into a critique of Erasmus and Luther's views on free will. While we
might have expected Rossi to read Arnald with an eye toward the possible
friction between astrology and free will, his list of expurgations suggests
that he was thinking about the issue of free will explicitly as a dangerous
confessional problem rather than a scientific one.

Rossi also took an interesting approach to his analysis of the authorship of Arnald's book in general. Rossi commanded:

I would also delete the name "Villanova" which is placed at the beginning of the dedicatory epistle; not only to make allowance for the

conscience of the pious, who have read no mention made of Villanova in the book's title, but also because in fact, although many things in this commentary have been excerpted from Villanova's commentaries on this school, the style is nonetheless so changed, so many things are heard, and for the most part differently explained, that it can be characterized otherwise than as a work of Villanova. For this reason, I would also remove the word "ancient" in the title of the book, because it is not the work of the ancient Arnald [of Villanova], but of some younger author, perhaps [Celio Secundo] Curione. The recent authors whom he cites, such as Erasmus, Fuchs, [Agostino] Mainardo, and others, are a clear proof of this.[53]

As a humanist reader, Rossi was attentive to issues of style and language and to traditions of commentary. His role as a censor trained him not only to read with the aim of noting confessional difference but also to consider carefully the meanings of authorship. The editor, Johannes Curio (d. 1561) was a professor at Erfurt, and his editing of Arnald, in Rossi's eyes, was so poorly accomplished that Arnald's name should no longer be attached to the work at all. In this case, Rossi was not advocating for the removal of Arnald's name because he was a heretic but rather the removal of his name because the text itself had been so corrupted from the original work that Arnald had written. The humanist culture of textual analysis and correction was central to Rossi's work as a censor.[54]

Pallantieri and Rossi had carefully and revealingly provided line-by-line expurgations of Arnald's work, but ultimately, and perhaps for political reasons, Guanzelli's expurgations in the Roman *Index Expurgatorius* reproduced exactly the entry from the Spanish Index of expurgations of 1584. The entry merely listed seven of Arnald's treatises that were prohibited. It therefore constituted an expurgation of Arnald's corpus rather than expurgations of his individual texts. Censoring medieval authors was part of a larger humanist process of reviewing and revising authors of the past with present concerns in mind. Girolamo Rossi's reading of Arnald of Villanova reveals the ways in which the Counter-Reformation humanist reader, with pen in hand, was constantly aware of both style and the omnipresent threat of heterodoxy. The fortunes of medieval authors like Lull and Arnald were also sites of political negotiation between scholars and rulers in Spain and Rome.[55] In the Counter-Reformation, the Middle Ages were read and evaluated through the lens of recent confessional conflict, to which we now turn.

THE LUTHERAN PHYSICIANS

As we saw in chapter 1, the Lutheran physicians Janus Cornarius and Leonhart Fuchs spent much of their careers disparaging each other in print. Of
the many German Protestant physicians banned on the Roman Indexes,
they were the only two for whom official expurgations appeared in the
1607 *Index Expurgatorius*. In the case of Cornarius, the expurgations are
copied exactly from the Spanish Index of 1584, and for Fuchs, the expurgations drew heavily, though not entirely, on the expurgatory Indexes of
Spain and Antwerp.[56] Efforts to correct Cornarius had begun in 1562 when
Camillo Paleotti proposed that the Tridentine Index allow Cornarius's editions and commentary on Galen and Dioscorides.[57] The Antwerp Index of
expurgations of 1571 took Paleotti's proposition a step further. The beginning of his entry in the Index stated simply, "In the commentaries of Janus
Cornarius there is little that offends."[58] The Antwerp Index then listed
each of his commentaries and noted in italics that they did not offend
or contradict the Catholic faith. The only corrections the Index required
were the removal of a preface referencing conversion and the expurgation
of a claim that fasting, especially in March and April, was bad for the
body. The censor clearly saw the latter as an attack on Catholic practices
surrounding Lent.[59] The censor for the Antwerp Index was familiar with
Cornarius's corpus, pointing out that Cornarius translated many other
authors, including Plato and Epiphanius of Salamis. He noted a number
of polemics against Leonhart Fuchs, describing, in the text of the Index,
that in one Cornarius excoriates Fuchs, calling him "Vulpecula" (little
vixen). "Whether in fact," the censor continued, "some of it dealt with
religion, I do not remember; for they [those works] have not yet been examined."[60] The Spanish Expurgatory Index of 1584 corrected two passages in
Cornarius's commentary on Galen's *De compositione pharmacorum* (*On
the Composition of Medicines*). The censors determined that the first passage unnecessarily drew attention to pubic hair, while the second passage
addressed the preparation of a medicine made by mixing wine with the
plant *Cynoglossum* (colloquially known as hound's-tongue), with instructions that it should be combined with the left hand before the sun rises.
The censors noted that since Cornarius had not rejected Galen's apparent
superstition, it was necessary for the reader to acknowledge the error by
adding in the margin that "this is vain and superstitious."[61]

The Neapolitan congregation of censors took a different approach to
Cornarius's books on medicine and on Hippocrates. Their short list of

expurgations required only that that a dedicatory epistle to Cornarius be removed and that Cornarius's name and that of his printer Oporinus be struck from the titles and title pages of works.[62] The Neapolitan expurgations of Cornarius represented an attempt distinguish the substance of his work from his confessional identity, thereby separating the author from the work. We should dwell for a moment on the paradox inherent in a system that nominally evaluated works based on the beliefs of their authors and then worked hard to remove traces of authorship to cleanse good texts of their references to those very authors. As we will see in chapter 6, however, readers were not meant to forget the author as they removed his name. Attention to the material processes of expurgation added a ritualized dimension to the process of Counter-Reformation reading. At the level of establishing a set of legally binding expurgations, the Neapolitan expurgations essentially added nothing to requirements already present in the rules of the Clementine Index, which since 1596 had required the removal of references to and praise of heretics.[63]

These three distinct approaches to Cornarius's works focused on identifying Cornarius's scholarship as medical and humanist, removing stray sexual or superstitious content, and finally expunging Cornarius himself from the prefatory material of his medical books. The Roman censors involved in compiling the 1607 Index drew on the Spanish and Antwerp Indexes but not the Neapolitan expurgations. The 1607 entry begins by exactly copying the expurgations in the Spanish edition. It then adds to Cornarius's bibliography his translations and commentaries on Plato, Constantine, and Synesius of Cyrene, noting at the end of each entry that they do not deal with religion or contain anything offensive to the Catholic religion. Each entry uses slightly different language to denote that the work is acceptable.[64] The lack of formula suggests that these entries were composed by different censors.

The rest of the entry for Cornarius exactly reproduces the Antwerp Index, but with two telling differences. The first change is to Cornarius's prefaces to Marcellus and Galen from 1536. The Antwerp Index determined that "it contains nothing [against religion], therefore it is allowed to all." The Roman *Index Expurgatorius*, which supplemented a robust system of licensing readers on the Italian peninsula, removed the clause "to all." In the Italian context, keeping and reading prohibited books technically still required a reading license even after the text was corrected, so no work by a prohibited author would be permitted "to all." The second change to the Antwerp entry removed the censor's first-person reflection on texts he had not yet examined. The Roman *Index Expurgatorius* pulled the work of the censor further from readers' view, masking the interpretive nature of expurgation within a normative legal context.

Just as Fuchs and Cornarius's works had been closely linked through polemic and controversy during their lives, the expurgations of their works in the Roman Index followed similar trajectories. Leonhart Fuchs's works were banned in 1559, though the ban was attenuated the next month to allow people to read them with permission after expurgating Fuchs's name. Fuchs's works were extremely popular and circulated widely on the Italian peninsula and throughout Catholic Europe. His *De historia stirpium* (*On the History of Plants*) had been translated into Spanish in 1557, though following the 1559 Pauline prohibition the printer Arnold Birckmann replaced the title page and letter to readers with new text that identified the work instead as *Historia de yervas, y plantas, sacada de Dioscoride Anazarbeo* (*The History of Plants according to Dioscorides*).[65] The Spanish Inquisition also required a few changes in content before the book could be printed in Spanish, removing Fuchs's reference to Dioscorides's description of the abortifacient properties of the cyclamen (*Cyclamen hederifolium*). This early expurgated translation, cleansed of pernicious content, dedicatory epistles to Protestant princes, and references to its own heretical author, was the kind of corrected reprint that Catholic officials in Italy initially imagined but which never came to fruition.

In Italy, editions of Fuchs circulated in Latin accompanied by licenses inscribed by local inquisitors and expurgations applied on an individual basis. It was clear within a month of Fuchs's total prohibition that his works merited an official set of expurgations. Alfonso Chacón tried to attenuate the total prohibition on Leonhart Fuchs's works in 1590–91, arguing that the corrections published in Spain in 1584 were largely sufficient. In the words of Giovanni Battista Porcelli, the inquisitor of Asti, in 1597, "Leonhart Fuchs was long ago prohibited in the first class, but despite this he is permitted almost everywhere, and many copies have been presented [to me] undersigned by inquisitors and it seems as if there is an expurgation that I do not have."[66] In Padua, the inquisitor Felice Pranzini's expurgations noted that the 1565 Vincenzo Valgrisi edition of Fuchs's *Institutiones medicinae* (*Principles of Medicine*) printed in Venice had already been expurgated, presumably in Venice though there are no surviving copies of these expurgations.[67]

Ultimately, the expurgations of Fuchs promulgated in 1607 were, yet again, primarily verbatim excerpts or slight rewordings from the expurgatory Indexes of Spain and Antwerp with few additions. An anonymous and undated draft of these final expurgations in the Archive of the Congregation for the Doctrine of the Faith (ACDF) in Vatican City shows the compiler at work copying from the Antwerp and Spanish Indexes, excerpting and recombining lists of corrections. The major changes added in Rome included removing language stating that works could be permitted and replacing it instead with formulations about the works not offending religion. Roman censors also changed the ways they referred to a few prohibited portions of texts. Phillip Melanchthon's last name does not appear in the 1607 *Index Expurgatorius*, which refers to him instead as "Phillipus etc." Other changes by the anonymous Roman censor, or perhaps his scribe, are more amateurish. Where the Spanish Index included Greek numbers, the Roman censor did not copy these out, leaving ellipses in their place. These ellipses are reproduced faithfully in the 1607 Index.[68]

The final 1607 expurgations focused primarily on erasing Fuchs's connections to the Protestant Reformation from his medical texts and secondarily on removing Fuchs's attacks on Catholicism and mockery of ecclesiastics. The work that warranted the most attention from censors was Fuchs's Latin translation and commentary on the Byzantine physician Myrepsus's *Medicamentorum opus* (*Work of Medicines*). The preface to the Senate of Nuremburg discussed Fuchs's approach to making medicines and then went on to excuse Myrepsus's superstitious inclinations that emerge in the work. Fuchs attributed the source of Myrepsus's superstition to the period in which the author lived, which Fuchs described as a time of false religion. This barb was clearly aimed at Catholics, and it hit its mark. While the Roman *Index Expurgatorius* followed Spain's wording exactly in removing only the sentence about false religion, Italian censors had repeatedly recommended removing even more of the preface, and the Neapolitan congregation suggested removing it in its entirety.[69] The fact that two anonymous expurgations from Perugia fail to mention the dedicatory epistle suggests that these censors were working from a copy of the work from which the preface had already been removed.

The expurgations that Italian censors submitted for Fuchs's works were not ultimately integrated into the official expurgations of 1607. Italian censors paid particular attention to insults aimed at Catholics, but they also noted repeatedly the importance and utility of Fuchs's work.[70] Girolamo Pallantieri and the team of censors in Padua submitted expurgations explaining that in Fuchs's *Methodus seu ratio compendiaria perveniendi* (*Method for Preparing Medicines*, 1548) the dedicatory letter could be removed because it did not deal with medicine and added nothing useful.[71] They repeated a version of this verdict in their expurgation

of the *Institutiones medicinae* (*Principles of Medicine*) from 1594, pointing to Fuchs's praise of the Lutheran jurist Ludwig Gremp von Freuden-
stein and suggesting that it be removed since "it contains nothing dealing
with the art of medicine."[72] This kind of reasoning and explanation was
foregrounded in the printed expurgations of 1607. The anonymous censor
who compiled the official expurgations took pains to remove these explanations from within the entries to make them headings. Where the Antwerp Index explained that "the invention of Medicine and plants should
be attributed to God, [and] it is not considered to be exposed to reproof,"
the Roman expurgations edited this slightly to begin the entry with an
explanation that Fuchs's *On the History of Plants* "has nothing pertaining to religion."[73] The Roman Index focused on distancing the material in
this medical and botanical text from its relationship to theological issues,
rather than stating that there was an appropriate relationship between the
religious and the botanical and medical content.

The cases of Cornarius and Fuchs raise several important conclusions
that carry over to other official expurgations promulgated in 1607. Books
written by Lutheran humanist physicians brought with them a web of references to and citations from other non-Catholics. One of the most important aspects of censoring their books was removing these references to
Reformation theologians, rulers, and colleagues. In a humanist culture of
citation with a distinctly early modern appetite for information, expurgating medical texts meant changing the ways that texts related to a world of
living and recently dead scholars and patrons.[74] This kind of expurgation
was not related to regulating the medical content of the work but to cir-
cumscribing the social context that created and validated medical knowledge.[75] The second major area of expurgation in Cornarius and Fuchs was
the magical or superstitious content, especially that which was gathered
from classical and medieval authors. Once again, we see that the medical
humanist enterprise of editing, compiling, and evaluating texts was made
explicitly religious and confessional in the hands of censors.

The expurgations of Cornarius and Fuchs push us to confront the notable absence of Conrad Gessner from the list of authors with official corrections in the 1607 *Index Expurgatorius*. In 1592 Gonzalez Ponce de Leon,
a consultor to the Congregation of the Index, expressed his opinion that
although Gessner was prohibited in Italy, in his view the Spaniards had
taken the better approach by expurgating the work, for Gessner was "a
most diligent author and wrote many useful things that could very easily
be expurgated." For example, he continued, "*On Animals* contains nothing offensive except citations of heretical authors."[76] Gessner was, even

more than Fuchs, a prolific author. One anonymous expurgation in the archives of the Congregation of the Index painstakingly identified a short list of superstitious and confessional problems in the first two books of Gessner's work on birds and then abandoned the task, confessing that "I have only seen 217 pages of Gessner."[77] The whole work numbered 779 pages, and this partial expurgation clearly represented the degree of fatigue that censors faced as they grappled with the laborious work of selectively censoring texts.

Although Theophrastus Paracelsus was not a Protestant, thematically, this is the place to briefly address the Catholic efforts to expurgate the so-called Luther of Medicine. Paracelsus's medical heterodoxy lay in his rejection of Galenic medicine, though Catholic censors were most concerned with his attacks on the lifestyles of ecclesiastics, Protestant themes in his works, and above all his approach to magic and demonology.[78] The 1607 Index included a set of corrections to Julius Caesar Scaliger's edition of the classical author Theophrastus of Eressos, after whom Paracelsus had styled himself.[79] This entry is followed by a brief correction of his *De chirurgia minore*. These corrections were copied exactly from the 1584 Spanish Index and did not include the influence of Italian censors.[80] Based on the very few corrections submitted to the Congregation of the Index, we might wrongly assume that Paracelsus was not widely read in Italy. However, information from reading licenses and the case of Girolamo Rossi suggest otherwise.[81] Rossi cited Paracelsus and discussed his works in his book *De destillatione (On Distillation)*, though he never submitted expurgations for Paracelsus.[82] Further, Rossi's index to his *On Distillation* does not describe Paracelsus as "lapsed" as it does for heretical authors since Paracelsus was neither a Protestant nor officially banned when Rossi wrote the book.

Rossi not only read Paracelsus, he also engaged with disputes about Paracelsus's philosophy through his reading and expurgation of the works of Thomas Erastus, another banned physician, though not one for whom corrections were published in 1607. Rossi's close reading and expurgation of Erastus's attacks on Paracelsus focused on details that reflected Erastus's Protestant theology. In a discussion of magic, Erastus argued that words are "articulated sounds" (*soni articulati*) and that they act only through their direct meaning. This interpretation then disqualified any number of magical uses of language for healing powers. Its corollary also had deep religious implications for Rossi since "the words of the sacrament, do indeed turn words into elements."[83] Rossi suggested that the passage be prefaced with the caveat "speaking naturally." Words used as incantations to alter natural things (with respect to medicine in this case) were considered superstitious, but the words of the sacrament could still theologically call the sacrament into being, turning the bread and wine into Christ's body and blood. Rossi's addition of "speaking naturally" attenuated the problematic nature of this passage, but the censor was clearly aware that Erastus's critique of magic was also a direct attack on the Catholic Church. In an era in which Protestants regarded Catholic ritual as superstition, the idea of superstition came to transcend its magical sense and to be leveraged as a polemical, confessional attack. The controversy that surrounded Paracelsus's medicine and philosophy was tied up in disputes about religion.

The Congregation of the Index turned its attention to Paracelsus more seriously following the 1603 edition of his works. In 1616, the German physician and member of the Academy of the Lincei Johannes Faber issued a set of corrections to Paracelsus's texts.[84] Faber's expurgations, though later than the rest in this chapter, echoed the established discourses of medical utility in the face of ecclesiastical censorship. Faber wrote, "Numerous other passages are in no way acceptable in Paracelsus, but many are sound, especially where he discusses cures of diseases and preparations of cures, which are especially useful for the medical faculty."[85] The pious physician could read and interpret the Luther of Medicine in ways that were useful, while actively rejecting those which were unacceptable.

The Reformation had created a network of physicians whose works were prohibited to Catholic readers. However, Catholic censors largely agreed that the utility of these works justified the work of correcting them, emending them to remove their confessional contexts, and repressing signs of superstition or accusations of Catholic superstition. Censors' attentiveness to the connections between and across these authors, which they mentioned in their expurgations, reveals that Counter-Reformation readers were acutely aware of the personal acrimony within the medical republic of letters, in addition to the religious debates in which they took part.

HOMETOWN HEROES AND HERETICS

Close analysis of the expurgations of medieval and Lutheran prohibited physicians on the Indexes of Antwerp, Spain, Naples, and Rome revealed that differing local concerns led to different solutions for correcting these texts. Nowhere does this become more apparent than in the corrections to the works of three physicians—Guglielmo Grataroli, Amatus Lusitanus, and Girolamo Cardano—who lived and practiced in Italy and who were simultaneously renowned for their important medical works and treated with caution due to their unorthodox, and even heretical, religious beliefs. None of these authors were expurgated in the Antwerp Index of 1571, and the expurgations in the Spanish Index of 1584 took up less than a page for each author, despite their prolific works. These hometown heroes and heretics simultaneously presented problems and opportunities for Catholic censors, who worked to keep the texts of these well-known physicians circulating while removing impious material from their medical texts.

Whereas Girolamo Rossi's expurgations of Arnald of Villanova were not included in the 1607 *Index Expurgatorius*, his expurgations of Guglielmo Grataroli were published almost verbatim. This is one of the rare examples where expurgations solicited from congregations of censors outside Rome were adopted as the Church's official corrections. Guglielmo Grataroli was an Italian physician who studied in Padua and Venice and worked in Bergamo.[86] He was arrested and tried for heresy and abjured before the Inquisition in Milan in 1544. In 1550 he was investigated again, this time by the Inquisition in Venice, which was concerned about reports that he kept prohibited books and that he held problematic views about indulgences, the Eucharist, the pope, free will, purgatory, and the effectiveness of good works and saints. In short, Grataroli was a Calvinist. He anticipated the Inquisition's next moves and fled to Basel, where he integrated into the community, rising eventually to the position of dean of the College of Physicians and saving the life of Girolamo Cardano by warning him against lodging in a plague infested hostelry, as Cardano remembered in his autobiography.[87] From exile in Switzerland, Grataroli wrote and edited a number of books, which were not formally prohibited on the Index until in 1590. Even when they were prohibited, it was with the provision "until emendations are brought forth."[88]

In winter 1598, Rossi sat down with his 1558 copy of Grataroli's *Opuscula*, a duodecimo volume containing a collection of his treatises printed in Lyon. Rossi dutifully noted passages that mentioned Erasmus so they could be removed from the text. Rossi also noted that when Grataroli quoted from the Bible, he was not using the Vulgate edition.[89] In fact, though Rossi did not know it, Grataroli was using an edition translated by Erasmus. In the section of Grataroli's work on physiognomy, Rossi objected to the author's description of people with long, malleable heads as being particularly circumspect and farsighted, since this was the head shape and set of character traits that Grataroli identified with the Turks. Rossi rejected Grataroli's implicit Turkish compliment and listed the pas-

sage for expurgation, explaining merely that "they were not." The praise
of the Turks may have especially upset Rossi, since one of his sons was
at that time in the eastern Hapsburg Empire fighting the Ottomans.[90] In
another part of the text on physiognomy, Grataroli discussed people with
cone-shaped or pyramidal heads. He wrote that the Genoese "have this
form and nature to the greatest degree, and many hooded ones—whom
they call religious, are particularly cowards and hypocrites."[91] Rossi rea-
sonably concluded that this passage was "damaging the reputation of the
clergy" and that the passage should therefore be deleted.[92]

While most of Rossi's expurgations related only tangentially to the
medical content of Grataroli's work, his concerns became more substan-
tive as he moved on to Grataroli's short text on forecasting the weather.
Rossi eliminated mention of the swan (here the constellation Cygnus) as
auguring happiness, since "indeed an auspice would not be considered by
a Christian man."[93] A few pages later, Rossi indicated that "demonic sickness" should be erased from the list of illnesses that occur when air is too
dry. Rossi reasoned that "demonic sickness" is not caused by the qual-
ity of the air, "as I explained in the expurgation of Cardano."[94] But what
role exactly did demons play in health and sickness? A good Catholic like
Rossi could not deny demonic illness because exorcism remained part of
Catholic priestly practice. However, Rossi believed that priestly practice
was exactly where ideas about demonic illness needed to stay—it was not
something that physicians could influence. Rossi's reasoning is more fully
explained in his expurgation of Girolamo Cardano's discussion of melancholy and demoniacs in his work on genitures.[95] Cardano, discussing
the geniture of a possessed man named Battista da Bergamo, attributed
his suffering to the alignment of the stars. Rossi argued that this was impossible "since demons, because they are naturally incorporeal, are not
subject to the stars, but since they abuse the humors existing in bodies,
they often bring this [suffering] about by means of melancholy. Therefore,
we watch exorcists remove them both through vomiting and separation."[96]
Cardano then went on to cite Avicenna, who suggested that demons could
be cured through natural remedies. Rossi continued to protest, "A real demon we say easily possesses those with melancholy, and abuses the melancholy, because that humor above all others is capable of causing insanity and desperation in people, since it causes the greatest evil, bad demons
choose it; finally, those [demons] who know the disposition of bodies and
humors enter those [people] made disposed to illnesses of this kind, and by
God's permission they impel them and induce them to [act]."[97] Rossi did
not deny the existence of demonic illness; in fact, he noted that demons

seek out men who are humorally predisposed to sickness and then lead them further into insanity and desperation. However, because the demons were incorporeal, the planet Mercury being in the ascendant could not influence them, though Rossi seemed willing to concede that Mercury's position could influence the humors, creating a set of circumstances that further opened the door for demons and demonic possession. In the end, only God, not the stars, could allow a demon to enter a person, and only through God (and therefore through priests) could the demon be removed.

Rossi's expurgations represent his systematic understanding of the boundaries of Catholic medicine at the turn of the seventeenth century because he connects ideas and expurgations across the works of various authors. While Rossi criticized Grataroli for mocking the clergy and praising the Turks and Erasmus, he we went further by connecting his ideas about medicine and healing beyond the bounds of the individual works by Grataroli or Cardano. Rossi's reading and expurgations were aimed at creating a consistent set of principles that guided Catholic medicine and which could be applied to a range of texts.

While Rossi shows consistency across texts, the Catholic system of dispersed congregations correcting texts undermined this approach. Neapolitan censors took a contrastingly superficial approach to expurgating Grataroli. Their expurgations focused on removing references to Erasmus, removing language suggesting certainty in relation to predictions, and deleting passages that mocked the clergy. The 1607 *Index Expurgatorius* adopted about half of Rossi's suggested expurgations. On the whole, its compilers were less concerned than Rossi was about superstitious content in Grataroli's work on physiognomy, and they did not share Rossi's concern about Grataroli's discussion of demons. The expurgations in the 1607 Index that were not composed by Rossi were submitted anonymously and all come from the same report. Of these five additional and anonymous corrections, only the passage about Genoese heads appears on the list, though in a shortened form that does not reproduce the offending text as Rossi had.[98] The other four passages required simple deletions of words and did not relate to medical content. Rossi and the anonymous censor's reports are integrated following the progression of the content in Grataroli's books, but they were clearly combined in haste and perhaps without even referencing Grataroli's texts. The end of the entry reproduces the anonymous censor's suggestion, "On line 1 of page 24, delete: 'and with pure knowledge, against the will of Satan and his members.'"[99] The next line of the Index adds Rossi's expurgation for page 244, instructing readers to correct "the beginning of the same page, where it reads 'Christ our servant,' it should read 'Christ our Savior' so that the profane new forms of utterance be avoided."[100] In fact, Rossi and the anonymous censor were referring to the same passage in Grataroli.[101] This sloppy oversight occurred in part because Rossi thought it would be remiss to repeat the "profane new forms of utterance" even in a list of expurgations bound for the Congregation of the Index.

Whereas Rossi's expurgations of Grataroli were integrated into the 1607 Index, his expurgations of Amatus Lusitanus were not. Lusitanus, a Portuguese crypto-Jew exiled to Italy and then Thessaloniki, wrote one of the most popular medical texts of the sixteenth century. His *Centuriae* were a collection of descriptions of cases he had attended in his medical practice. The work swelled with ever more examples in successive editions and was widely read. Several of the cases were accounts of priests, friars, and nuns, with medical conditions that indicated behavior contrary to the religious principles they supposedly upheld, such as excessive food consumption and sexual activity. Protestant polemicists regularly leveled criticism of this kind at Catholic clergy, but it was not immediately clear for censors that these medical cases qualified as prohibited "obscene narratives," a prohibited category meant to apply to literary and not medical narrative. Indeed, as Baldini and Spruit have pointed out, it was legally dubious to take this approach to medical texts since "obscenity regarded literary works, not medical practice," though in effect this line was regularly blurred.[102]

While the law was unclear, Rossi and other censors were in agreement that Lusitanus's medical cases cast the clergy in a negative light. For example, Lusitanus steadfastly denied that a nun could become pregnant from taking a bath in water that contained semen. Rossi's expurgations suggested that "a certain nun" be replaced by the less precise phrase "a certain woman," thereby removing associations with the clergy that cast aspirations on their celibacy.[103] In general, censors were careful to remove discussion of the maladies of monks, nuns, and monasteries from medical treatises because these cases were often related to sexual conduct. Like Rossi, Girolamo Pallantieri identified sexual content in his expurgations of Arnald, and he removed "reference to priests, monks, and the convent."[104] Removing references to the clergy effectively decontextualized cases in a way that appeased censors, but it also erased the descriptive content that made case histories a unique and important genre of medical literature.[105] On occasion, to the accounts of patients in medical texts crossed the line from being damaging to the clergy to being purely titillating. Pallantieri concluded that Arnald's description of masturbating nuns needed to be removed in its entirety, not merely removed from the context of the convent.[106] While jokes about the sexuality of the supposedly celibate religious may have landed well in the Middle Ages when Arnald was writing, the heightened sensitivities to attacks on the clergy in the long wake of the Reformation gave these jokes an added and dangerous barb. These passages were, without fail, removed or neutered by changing their references from religious to laypeople. Counter-Reformation censors understood that their own contexts required manifesting a certain sensitivity in print toward any issue that touched upon the clergy.

Girolamo Cardano, by contrast, was both acutely aware of the turbulence of his times and largely unwilling to fundamentally circumscribe his philosophies. Combined with his prolific book production, these traits turned the censorship of his works into one of the most complex cases of the sixteenth century. The ecclesiastical efforts to reform both author and texts produced an astonishing number of archival documents.[107] Cardano was a prolific author with eclectic interests, and his works were widespread and widely debated. He was alive during the early expurgations of his work and was invited to partake in correcting his own errors. Cardano's response to these invitations to self-censorship were cursory and often resulted in him adding material to his books rather than removing offending passages. When the Roman Inquisition banned all of Cardano's nonmedical works in 1572, parts of his corpus had already been prohibited in Paris, Spain, and Portugal. The Roman prohibition was repeated in 1590, 1593, and 1596.[108] However, the line between the medical and the nonmedical in Cardano's works was difficult to define, even in its own time, and texts which were not necessarily medical were nevertheless useful for medicine. In 1572, the consultor Alfonso Chacón argued against the full prohibition of *De subtilitate* and *De varietate*. At the end of his thirty-two-page expurgation, he concluded, "These books of Girolamo Cardano, if expurgated, will be most useful for the work of all philosophers, physicians, mathematicians, astronomers, architects, farmers, sailors, for the care of family matters, and finally for all artisans."[109] Despite Cardano's major deviances from Catholic orthodoxy, Chacón was adamant that there was much worth saving in the works of the Milanese doctor.

As we have seen, Girolamo Rossi admired Girolamo Cardano. It is likely that as he wrote his own work *On Distillation* he had the 1554 octavo Lyon edition of Cardano's *De subtilitate* by his side, the very same volume he would later expurgate for the Congregation of the Index.[110] Rossi knew that Cardano was a problematic character, and he had censored his praise of the heterodox physician in his own writings, appended the adjective "lapsus" (lapsed) next to his name in the printed index of his book, and composed formal expurgations of Cardano's work, which he sent to the Congregation of the Index in Rome. Rossi had thought through Cardano's unorthodoxy as part of his context of reading and writing, and he linked it to a broader set of questions about Catholic medicine in his expurgations of Grataroli. Rossi expounded on the connections between astrology, demonology, and illness in his expurgations of Cardano's astrological works. His expurgations repeatedly mitigated the degree of certainty and necessity that could be derived from astrological predictions. Rossi was also careful to delineate that God's will could never be circumscribed by the stars. However, his expurgations show him to have been deeply preoccupied by Cardano's discussions of demons. In these sections, Rossi took the time not only to mark the sections that needed to be deleted but especially to explain and repudiate Cardano's thinking. In his expurgation of Cardano's commentary on Ptolemy's *Quadripartitum*, Rossi explained why he removed a section of Cardano's text related to demonic possession and illness:

> I wanted to say this, to show how pertinent it is to this point, how astrologers and physicians are able to speak about people possessed by demons, but with great seriousness, lest it be extended to all possessed people, and someone think that what comes from a demon derives from a humor; therefore it is better to remove such material, especially since the fact that someone is possessed by a demon, though they are melancholic, is not because of the stars but from God and by his permission.[111]

Rossi consistently identified the problem in Cardano's approach to demons as deriving from the way a doctor may or may not speak about an issue on the boundary of his professional expertise (illness caused by demons). Demons, in Rossi's view, were very much part of the medical realm, but they must be dealt with carefully and seriously, and ultimately demons acted by God's will, not according to astrology. In his expurgation of *De subtilitate*, Rossi condemned passages about sorcery and demons and responded with citations of approved Catholic sources including Nicholas Eymerich's *Malleum maleficarum* (*Hammer of Witches*) and works by Paolo Grillandi and Bartolomeo Fumo.[112] Rossi's expurgations, especially those dealing with demons, were more than lists identifying passages to be removed. They were often learned, Catholic refutations of heterodox positions that lay at the intersections of medicine and religion.

When expurgations of Cardano appeared in the *Index Expurgatorius* of 1607, the compilers of the Index acknowledged the complex process of expurgating an author as complicated as Cardano, addressing the reader of the Index directly:[113]

> Greetings, reader. In order to make it easier for you, dear reader, we have invented a way for you to more easily find that which must be expurgated in the works of Cardano, since indeed his writing extends to many books without any chapter divisions. Therefore to individual citations of passages we have added particular words to seek from the index of the same book that should be corrected. On every page where they are found, these words will fully indicate, as though pointing with a finger, the passages which must be emended by order of this censorship. We also took care that these correspond to all editions (as much as it could be done). Farewell, and be pleased with this diligent concern for the good on our part.[114]

Cardano represented a particular problem in that he and his works were prolific, widespread, and deeply integrated into scholarship in Italy. To correct them, censors needed to help readers navigate across multiple editions of the text, which contained different page numbers and even different content. This complicated task was worth so much time and energy because of the widely expressed view of the utility of Cardano's corpus.[115] Cardano was difficult for censors to pin down, but his omnipresence on the bookshelves of Italian readers necessitated the difficult task. While his correction was a task largely avoided by censors in Antwerp and Spain, it was one that censors like Rossi tackled head-on, searching for a consistent and reasonable approach to this popular author.

In addition to popular Italian authors, Guanzelli's 1607 Index includes expurgations for two important Catholic authors from Spain and the Netherlands who had related intellectual projects that came under the scrutiny of censors. Both Francisco Vallés and Levinus Lemnius were devout Catholics and vocal members of the medical community.[116] Vallés studied at the University of Alcalá, was the physician to Philip II, and published a number of books, the most important and influential of which was the *De sacra philosophia* (*On Sacred Philosophy*). Vallés's work systematically moves through the entire Bible, explaining natural philosophical phenomena described in it and arguing for the particular importance of the Bible as a source of knowledge about the natural world.[117] Vallés's project was intended as a pious, Catholic project, but in December 1597, the inquisitor of Bologna wrote to the Congregation of the Index to denounce superstitious passages in the text.[118] Two years later, the Congregation in Rome commissioned an expurgation of this work from the Spanish Dominican Luis Ystella, who had taught biblical exegesis in Valencia before being dispatched to Rome as a political and religious diplomat and who would go on to become Master of the Sacred Palace after Guanzelli. Ystella offered praise of Vallés's approach to astrology and other divinatory arts, of which, wrote Ystella, "he speaks Catholically."[119] He raised questions about some of the Spanish physician's natural accounts of miraculous events and especially called into question his natural philosophical account of the Creation, which assigned spiritual qualities to the four elements.[120] Of Ystella's twenty proposed corrections to Valles's *De sacra philosophia*, Guanzelli's Index reproduced fifteen verbatim and one without Ystella's explanation. Five of Ystella's recommendations were omitted entirely.[121] Ystella concluded his expurgations by suggesting that the expurgations were relatively light, that the author was renowned for his Catholicism, and that, ultimately, there were many things that a reader would enjoy in Vallés's works.[122]

Levinus Lemnius held medical degrees from the Universities of Louvain and Pisa, and he published four books dealing with the intersections of medicine, natural philosophy, and the Bible that drew scrutiny from the Congregation of the Index.[123] Lemnius's *Occulta naturae miracula* (*Secret Miracles of Nature*) was initially published in Latin in Antwerp in 1559, immediately translated into Italian and reprinted in Venice by Lodovico Avanzi in 1560, and repeatedly reprinted in both languages thereafter. Lemnius, who had himself studied in Italy, was widely popular beyond the Latin-reading audience. The first editions contained two books with 74 chapters, but an expanded version published in Antwerp in 1564 swelled to four books and 104 chapters. Lemnius's text fits into the genre of books of secrets, dealing with a range of subjects loosely related to medicine, from conception and corpses to remedies and souls.

Like Vallés, Lemnius was not listed among authors in the first class since, according to Alfonso Chacón, he was "indeed a pious and Catholic man . . . and after this his [book about] natural miracles can be made available to Catholics for expurgation."[124] This task was assigned to the congregations of censors in Padua and Pisa, though the expurgations held in the archives of the Congregation of the Index were written instead by the consultor Ambrogio da Asola and undersigned by Pranzini. The lists of expurgations they presented to the Congregation of the Index in 1598 addressed problems in the 1588 Lyon edition of *Herbarum atque arborum quae in*

Bibliis passim obviae sunt (*An Herbal for the Bible*) and the 1593 Frankfurt edition of the *Secret Miracles of Nature*. Da Asola's first problem with Lemnius's text was the dedicatory letter to the Lutheran king of Sweden Erik XIV Vasa enumerating the king's many virtues. Da Asola also corrected a historical error conflating Saint Paul the Hermit with Saint Anthony of Egypt and deleted praise of Conrad Gessner.[125] His expurgations included corrections to both the text and the printed marginalia. Since Lemnius's text dealt with procreation, da Asola suggested more pious rewordings that made God and the sacrament of marriage central to human reproduction.[126] Lemnius's analysis of procreation contended that parents' diseases and personality traits were passed on to children. Da Asola objected to much of this description as "against good morals" and suggested instead reading the Wisdom of Sirach in Ecclesiastes, which had been confirmed as part of the Catholic canon at the Council of Trent.[127]

Da Asola's reading of Lemnius revealed an intellectual world filled with heretics such as Henry XIV and Gessner and also uncouth patients with dubious morals. His expurgations proposed replacing this network with a Catholic intertextuality, connecting Lemnius's observations and suggestions instead to a post-Tridentine Catholic canon, taking pains to redirect readers to the Vulgate translation of the Bible and to the term "demon" rather than "evil spirits."[128] Da Asola's corrections and additions to the text repeatedly highlighted the sacramental nature of marriage, baptism, God's grace, and forgiveness.[129] As the censor read, he took pains to insert text differentiating between Jews and Catholics and to identify parts of Lemnius's word choice that were potentially influenced by Calvinism.[130]

Ultimately, only Lemnius's *Secret Miracles of Nature* appeared with formal expurgations in Guanzelli's 1607 Index. The expurgations required by Roman authorities were condensed onto a single page, and they were taken directly from the Spanish Expurgatory Index of 1584.[131] The Congregation of the Index in Rome had requested a copy of these Spanish expurgations sometime between 1593 and 1596.[132] Despite Ambrogio da Asola's careful reading and eleven pages of expurgations to Lemnius's text, Rome instead followed the Spanish censors, which focused on only a half-dozen errors in Lemnius's wide-ranging book. All of these passages were highlighted by da Asola in his extensive expurgation, but the adopted expurgations included no explanation of the necessary changes. Given the length and breadth of Lemnius's text and the extensive nature of da Asola's complaints, the final list of expurgations were quite minimal. The six passages to be removed dealt with the relationship between human and angelic minds, salvation, free will, and astrology. The two chapters to be removed in their entirety included discussions of superstitious traditions (book 3, chapter 8), and the relationship between the state of one's conscience and one's physical health (book 4, chapter 21). As da Asola had explained, Lemnius's description of what stimulates the conscience sounded Calvinist, and further, the sacrament of confession conferred divine grace. Though da Asola did not directly critique the part of the deleted chapter that dealt with Lemnius's discussion of the roles of doctors and ecclesiastics, it is clear from Lemnius's description that a healthy mind and body required both priests and doctors.[133]

THE UTILITY OF MEDICINE

The corrections required for medical works listed on the *Index Expurgatorius* of 1607 specified how to alter printed books in order to preserve their utility for medical practitioners while removing content that could be considered harmful to Catholic faith and morals. For the physician Girolamo Rossi, the process of reading to expurgate repurposed his humanist reading skills and became ingrained in the way that he approached texts. Catholic censors, like Rossi, carefully addressed those portions of medical texts that were potentially heterodox, dwelling above all on maintaining a community of Catholics free of ridicule and absent the (explicitly announced) voices of Protestant scholars. They also grappled with the influences of God and nature on the human body. The remainder of this chapter returns to Girolamo Rossi to explore how his expurgations reflected his views on the practice and experience of a working physician.

Across his expurgations, Rossi repeatedly addressed the useful work that physicians did, in addition to the knowledge they had. For Rossi, the utility of medicine was not just an excuse to keep books, it was about keeping books in order to practice medicine. In the dedicatory epistle at the beginning of Guglielmo Grataroli's treatise on maintaining health, Grataroli addressed Francesco Grataroli, his young relative. The elder physician counseled, "And most widespread is the popular piece of poetry: one should wish for a healthy mind in a healthy body, and indeed there is no joy in money, or in children or kingdoms, and in the end virtues cannot be useful or beneficial without health."[134] Girolamo Rossi took issue with this passage and requested that the last part of the phrase be removed. He explained his thinking in terms that were later inserted directly into the 1607 Roman *Index Expurgatorius*: "Take out [this sentence] because temperance, patience and other [virtues] are useful even in a sick person."[135]

Whether or not Rossi realized that Grataroli was quoting Juvenal, he was immediately concerned that the passage downplayed the importance of Catholic virtues. Catholic piety was central to Rossi's work as a physician and also to those suffering from illness.

Rossi was the ideal physician-censor, because while he read like a censor, he repeatedly foregrounded his perspective as a physician in his expurgations. The practice of medicine was useful and relied on particular skills combined with piety. The priesthood relied on its own rules and expertise, and according to Rossi, the two professions coexisted symbiotically, a belief he described in his expurgations of Arnald of Villanova. Rossi drew attention to a passage which began, "The rule of the clergy orders that this be regarded as law." Rossi refused to reproduce in his expurgations the rhyming rule that had followed and which he considered to be offensive to the clergy (that eggs are good when they are pale, long, and fresh). Rossi identified this passage as unfairly mocking the clergy. "Though they are priests precisely in that they are not doctors," he wrote, "so as to be able to deliberate about these matters for the sake of good health, they seem to be reproached by this verse as if they were devoted to their palates and kept their souls in their bowls." But Rossi offered a solution, an expurgation or emendation that would preserve the position of the priest and draw attention to the work of the doctor instead. "One could replace it on the other hand perhaps with: 'The rule of the doctor orders that this be regarded as law.'[136] Rossi's emendation put dietary regulations under the authority of physicians, in this case both to highlight their authority in these matters and moreover to remove the possibility that clergy would be ridiculed for thinking too much about food and not enough about souls. With their separate realm of expertise, physicians were useful to the Catholic faith through their content knowledge, and that expertise shifted a possible source of scorn away from the clergy.

A life studying medicine was a route to piety for Rossi. In Cardano's *De subtilitate*, Rossi drew attention to a passage in which Cardano identified wisdom as the "greatest happiness that God could or wanted to bestow upon man."[137] In the paragraph that follows, Cardano admonished that "to achieve this wisdom, one must delight in study," and he proceeded with a list of authors beginning with Euclid and Al-Kindi, progressing through a list of classical and medieval authors, and arriving at Ptolemy and Vitruvius. "From there we turn to the art we want to profess, such as medicine, jurisprudence, or theology. And this is the order of the disciplines," Cardano concluded.[138] Rossi, the pious doctor, took aim at this passage, because for him the highest good must always be God, and human wisdom would always be inferior to that of God. Rossi further protested Cardano's suggestions that one must delight in study to achieve wisdom, and Cardano's order of the human disciplines, which ought to acknowledge that theology has greater virtue than law or medicine.[139] Instead, according to Rossi, readers should consider the double nature of beatitude (*beatitudo*): beatitude of the Father (*Patris*) and beatitude of the Way (*Viae*). Following the path of beatitude (*beatitudinis Via*) is inherently imperfect and requires sacrifice. It is therefore not the greatest happiness that God could bestow, nor want to bestow on humans, as Cardano had written.

Rossi then turned to Aristotle's *Nichomachean Ethics* to question yet another level of Cardano's discussion of happiness. Does the greatest happiness come from contemplation or from actions? Cardano, he affirmed, argued in favor of contemplation, but Cardano's example of medicine, which Rossi practiced, was an example of happiness that derives from actions.[140] Rossi protested that Cardano neglected the teaching of theologians and that he thought contemplation alone would give him knowledge of the divine and eternal life. To counter Cardano's irreverent and superstitious approach to knowledge, Rossi turned to the Bible. He pointed to passages in John 17:3 ("Now this is eternal life: That they may know thee, the only true God, and Jesus Christ, whom thou hast sent.") and Matthew 5:3 ("Blessed are the poor in spirit: for theirs is the kingdom of heaven."). Where Cardano praised wisdom as God's greatest gift and delight in the study of both pagan and Christian authors as the means of achieving wisdom, Rossi countered by emphasizing the practical nature of medicine and the necessity of faith and humility. "I wanted to add this here, so as to confirm what I stated above," concluded Rossi, the physician turned censor, "True philosophy does not oppose faith."[141] Rossi's expurgations are a personal testimony about the nature of one man's medical practice, his conception of the utility of medicine, and the faith that gave meaning to his medical work. For Rossi, piety and utility were one, and books by prohibited authors could be revised to enhance medical knowledge and practice without endangering faith. As we have seen, in Rossi's view, no true science would oppose faith, and faith informed the practice of medicine.

CONCLUSION

The 1607 *Index Expurgatorius* is a strange document on many fronts. It draws haphazardly from numerous sources. It was compiled inattentively and is filled with errors. It did not establish the legal precedent for which ecclesiastical officials had hoped. But from the perspective of its treatment of medical authors, the most perplexing aspect of the *Index Expurgatorius* might be that after nearly fifty years of unsuccessful Roman efforts to "correct" certain medical texts, there were still people like Rossi, Pallantieri, and Chacón willing to do the painstaking work of converting these texts into objects of piety in addition to objects of utility.

Catholic readers extensively used Guanzelli's 1607 *Index Expurgatorius* to expurgate their own copies of suspended books in order to comply with their reading licenses. In 1610, the printer Giovanni Antonio Seghino calculated that among physicians there was an untapped market for an Index of expurgations of exclusively medical books. At his press in Turin, Seghino compiled the expurgations of medical books with a prominent list on the front of the prohibited authors for whom he provided corrections: Lusitanus, Arnald, Grataroli, Cardano, Cornarius, Fuchs, Lemnius, and Vallés.[142] However, when Seghino arrived at the layout of the final folios E1 and E2, he realized that he had extra space in the remaining pages. To fill this space, he added the expurgations of Plato, though he made no change to the title page to signal this addition. All the expurgations within the volume were taken exactly from Guanzelli's Index. Seghino's sloppy printing introduced a number of minor errors and typos to the text, including repetitions of lines of expurgations, several omissions of the capital versus lowercase letter that differentiated between editions of Cardano, and occasional missing expurgations, including one on the final page where he (or the typesetter) must have realized there were more lines of expurgations than would fit on the page. With Seghino's 1610 Index of medical expurgations we see an enterprising printer betting on the continued importance of the works of these eight prohibited medical authors (minus the confusion surrounding Theophrastus and plus Plato, since there was space). His compilation of expurgations was an attempt to make money off the regulations that censorship had imposed. After all, physicians made up a professional group that was well known for amassing large libraries and reading prohibited books. The standard canon of prohibited medical texts was so well known, it was even being commodified.[143]

In 1627, the Anglican clergyman Thomas James, the first librarian of the Bodleian Library in Oxford, famously described the 1607 *Index Expurgatorius* as helpful for librarians who were assembling collections.[144] James's statement was not just an opportunity to mock the "idiocy" of the "Papists" who censored great books but also a recognition of the fact that the books corrected in the 1607 Index were works that were so desirable, even in Catholic Italy, that they were corrected, not destroyed. James's reflection on the importance of the 1607 Index was a backhanded compliment. Catholic censorship was changing the ways that people far beyond Italy approached reading books and assembling libraries. If Catholic Italy could not do without the books listed in the *Index Expurgatorius*, neither could the Bodleian Library.

CHAPTER FIVE

Prohibited Medical Books and Licensed Readers

On the eve of the publication of the Pauline Index of 1559, Ulisse Aldrovandi and Ippolito Salviani discussed how to obtain licenses to read books that were soon to be prohibited by the Catholic Church. The two physicians were taking part in a system of Church-run licensing that was still in its infancy. Over the course of the next hundred years, this kind of strategic planning to obtain reading licenses would take place countless times among learned physicians across Italy. While conversations among intellectuals, such as the exchange between Aldrovandi and Salviani, have left only scattered archival traces, the official records regarding reading licenses in the archives of the Roman Inquisition and the Congregation of the Index are extensive, though little studied.[1] Due to the complicated and shifting bureaucratic structure of these two congregations, the records of licenses are scattered across several archival series.[2] While none of these collections are complete, it is immediately clear that Aldrovandi and Salviani's effort to obtain official permission to read prohibited books was widely adopted. For the period between 1559 and 1664, I have identified 5,211 requests for reading licenses, at least 428 of which were from medical professionals. I estimate that the total number of recorded license requests for the period between 1550 and 1700 is close to 10,000 licenses. The larger number (compared with the 5,211 licenses discussed in this chapter) reflects many licenses granted to ecclesiastics, which are archived separately from the series that include licenses for both lay readers and some ecclesiastics. To focus on medical readership, I have limited myself to the relevant archival series and bounded my search by the Pauline Index of 1559 and the Alexandrine Index of 1664. Of the 5,211 licenses, approximately 8 percent of license requests directed to the Congregations of the Inquisition and Index were sought by people who identified themselves as having medical qualifications. While medical professionals may have been only a small part of this whole, they left comparatively extensive and eloquent evidence in their requests describing how they could be trusted to take only what was useful and safe from books written by prohibited authors.

The extensive data that can be gleaned from these license requests provide an opportunity to approach censorship from the difficult to access perspective of readers. The applications for access to prohibited books document physicians as readers and collectors who read a wide range of prohibited books and justified this activity as useful and necessary for their work. These new archival data also enable us to trace the reception histories of prohibited medical and scientific books on the Italian peninsula through their licensed readers. Medical professionals' requests for licenses to read medical books constituted a forum in which physicians repeatedly articulated their medical practice as useful for society, and which in turn shaped the Catholic Church's evolving position on censorship.

The licensing of readers to read prohibited books also raises fundamental questions about the effectiveness of censorship in the sixteenth and seventeenth centuries. Despite the broad bans imposed by Catholic authorities, the regulation of texts through licensing readers explicitly allowed for the extensive, licit use of prohibited books. While this system originally arose to mitigate damage to venerable libraries and elites with close ecclesiastical ties (such as Salviani, the papal physician), the process of expurgation extended the privilege of reading prohibited books more broadly. By the seventeenth century, with the standardization of many expurgations delineated in the 1607 *Index Expurgatorius*, ever more readers applied to authorities in Rome to read banned books. Reading licenses became an institutionalized negotiation between Catholic authorities and the needs of pious, professional readers.

Previous chapters of this book described the process of expurgation that made it theoretically possible for readers to keep and study prohibited medical books. This chapter describes how physicians applied for, and often received, the reading licenses that made it permissible to "keep and read" prohibited books in the sixteenth and seventeenth centuries. Data from early modern reading licenses allow us to move between the scales of individual readers and the reading communities of physicians and other collectors of medical books. Physicians were precocious book collectors with large and valuable libraries that they took steps to preserve both during their lives with reading licenses and after their deaths through their wills and testaments. The care that physicians took to preserve and document their libraries by seeking licenses provides us with remarkable insight into their reading and collecting practices and into the changes in popular scientific content during a period of scientific transition. The chapter concludes by considering how medical professionals used reading license applications to emphasize the utility of their scholarly expertise to the Catholic Church.

REQUESTING A READING LICENSE

Throughout its history, the Catholic Church provided a means by which the pious could take part in certain activities that were otherwise prohibited without violating Catholic doctrine. Records of requests for reading licenses from the archives of the Holy Office appear alongside, though much more frequently than, licenses for other regulated activities, such as licenses to carry weapons, to eat meat on fast days, and to receive medical treatment from Jewish doctors.[3] The Roman Inquisition was charged with regulating a wide range of behaviors, especially practices pertaining to heresy. Similarly, reading licenses granted permissions to deviate from the normal rules of Catholic postpublication censorship. Catholics from the sixteenth through the twentieth centuries could legally access books written by authors condemned on the Index of Prohibited Books, provided they first applied for and obtained a license.[4]

An individual interested in reading a prohibited book could apply to a range of authorities to obtain permission. The Congregation of the Holy Office of the Inquisition, the Master of the Sacred Palace, the Congregation of the Index of Prohibited Books, and the pope all granted reading licenses from Rome. Licenses found in books and those referenced in inquisition trials also indicate that local bishops, inquisitors, and at times even parish priests or personal confessors granted licenses to local readers. Over the course of the sixteenth and seventeenth centuries, frequent bulls and edicts from Rome point to repeated efforts to control and standardize the granting of licenses, and especially to prevent local bishops and inquisitors from granting licenses.[5] When Galileo wrote to Fulgenzio Micanzio in June 1636, he complained that he was especially upset about the banning of his *Dialogue* since "getting a license to read it had been reduced strictly such that only the Pope himself reserved [the privilege to grant licenses to read it]."[6] In this instance, the disgraced astronomer may have been giving too much credit to the Catholic censorship apparatus. While his case was remarkably high profile and unlikely to slip through any cracks, the jurisdictions covered by granting authorities were shifting and porous, often leaving potential readers with several avenues for the pursuit of a reading license.

Twenty years before Galileo's comment, in the wake of the Congregation of the Index's 1616 condemnation of Copernicanism, Galileo's friend and fellow member of the Academy of the Lincei Cosimo Ridolfi applied for a renewal of his license to read the prohibited works of Paracelsus (see figure 5.1). Dated March 15, 1617, and undersigned by the notary of the Roman Inquisition, Andrea Pettini da Forlì, the text is the standard formula for a reading license, with Ridolfi's details and dates inserted.[7]

> In the General Congregation of the Holy Office of the Roman and Universal Inquisition, customarily assembled in the Palace [of the Holy Office], [and today assembled] in the House of the very illustrious and reverend Cardinal Aldobrandini in the Trevi neighborhood, in the presence of the very illustrious and reverend Cardinals specially deputized by the Pope as general inquisitors to prosecute against heretical depravity in all Christendom.
>
> A memorial regarding Cosimo Ridolfi, nobleman from Florence, was read by the very illustrious and reverend Cardinals. The present general inquisitors renewed for another three years the license conceded to him on November 6, 1613, to keep and read the works of Theophrastus Paracelsus, with the condition that he expurgate them according to the corrections in the Index Expurgatorius published in Rome in 1607, and at the end of the three years he should consign them to the Inquisitor of Florence.[8]

This license is unique only in its survival. It gives the date, the location of the Congregation's assembly, and the status of the cardinal inquisitors. Ridolfi was not a physician, but he was requesting the works of the increasingly popular medical author Paracelsus. The license stipulated that Ridolfi could "keep and read" these works; he presumably owned them already. Reading licenses only very rarely deviated from this form to explain that readers could purchase the texts requested.[9] Ridolfi's license also indicates the widespread use of the 1607 Roman *Index Expurgatorius* as a guide to expurgating books for personal, licensed use. It is likely that Ridolfi expurgated his copies of Paracelsus himself.

Given Ridolfi's status as a member of the nobility, it is less likely that his books were consigned to the inquisitor of Florence two years later as the license stipulated (and in this case Ridolfi died before his license expired). Occasionally, cardinals in Rome did follow up on licenses granted

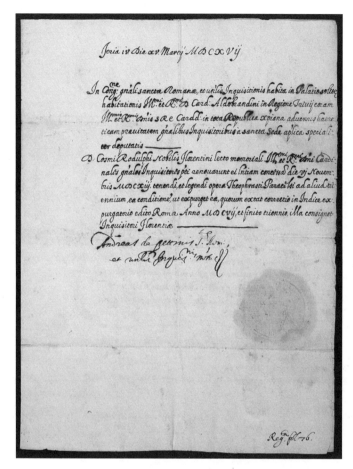

Fig. 5.1. Reading license granted to Cosimo Ridolfi, a Florentine nobleman and friend of Galileo, allowing him to read the works of the prohibited author Paracelsus. MSS gen 25, item 7, Thomas Fisher Rare Book Library, University of Toronto.

to residents of other cities. In Florence in 1613, a bookseller, Signore Bracciolino, was granted a license to read Conrad Gessner's *Bibliotheca universalis* and the anonymous compilation of political tracts known as the *Tesoro Politico*. However, Cardinal Giovanni Garzia Mellini in Rome wrote to the inquisitor in Florence that Bracciolino had also requested five other books, including an uncensored edition of Boccaccio. Mellini warned that since these other requests were denied, the inquisitor ought to be sure that the books were consigned, and he should keep an eye on Bracciolino to be sure that he obeyed.[10] In another case in Faenza, a reading license granted to Doctor Giovanni Fontana of Modigliana on September 24, 1631, was

exhibited to the Dominican inquisitor general of Romagna, Tommaso No-
varri da Tabia, and transcribed into the archive on January 21, 1632. The
inquisitor "verbally registered" this license, and it was recorded by the
local notary, a certain Fra Ippolito.[11] Given the distance from Rome and
the vast number of reading licenses, it is tempting to assume that no one
abided by the rules of this system. However, it is clear from widespread archival evidence that both petitioners and Roman authorities were surprisingly conscientious about following these procedures. Reading licenses
created a trail of paperwork that crisscrossed the Italian peninsula and can
be cross-referenced against many archival sources.

The variety of sources of information about reading licenses also indicates the complicated and changing status of this system over the
course of the sixteenth and seventeenth centuries. Bureaucratic systems
for granting reading licenses reveal tensions between the centers and pe-
ripheries of Italian Catholicism, as well as competing jurisdictions among
authorities within Rome about the control of book circulation. With both
the Congregation of the Inquisition and the Congregation of the Index of
Prohibited Books issuing licenses as well as the pope and the Master of the
Sacred Palace, regulations surrounding reading licenses were subject to
multiple and sometimes conflicting authorities throughout the sixteenth
and seventeenth centuries. As a result, the process of seeking, receiving,
and renewing licenses could be as confusing for petitioners as it is for historians. Some readers of prohibited books certainly took advantage of this
confusion to request from the Inquisition a book denied by the Index, but
the dynamic that emerges most clearly from the jurisdictional overlap is
the multiple authorities working to make prohibited books available to
qualified readers.

LICENSE REQUESTS FROM MEDICAL PROFESSIONALS

Reading licenses may have been widespread in the sixteenth and seventeenth centuries, but their survival has proved incredibly ephemeral. Occasionally a reading license (or more often a copy) is found in the personal
archives of physicians and scholars. For example, the archive of the Caimo
family in Udine contains multiple notarized copies of reading licenses
granted to members of the family, especially Giovan Battista Caimo.
These licenses to read copies of books, including Conrad Gessner's *Historiae animalium*, note that the license was also inscribed by the local
inquisitor on the first page of the folio volume. Indeed, rather than actual
licenses or even copies, evidence of licenses most often exists in the form

of an inscription in the early pages of prohibited medical books, as Caimo's license indicates, but without the documentation in personal archives. A copy of Leonhart Fuchs's *De historia stirpium* at Harvard's Francis A. Countway Library was owned by Antonio Cappelli, a physician born in Montepulciano who studied in Pisa at the end of the sixteenth century (figure 5.2). The inquisitor of Pisa undersigned the expurgated volume,

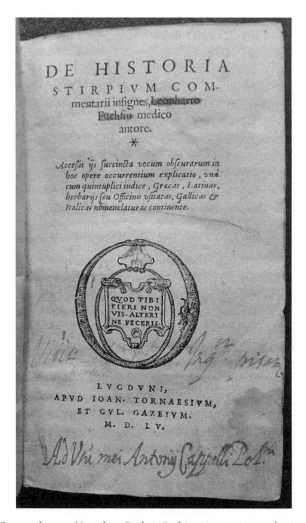

Fig. 5.2. Censored copy of Leonhart Fuchs's *De historia stirpium* undersigned by the Inquisitor of Pisa and granted to Antonio Cappelli. In addition to the note recorded on the title page of this volume, there is a record of Cappelli's license in the Archive of the Congregation for the Doctrine of the Faith. Leonhart Fuchs, *De historia stirpium* (Lyon, 1555), Francis A. Countway Library of Medicine, Harvard University.

verifying that the copy was appropriately expurgated and legally granted "for the use of Antonio Cappelli."[12]

While personal archives only rarely contain reading licenses and licenses were inconsistently inscribed into books, records in the archives of the Congregations of the Inquisition and Index of Prohibited Books testify to the thousands of licenses sought in the sixteenth and seventeenth centuries, hundreds of which were granted to physicians. This new information supplements license information from other sources, revealing in Antonio Cappelli's case that his license to read Fuchs was granted by the Congregation of the Index in September 1599, and it additionally gave him permission to read works by Arnald of Villanova, Conrad Gessner, Otto Brunfels, Johannes Lange, Johann Winter, and Paracelsus.[13]

The records of reading licenses sought and granted are a centralized, though still incomplete, repository of information about readership across Italy. This vast account of readers and books is paradoxically documented and visible because the texts were prohibited and the Counter-Reformation Church sought to centralize the administration of these privileges over the course of the seventeenth century. The records of licenses in the archives of the Roman Inquisition and Index survive in the form of letters, in registers of licenses sought or granted, and in the minutes of the meetings of the congregations.[14] The form and content of the various license requests is varied. However, each request included the name of the petitioner and usually included the date of the request. While nearly half of the requests do not indicate the qualifications of the petitioner, approximately 8 percent of petitioners described their medical qualifications in support of their petition. Many of the records include a list of prohibited books that the petitioner sought to acquire or read. Like many historical data sets, these license requests have too many gaps to lend themselves to robust statistical analysis. As is obvious in figure 5.3, there are many holes in the archival records of reading licenses, and relatively consistent records exist only for the late 1610s through about 1635—though this period, also has a glaring lacuna from the late 1620s to early 1630s. Even in the 1650s through the end of my sampling period in 1667, there are many license requests, but these licenses include no information about the books requested and little aside from the name of the petitioner. Despite these caveats, the 428 licenses sought by medical professionals, and still more licenses requesting medical books, provide a fascinating window into the reception history of prohibited medicine on the Italian peninsula.

The geographical information in reading license requests can be used to draw a provisional map of prohibited medical reading in Italy (see

Fig. 5.3. Total License Requests, 1559–1664

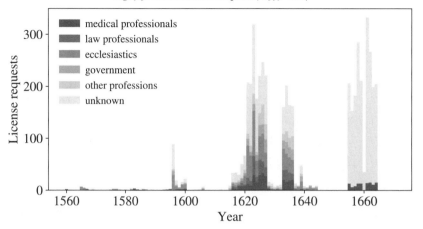

Graph showing the number of reading license requests over time. From this image the gaps and changes in the archival records (1601–1614, 1628–32, 1645–54) are immediately obvious. The graph shows only license requests for which there is a clear date; 1,320 requests do not have a clear date and are therefore not shown here at all. Shading in the bars represents specific qualifications of the petitioner, when known, though the majority of license requests do not provide information about the qualifications of the applicant.

figure 5.4). Physicians from across the Italian peninsula sent petitions for reading licenses to authorities in Rome. Though we must bear in mind that only about half (228) of the licenses contain information about the physicians' location, even these incomplete data are revealing. The first important observation from this geographical information is that reading prohibited medical books was not an activity confined to Italy's cultural centers but was instead widespread and dispersed, including medical practitioners in many small towns. We need only think of Girolamo Rossi and Giovan Battista Codronchi, the active censors of medical books at the turn of the seventeenth century, who undertook their work not from the university cities of Padua and Bologna but instead from the small towns of Ravenna and Imola. Although I have not found records of either of these individuals' reading licenses, they certainly would have had them. The cases of missing licenses for important censors is yet another reminder that while the surviving data about reading licenses is vast, it is also fragmentary. Across Italy, even outside the great centers of learning, physicians understood the benefit of reading prohibited books to their professional work. The only other professional status with a similar degree of geographic distribution is that of ecclesiastics—whose comparative presence

Fig. 5.4. License Request Locations by Qualification

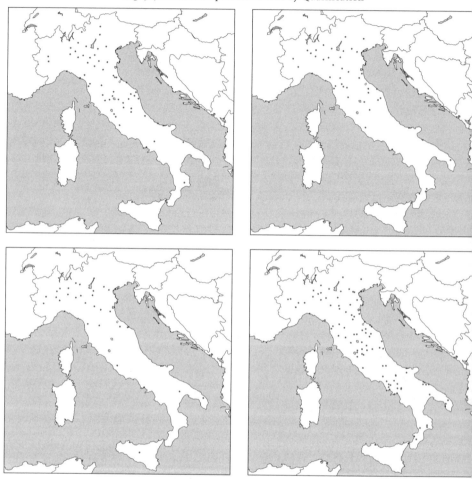

Top left: Medical professionals; top right: Legal professionals;
bottom left: Government/local nobles; bottom right: Monks and ecclesiastics
Maps of Italy showing locations from which people with specific qualifications
requested reading licenses. To emphasize the geographic spread of requests
by various professional groups, the points indicate a license request
from that location but do not represent the number of requests.
Only requests from monks and ecclesiastics (bottom right) rival
the geographic spread of requests by physicians (top left).

was more pronounced in the provinces of southern Italy in present-day Campagna, Calabria, Basilicata, and Puglia. Requests from nobility and secular officials and men trained in law came more often from the cities of northern Italy.

The cities with the most requests from physicians for reading licenses are Naples, with seventeen requests from ten different physicians, and Bologna, with fifteen requests from nine different physicians (see figure 5.5). The number of requests from Bologna is not surprising given the strong medical faculty at the university and its prominent location in the Papal States. The number of requests from Naples is more unexpected,

Fig. 5.5. Licenses Requested by Medical Professionals

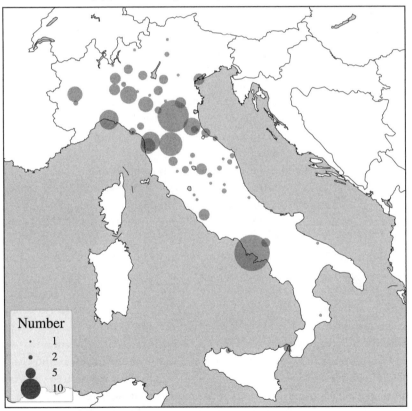

Map showing requests from medical professionals with points sized to correspond to the number of requests from a particular location. The largest points are Naples (17), Bologna (15), Florence (11), Lucca and Genoa (9), and Piacenza and Forlì (8). The legend indicates the scaling of point size to number of counts at city.

especially because of the complicated nature of how the Inquisition operated in this Spanish-ruled city, though it is clear from the requests that learned Neapolitans turned to authorities in Rome for permission to read prohibited books.[15] There are also seven medical license requests from Pisa, eleven from Florence, and a number from scattered small towns across Tuscany including Montepulciano, Fivizzano, and Montalcino. Five different physicians requested licenses, sometimes repeatedly, from the small city of Lucca, which had a total of nine requests. Francesco Maria Fiorentini, reported (in the third person) in his own license petition that "he assures you that he does not intend to use them [prohibited books] to study pernicious or vain doctrines, but only in as much as it concerns his profession of medicine."[16] Fiorentini was a pious Catholic reader who could be counted on to read and interpret selectively, paying attention to medical content while ignoring or even expurgating "pernicious or vain doctrines." He may have believed that the explanation in his request was necessary because his city, Lucca, had produced a number of prominent physicians in the sixteenth century who had been accused of heresy, including Donato Ori, Simone Simoni, Giovanni Battista Donati, and Michelangelo Bertolini.[17] Lucca may not have been a center of medical learning, but it was a place where prohibited books were widespread and physicians sought permission to read the prohibited works that were relevant to their profession.[18]

Although the license data are fragmentary, the Roman records for the years around 1630 are fairly consistent, which invites some comparison between requests for reading licenses and the lists of active physicians in 1630 that Carlo Cipolla compiled for the region of Tuscany.[19] Four of the twelve members of the Florentine college of physicians—Giovanni Battista Aggiunti, Cristofano dell'Ottonaio, Benedetto Punta, and Giovanni Ronconi—sought and received licenses from Roman authorities.[20] Similarly, one of three physicians in Montepulciano in 1630 also held a reading license.[21] However, it would be misleading to extrapolate that a third of a city's physicians held reading licenses. Of the other twenty-one physicians Cipolla cites as practicing in Florence, I have located licenses for only two of them (Giovanni Nardi, who requested the works of Paracelsus and Gessner, and Gian Vittorio Rossi). For the twelve physicians practicing in Pisa in 1630, I have located only one license request that appeared in the records of the Index and Holy Office (Giovan Battista Ruschi). Records of reading licenses reveal thousands of readers with permissions to read prohibited books, though we should not overestimate these cases as a significant number of the population, even among elite professionals.

Italy's most famous center of medical learning, Padua, is strikingly absent from the map of license requests. In general, the cities of the Veneto are likewise underrepresented among petitioners for reading licenses. This is not an accident of the archive nor an indication that physicians in the Veneto did not read prohibited books but rather a reminder that physicians in the Veneto did not look to authorities in Rome for permission to read prohibited books. For example, Gian Vincenzo Pinelli, the owner of Padua's most famous private library, had many ecclesiastical connections, including a close friendship with the inquisitor of Venice who allowed Pinelli to read any books that did not pertain directly to heresy.[22] However, there is no evidence that Pinelli ever possessed a license from Roman authorities to keep the many prohibited books in his collection.[23] Pinelli's contacts with heretical ideas were well known in ecclesiastical circles, though his own piety was never called into doubt. Indeed, Cardinal Francesco Maria del Monte contacted Pinelli for help finding copies of Girodano Bruno's works during Bruno's trial.[24] Pinelli's library was a collection of books and also a place for scholars in Padua to congregate. When Galileo moved to Padua, he lived in Pinelli's home on the Via del Santo, as did Antonio Querenghi, Lorenzo Pignoria, and Claude-Nicolas Fabri de Peiresc, and Bruno himself may have passed through in 1591.[25] In the years after Pinelli's death, the Pinelli collection would make its way, partially, to the newly founded Biblioteca Ambrosiana in Milan, where its prohibited books were properly licensed.[26] The scholars and physicians of Venice and Padua were located at one of the centers of the European book trade and had access to itinerant populations of students and merchants and to many prohibited books, which they read liberally and usually without permission from Rome.[27]

Roman physicians are also underrepresented among requests for reading licenses; however, this is a representation of archival loss. Within the city of Rome, readers seeking licenses applied to the Master of the Sacred Palace, who was responsible for censorship within the city, rather than to the Holy Office or Congregation of the Index. The archives of the Master of the Sacred Palace are long missing and with them, we must assume, the vast majority of reading license requests from Roman physicians.

As the cases of Donzellini and Aldrovandi remind us, physicians initially sought reading licenses from local inquisitors, confessors, and even parish priests. Over the course of the sixteenth and seventeenth centuries, requests were increasingly addressed directly to authorities in Rome or forwarded by local authorities to the Roman Congregations of the Index and Holy Office. When Sebastiano Pardini, a physician from Lucca, petitioned the Holy Office for a license to read astrological texts in 1625, he applied with a local nobleman.[28] The request contains the letter from the physician and a note on the envelope stating that the Holy Office had granted the request. It also contains another note, perhaps written by an intermediary in Lucca, either the bishop or the inquisitor, suggesting that if the Roman authorities were not willing to grant the license to both Pardini and the gentleman, they should at the very least give it to the doctor. This unidentified hand continued, "if you will not grant it all, grant it at least in part."[29] The intermediary was perhaps reflecting on the long list of controversial astrological titles inside, including Cardano's commentary on the *Tetrabiblos* and Jean Taxil on physiognomy. We see in Pardini's request a network of individuals locally and at a distance that physicians leveraged to obtain reading licenses.

While the Congregations of the Holy Office and the Index granted or denied the vast majority of license requests, five of the licenses requested by medical professionals were granted directly by the pope, and four of those by Urban VIII.[30] The licenses that Urban granted to physicians were all general licenses—that is, the pope used his papal prerogative to allow approved physicians to read prohibited books without naming the specific prohibited authors. Benedetto Averino, the personal physician to Cardinal Carlo Gaudenzio Madruzzo, limited his request to "prohibited books dealing with medicine," as did Niccolò Bevilacqua.[31] Urban granted their licenses for three and five years, respectively, specifying in Averino's case that the books be corrected according to the *Index Expurgatorius* of 1607 and that the names of heretics be removed from them. In February 1623, Pope Gregory XV granted a reading license to Demetrio Canevari, who was widely known in Rome as a consummate bibliophile and had amassed a great library in the early seventeenth century.[32] Canevari's license reflected his broad interests in important medical authors such as Fuchs, Gessner, Cardano, Erastus, and Brunfels, and in humanist commentaries by Erasmus and Melanchthon and the controversial astronomers of his day, including works by William Gilbert and an expurgated copy of *"De revolutionibus* by Nicolaus Copernicus, without the chapters in which he teaches, following ancient thinkers, that the earth moves."[33] Popes used their authority to bestow exceptional privileges upon trusted physicians.[34]

Circles of intellectuals in and around Rome also facilitated their colleagues' petitions for reading licenses.[35] Johannes Faber, a Protestant-born and Catholic-educated physician from Bamberg and member of the Academy of the Lincei, received assistance in crafting his petitions for reading licenses. Faber was close to Antonio Bucci, a physician from Faenza and consultor to the Congregation of the Index. Bucci knew the Master of the Sacred Palace and helped ensure that he granted Faber reading licenses.[36] In turn, Faber served as mediator between his own intellectual circle and the curia.[37] Faber attended to his colleagues' reading practices through correspondence, but his archive also includes copies of reading licenses granted to his peers. The newly declared Pope Urban VIII granted one such license to Faber's mentor and fellow Bavarian convert Kaspar Schoppe in November 1626.[38] Ultramontane scholars and converts occupied a special position in the intellectual world of seventeenth-century Rome, facilitating the movement of knowledge across Europe despite the impediments of ecclesiastical censorship.[39]

The patronage of spiritual and temporal rulers was a considerable asset in obtaining reading licenses. In spring 1623, the Holy Office granted to Gian Pietro Rasselli, papal physician to the ailing Pope Gregory XV, a license to read a remarkable list of astrological texts, perhaps in an attempt to better understand and preserve the health of the fading pope.[40] Similarly, the court physicians Giovanni Comiti in Parma and Pier Antonio Caballo in Mantova specifically emphasized their relationships to the dukes in their requests for reading licenses.[41] In his request of 1625, Caballo certainly found that his relationship to the Gonzaga dukes helped him secure a license to read prohibited books by thirty-seven authors, with titles pertaining primarily to medicine, mathematics, astronomy, and astrology.[42] The next year, a new author had come to Caballo's attention, and he separately sought and received permission to read the English physician Robert Fludd's mystical *Utriusque cosmi Maioris scilicet et Minoris, metaphysica, physica atque technica historia* (*The Metaphysical, Physical, and Technical History of the Two Worlds, Namely the Greater and the Lesser,* 1617–21) that had been recently published in Frankfurt.[43] Despite the Italian Counter-Reformation culture of censorship, medical credentials and proximity to secular rulers allowed Caballo to stay up-to-date with contemporary scientific authors.

The pious Gian Vittorio Rossi was well connected in medical circles in Padua at the beginning of the seventeenth century and later received some of the most extensive reading licenses granted over the course of the seventeenth century.[44] Rossi sought licenses from the Congregations of the Index and the Holy Office on at least six occasions between June 1616 and March 1626.[45] The requests, which were repeatedly granted and renewed, present Rossi as exceptionally well connected. He had inherited part of his impressive library from his teacher in Padua, the famous anatomist Girolamo Fabrici d'Acquapendente.[46] Rossi's reading licenses covered a range of books including editions of classical texts edited by Erasmus and prominent Protestant authors (Fuchs, Cornarius, and Melanchthon), medical treatises (by Fuchs, Brunfels, Arnald, and Lusitanus), and an assortment of ephemerides and works pertaining to medical astrology.

Rossi's license from 1618 took the unusual step of explaining extensively how he was expected to read these prohibited books. While the Congregation of the Index stipulated in a fairly standard form that his license was granted "under the condition that he correct those things that are corrected in the Roman Index Expurgatorius of 1607 and delete the names of heretics," it continued on, requesting that he also "delete and make note of anything while reading that goes against good morals and Catholic truth, and refer the Holy Office to these passages."[47] A story retold by Rossi and published under his pen name Ianus Nicius Erythraeus illustrates the contemporary connections between publication, erasure, and repentance. Rossi recounted the famous musician Luca Marenzio's deathbed confession to the physician-turned-priest Father Giovenale Ancina. "If only I had not published my music," he supposedly pleaded. "If only I could erase it with my blood and leave no trace behind!"[48] Rossi's story reveals how contemporaries contemplated repentance and forgiveness in terms of writing and erasure, noting that although sins could not be revoked, God was able to pardon the repentant. For the learned Catholic, reading licenses and the alteration of books were integral parts of a broader Counter-Reformation piety that seamlessly integrated the physical and spiritual acts of expurgation.

While Rossi left a particularly rich trail of reading licenses and patronage, medical professionals outside of Rome and with fewer connections could obtain permissions to read prohibited books through their status as physicians or even merely as a "doctor of arts and medicine." Baronio Vincenzi, a physician from Spoleto, received permission to read numerous prohibited medical texts without reference to his patrons or specific qualifications beyond "medicus," as did the Calabrian physician Giovanni Battista Regolino.[49] We might suspect that there is a longer backstory to Giovanni Maria Riccio's 1624 license that quite unusually granted the "physician from Genoa" permission to read "prohibited books related to medicine" for five years, though in this case the record of the license provides only these sparse details.[50] However, the opportunity to secure a reading license often led medical professionals to be quite specific about their qualifications to read prohibited medical books. Pietro Paolo Pisano, the *protomedico* (public health official) in Messina, Sicily, applied for a license that would last ten years rather than the usual three. His request lists his accomplishments as doctor in arts and medicine, professor (*cathedratico*) in the same faculty in Sicily, and *protomedico* "many times over."[51] As medical authorities appointed by civic governments, *protomedici* held an elite position in society that facilitated their special privileges.

In addition to applicants with prestigious credentials, three surgeons are among the medical professionals who applied for reading licenses from the congregations in Rome. Francesco del Pezzo Cornetano, a surgeon "in Urbe" (Rome), received a license from the Holy Office in 1633 to read the *Examen ingeniorum* (*Examination of Wits*) by Juan Huarte.[52] Vincentino de Calofilippo and Tomaso Squilace, the latter a surgeon from Naples, requested to read the works of Paracelsus. Lest we think they were only interested in the iconoclastic physician's works on surgery, Calofilippo's request specified that he sought both the "medical and chemical" works of the prohibited German author.[53] Although none of the licenses requested by medical professionals identified the petitioner as an apothecary, it is reasonable to conclude that physicians shared prohibited texts with apothecaries in their shops, which were important sites for the exchange of goods and ideas in early modern Italian cities.[54]

Professors of medicine and philosophy from Bologna to Naples requested lists of prohibited books that were important to their work. Francesco Rolando, a physician and professor of mathematics at the University of Turin, received at least five licenses to read the works of prohibited authors between 1618 and 1636. His long lists of books in 1618 included the usual physician's requests for Fuchs, Lusitanus, and Cardano and also reveal his deep interest in mathematics, astrology, and astronomy with licenses to read the works of Cyprián Karásek Lvovický, Johannes Kepler, and Nicolaus Copernicus.[55] By 1624, Rolando had also developed an interest in and library of books related to occult topics and chemical medicine, receiving licenses to read Paracelsus, William Gilbert, and Robert Fludd.[56] In 1636, he added the works of Ramon Lull and the edited collection of alchemical texts *Theatrum chimicum* and *Turba philosophorum* to his license, but the Roman Inquisition drew a line at—indeed, a line through—his request to read a book on palmistry (*De manus inspectione libri tres*).[57]

Rolando's ambitious procurement of reading licenses and perhaps even prohibited books came through local expertise seeking and obtaining licenses in Turin—Rolando was not, after all, the first professor at the University of Turin with reading licenses. Orlando Fresia, the local *protomedico* and professor of medicine, had been granted a license in 1595 (which he requested through the local cardinal) to read works by Gessner, Cardano, Brunfels, Camerarius, Petrus Ramus, Paracelsus, and Achilles

Pirmin Gasser.[58] In addition to those named in his license, he also owned copies of works by other prohibited authors including Fuchs, Arnald, Brunfels, Erastus, Wecker, and Mizauld, which he had donated to the university library in Turin.[59] Fresia's books were prized in their own time for his "many annotations," and we might consider that Rolando, a few decades later in the same library, used his reading license to consult copies of prohibited books that his predecessor Fresia had owned.[60]

Three applicants for licenses identified themselves as physicians for their local offices of the inquisition. Aurelio Bussolo taught practical medicine for twenty-eight years at the university of Pavia. He listed his credentials for a reading license in 1626 and again in 1633 as "physician of the Pavia Holy Office."[61] Similarly, Costanzo Scotto lectured in logic, medicine, anatomy, and surgery at the University of Bologna from 1626 until his death in 1652. Scotto's request for a license in 1633 described him as "physician of the Bologna Holy Office."[62] Despite their overlapping credentials, these two pious physicians requested licenses for completely different lists of prohibited medical texts.[63] Bussolo sought and received permission for works by Thomas Erastus, Otto Brunfels, Joachim Camerarius, Hadrianus Junius, and Johannes Thomas Freig. Scotto's earliest recorded license request included several extremely controversial titles including Cardano's geniture of Christ and a work Scotto described as the "Centuriae del Niolano," which was probably the *Centum et viginti articuli de natura et mundo adversus peripateticos* by Giordano Bruno, also known as the Nolan.[64] Neither of these two requests were granted. By the time of the requests of 1633 and 1636 he had evidently accepted some degree of compromise and was reading the works of Cardano (except the *Tetrabiblos*), the works of Paracelsus, the *Secrets* of Wecker, and the works of Fuchs. The books considered necessary to the work of physicians were a matter of personal preference and opinion, even if their credentials for seeking them were the same.

The third physician for the Holy Office to seek licenses was Francesco de Curtis, a nobleman and physician in Naples who was born in 1592 in Cava. In 1678, an account of learned men in Naples described de Curtis as a "famous philosopher, physician, and astrologer," though the trail of his early reading licenses shows that de Curtis's astrological reading was not a straightforward process.[65] The first and second of de Curtis's recorded licenses were granted when he was twenty-six years old, in 1618 and 1619, and processed for renewal by the Congregation of the Index in 1628 and 1629.[66] A short series of letters are preserved by the Congregation of the

Index which include additions to these original licenses as well as information about what volumes were rejected. De Curtis was granted a license to read a number of medical books including works by Fuchs, Lusitanus, Lemnius, Erastus, and Mizauld. Next to the entry "works by Cardano" another hand has noted "except astrological." The list was then heavily edited as the official from the Congregation of the Index further emphasized the rejection of de Curtis's astrological requests, striking works by Johannes Schöner, Julius Firmicus Maternus, Alchabitius, Heinrich Rantzau, Francesco Giuntini, Guido Bonatti, David Origanus, and Luca Gaurico. With the renewals in 1628 and 1629, de Curtis requested the addition of works by Paracelsus and the humanist scholars Joachim Camerarius and Franciscus Vallesius. Creeping ever closer to a sanctioned astrological library, de Curtis successfully petitioned for the works of Lucio Bellanti, who had mounted a printed attack on Pico della Mirandola's take on astrology and argued instead for a form of astrology compatible with Christianity.[67] Annotations on this collection of letters show that the license was renewed in 1631. In 1635 de Curtis was granted licenses yet again, this time by the Holy Office.[68] The works of yet another astrologer entered his list by 1635, Joachim Fortius (or Joachim Sterck van Ringelbergh), the Flemish humanist and astrologer. By April 1636, de Curtis sought and received permission from the Holy Office to read the works of Girolamo Cardano. Whereas his previous license had noted that he could not read any of the astrological works, this request was granted with a more lenient stipulation, allowing all works except his commentary on Ptolemy's Tetrabiblos.[69] By 1635, when de Curtis presented himself as a physician for the Holy Office in Naples, not just a curious young doctor, he was finally making strides toward legally reading the prohibited astrological texts in which he had been interested for most of his life.

Reading licenses are a tantalizing window into a world of licit, prohibited reading among a group of medical professionals. The number of medical applicants is likely much larger than the 428 license requests reflect, since many licenses do not provide any information on applicant qualifications. Similarly, evidence of the geography of licensed medical reading is far from complete, but the available data reveal the depth and breadth of medical learning and patronage across the Italian peninsula. There were, of course, many more readers of prohibited books than even these licenses indicate, since there were undoubtedly medical professionals who read illicitly and those for whom license information is lost. We should read these licenses as individual historical artifacts and collectively as evidence of social and intellectual trends that, despite their limits as statistical data, provide a revealing perspective on the practice of pious reading of forbidden knowledge.

PROHIBITED BOOKS IN THE LIBRARY OF GIROLAMO AND STEFANO COLI

From the many petitioners for medical licenses, we turn now to a particular pair of licensed readers, shown in figure 5.6, and to their library in Lucca as an example of how reading licenses can be used to create intellectual portraits of readers in the seventeenth century. Painted by the Luccese painter Pietro Paolini around 1640, this portrait of Girolamo and Stefano Coli depicts a younger man and an older man surrounded by and interacting with books.[70] The letter on the desk in front of the younger man is addressed to Stefano Coli at his home. Stefano is on the left, depicted around 1640 as a beardless man between the age of twenty and thirty. The older man on the right is almost certainly Stefano's father, Girolamo Coli (d. 1644), a prominent physician and citizen in the city of Lucca.[71] The portrait emphasizes the importance of books to these two men's understanding of themselves and of their professional role as physicians. Like many other physicians, the Colis applied multiple times for reading licenses, and these requests allow us to piece together a story of their reading over the course of many years and even decades.

Though the words on the pages and titles on the spines are illegible, it is possible, even likely, that some of these books were volumes that the Catholic Church had prohibited in the sixteenth century. On September 10, 1636, the Holy Office of the Inquisition in Rome granted Girolamo and Stefano Coli, father and son, a joint license to read a long list of prohibited books for three years under three conditions: (1) that they correct (that is to say, censor) these works according to the *Index Expurgatorius* published in 1607; (2) that they show their reading license to the local inquisitor; and (3) that they delete the names of heretics mentioned in the books, including, but not limited to, the names of the authors. The list of books that they were allowed to read was extensive, including the complete works of Paracelsus, the medical and botanical works by Leonhart Fuchs, Otto Brunfels, and Conrad Gessner, and all of the works of Girolamo Cardano and Arnald of Villanova except those dealing directly with astrology.[72]

When Girolamo Coli, the elder subject of Lucca Paolini's portrait, applied to the Roman Inquisition for what may have been his first reading license on August 13, 1625, he described himself as Lucchese and as a "doc-

Fig. 5.6. Stefano Coli (*left*) and his father, Girolamo Coli, posing with the books in their library in Lucca, Italy. Books were an essential part of a physician's professional identity, and prohibited books were often an important part of a physician's library. The Colis applied for several licenses to keep prohibited books, which we might imagine on the shelves alongside the volumes in this portrait. Pietro Paolini, *Doppio ritratto con Stefano Coli*, Marco Voena collection, Turin.

tor of arts and medicine." He initially requested a modest list of books that included the works of eight prohibited authors.[73] Coli requested permission to read the complete works of Paracelsus, Arnald of Villanova, and Girolamo Cardano. He specified that he wanted to read only the medical works of Paracelsus's French Calvinist follower Joseph Duchesne and added that he wanted Caspar Schwenckfeld's books on the plants and fossils of Silesia, Otto Brunfels's history of plants, Levinus Lemnius's book on secret miracles, and Antoine Mizauld's medical *Centuriae*. Coli was instructed to take his license to the inquisitor in Lucca, and when the three-year period of his privilege concluded, he was to consign the books to the inquisitor. Prohibited books were not licensed to readers indefinitely, and Girolamo Coli, like all other readers, was expected to deliver prohibited books for which he had no license to proper authorities, in this case the local inquisitor.

The next license request from Girolamo Coli is recorded in the register

of the Congregation of the Index, and it suggests a more ambitious reflection of Coli as a professional and a reader (see figure 5.7).[74] The record of
the request does not contain a date, although it is likely from either 1628,
three years after the last license, or 1630, three years before the next record. The record also does not explicitly note whether the request was
granted; however, since the list has several items expressly struck from
the list (a particularity of the license requests in this archival series), we
can assume that it was granted with the exception of the deleted items. In
this second license request Coli strikingly presented himself as the "principal physician of the city of Lucca," and he proceeded to list the names of
twenty-five prohibited authors that he sought permission to read. No longer a man with only a degree in arts and medicine, Coli was now promoting his public role and requesting permission for a library that, at least in
terms of its prohibited books, had been considerably expanded during the
previous three to five years.

While some of Coli's requests, such as the three rejected astrological
works by Alchabitius, Francesco Giuntini, and Ptolemy (the *Tetrabiblos*,
of course), were certainly riskier requests than Brunfels's book of plants,
the texts reflect a broader portrait of Coli as a reader, rather than a hidden
trove of previously sequestered texts. Coli's prohibited medical interests
still included works of chemical medicine, but in this request Coli also
listed the Englishman John Caius's *Method of Healing*, the German po-
lemicist Ulrich von Hutten's firsthand account of syphilis, and the French
theologian Sebastian Castellio's Greek and Latin edition of the Sibylline
Oracles.[75] Coli's request to read the complete works of Leonhart Fuchs was
not a daring or contentious line item, and we might question why it was
that Coli did not have works by Fuchs on his list in 1625. With this request
for Fuchs's corpus, we see Coli coming into his own as a book collector
and public intellectual. The works of Fuchs were an established part of
any seventeenth-century medical library, and while it was not difficult
to get permission, the works may have been expensive to acquire. By the
time of his second license request, Coli's library had grown to include prohibited medical, humanist, and astronomical texts that represented the
breadth of learning essential for the self-proclaimed principal (though by
no means only) physician of his city.[76]

Coli's third license, granted in 1633 by the Holy Office under the same
conditions as his first, contains many of the same professional books and
authors as Coli's first license.[77] However, the "medical physician of the
city of Lucca" also introduced new prohibited religious and literary works
to his request. In addition to the copy of Pietro Aretino's translation of the

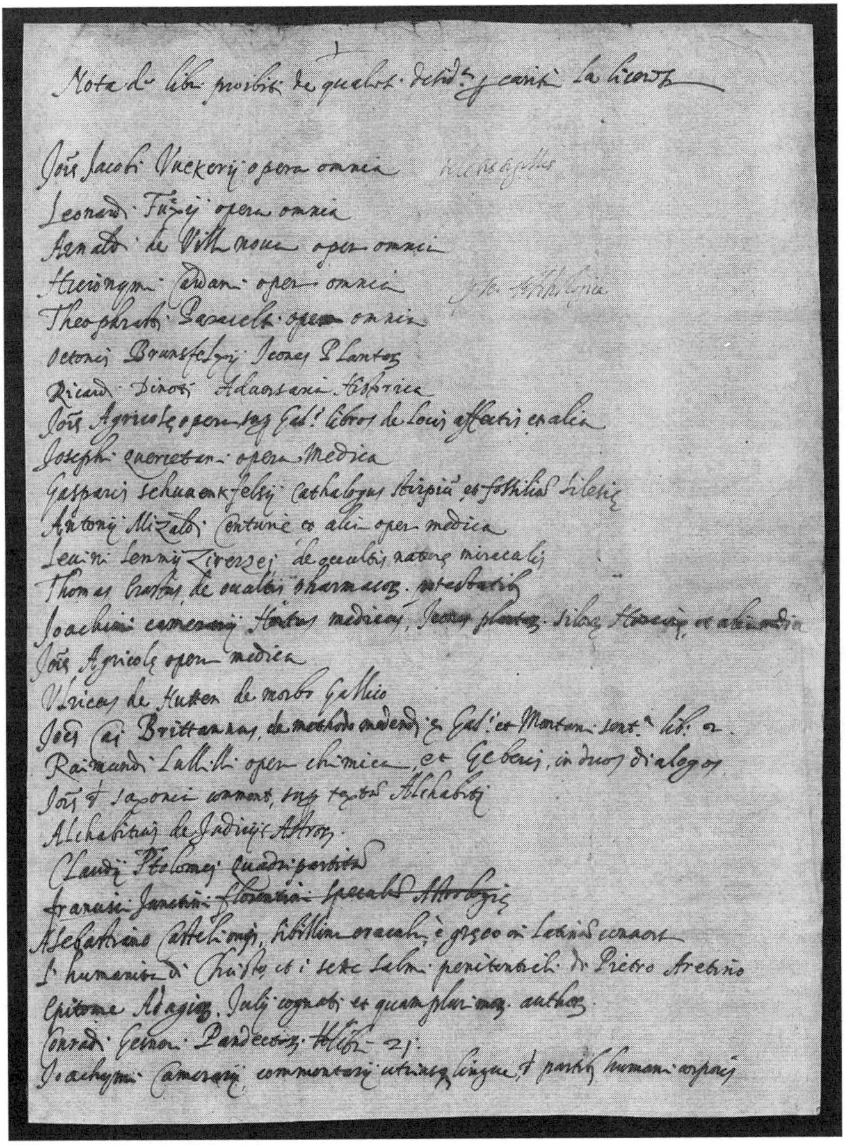

Fig. 5.7. Girolamo Coli's request to the Congregation of the Index of Prohibited Books for a renewal of his reading license. ACDF, Index IX, f. 344–45. Reproduced with permission from the Archive of the Congregation for the Doctrine of the Faith, Vatican City.

penitential psalms that he had been granted in his second request, Coli added Aretino's *L'umanità di Cristo* (*Humanity of Christ*) and *La vita di Santa Caterina vergine e martire* (*Life of Saint Catherine*). Coli also added, as the final item on his list, the corrected *Satire* of Ludovico Ariosto.[78] A passing remark in a nineteenth-century study of literary culture in Lucca suggests that Girolamo Coli's son Stefano grew to be locally recognized for his skill in Italian poetry, and he was listed as a member of the Accademia degli Oscuri in 1643.[79] We know now that he was raised with access to a rich and elite collection of books, many of which were prohibited. The Colis are yet another example of physicians with broad reading interests and admired expertise in subjects that defy narrow definitions of the medical profession.

The final license that the Holy Office granted to Girolamo Coli is dated September 10, 1636, and includes in the license provision both for the "medical physician" and for "Stefano his son."[80] From this license, we can imagine Girolamo and Stefano as they are depicted in the Paolini portrait, in their study and surrounded by great folio volumes, perhaps even consulting together works that were prohibited. It is likely that Girolamo Coli included his son, then in his late teens or early twenties, on the license for the posterity of his library. Reading licenses could not be passed on after death, so by including Stefano in his license, Girolamo Coli set him up to renew his own license in the event his father died.[81] By applying for a license with his son, Girolamo Coli was taking steps to ensure that the library he had worked to assemble would remain intact in the hands of his son following his own death.[82] The library of the "principal physician of Lucca" would be conferred upon the young physician, Stefano, who would go on to have a career as both a physician and a poet.

The list of books that Girolamo and Stefano submitted in 1636 is, with the exception of one book, identical to that of 1633. This does not mean that their library did not grow during this period, only that Girolamo Coli did not add much to his collection of prohibited books. The single prohibited book that Girolamo Coli added to his license must have been one that he bought used or already censored. Coli's request describes the book as a "dictionary of three languages, Latin, Greek, and Hebrew, whose name is now erased such that it can be found nowhere."[83] If this were a book Coli had censored himself, he would likely still have known the title and author. The problem of reporting specific prohibited volumes that had expurgated bibliographical information appears regularly in license requests. Sometimes owners still knew the author or title of a work, as in the case of a Toldo Constantini, a lawyer and poet from the Veneto, who in 1634 requested a license to read a copy of Conrad Gessner's *Biblio-theca universalis* "from which his name was rubbed out, the preface torn out, and many locations had thus far been corrected."[84] Ascanio Bulgarini from Siena requested a long list of classical and humanistic texts edited and printed in Protestant Europe, for which he provided detailed physi-cal descriptions of each book. His list included twelve volumes that were missing crucial bibliographical details. He recorded the entry for his 1541 copy of Julius Pollux's *Onomasticon* by copying the title page and writ-ing, "this is a very instructive book and dictionary of synonyms, now put to Latin by . . . ," filling in the rest of the entry with ellipses. Bulgarini's description then switched into Italian, "Here the name has been removed and blank paper has been put there."[85] The missing name was that of Ru-dolf Gwalther, a Reformed Protestant pastor and translator. Early modern readers were aware of earlier licenses for works in addition to expurgations in them. The rector of the Servites in Tuscany noted that on the first page of the Venetian 1525 folio edition of Francesco Giorgi's *Harmonia mundi* in his monastery, "there is a license from the Father Inquisitor, written in his hand in 1587."[86] Coli's license requests and those of his peers remind us that expurgated books circulated on secondhand book markets and were sufficiently valuable to readers that they purchased them even in what might be considered a damaged state.

Licenses also mentioned previous owners of books that were then resold and remained in circulation with expurgations, occasionally re-marking also on books that were corrected previously by theologians and inquisitors. Gian Vittorio Rossi submitted a long list of books to the Con-gregation of the Index around 1630. He noted in his request that the books had been granted to him in the past but with the recent revocation of li-censes he needed to apply for them again. He added that "a couple of weeks ago they were revised and corrected by the Inquisitor of Florence."[87] Read-ers even occasionally reported books that were corrected by the Spanish Inquisition. Francesco Cennini de' Salamandri, the Italian titular cardinal of San Marciano, requested a license for a book on German history that was written "by different authors in six folio tomes" and for the physician Zwinger's *Theatrum vitae humanae,* both of which he described as having been corrected by the Spanish Inquisition.[88] The used book trade spanned Europe and reflected the licit reading of licensed readers in addition to a heterodox secondhand market.

Stefano and Girolamo Coli are highly visible readers because of their unusually large number of petitions and because their portrait allows us to see these men among their books. But how did their reading practices and interests compare with those of other readers of prohibited medical books across Italy? In what ways were the Colis exceptional readers, and in what ways were their requests for prohibited books typical for physicians? By comparing the general trends in medical reading with the particular interests and requests of individual readers, we can begin to access a broader cultural realm that bridges, in the words of Andrea Ottone, "prescribed readings and personal preferences" and "professional duties and personal idiosyncrasies."[89]

MEDICAL READING IN CONTEXT

Quantifying and graphing reading licenses once again reveals the evolving history of reading licenses as a source base, just as any account of historical data will always be a story about archives and survival.[90] Bearing these archival gaps in mind, we can nevertheless probe the changes in subjects requested by petitioners over time (see figure 5.8). The subjects of the books I describe here are not inherent or obvious categories but instead are my subjective assessment of the works of particular authors. For example, I characterize the works of Leonhart Fuchs as medicine, Francesco Giuntini as astrology, Johannes Jacob Wecker as secrets, and Andreas Libavius as chemistry. Much of the narrative revealed by these larger-scale quantifications aligns well with trends described by historians of science and medicine. Over the course of the seventeenth century, we see a marked

Fig. 5.8. Subjects Requested in Physicians' Licenses

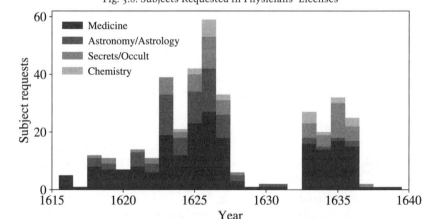

Graph showing requests for licenses by medical professionals, with gradations differentiating among several subject areas frequently requested.

decrease in the number of requests by physicians for prohibited astrology and astronomy texts. This decline stands in contrast to the increased interest in prohibited books of chemistry by authors such as Theophrastus Paracelsus, Andreas Libavius, and Oswald Croll. The increasing number of requests for secrets is likely a reflection of the extreme popularity of Wecker's and Levinus Lemnius's books of secrets, which were republished and translated many times in the sixteenth and seventeenth centuries.[91]

Paracelsianism has long been a touchstone for thinking about the changes in medical thought in early modern Europe. It is also a "slippery" term, full of contradictions and without a cohesive intellectual program.[92] Paracelsus iconoclastically turned his back on Galenic medicine, burned his copies of Galen's texts, and instead promoted cures made from chemicals and minerals. Historians have attributed the slow adoption of Paracelsianism in Italy to the strong predominance of Galenic medicine in the universities and also to the prohibition of chemical medical texts including those by Paracelsus, Croll, and Libavius.[93]

While prohibition of a text used to indicate a nearly invisible reception history or a reception at odds with Catholic orthodoxy, evidence from reading licenses reveals the ways that prohibited thought and Catholic faith coexisted for certain readers. Indeed, readers requested works of chemistry and chemical medicine in early modern Italy.[94] Among the well-known scholars interested in chemical medicine, Pietro Castelli, Raffaelo Gualterotti, and Benedetto Punta all held licenses to read prohibited books. Pietro Castelli's 1635 license, obtained after he moved to Messina to run the botanic garden there, contained a list of what we might expect from a physician working at the intersections of botanical and chemical remedies. His request included works by Lusitanus, Fuchs, and Gessner alongside Paracelsus, his critic Erastus, and Libavius, Croll, and Bernard Gilles Penot. Castelli's interest in heterodox subjects also included the *De rerum natura* (*On the Nature of Things*) of the Calabrian philosopher Bernardino Telesio.[95] The Florentine Punta's licenses were more restrained. The first, granted in 1616, requested only Paracelsus, Croll, and the works of Cardano (with the exception of his commentary on Ptolemy's *Tetrabiblos*).[96] Perhaps emboldened by his previous success, his request from 1623 granted him permission to additionally read Arnald of Villanova's works and Juan Huarte's *Examen ingeniorum*.[97] It seems that these two famous Italian Paracelsians pursued their interests openly and with the consent of the Roman Inquisition. Interestingly, although the Florentine Raffaelo Gualterotti, Galileo's fellow member of the Academy of the Crusca, embraced Paracelsian philosophies, his reading license from the Roman

Inquisition in 1618 does not include any prohibited books related to this subject. Instead, the poet's requests tended more toward the literary, including the works of Cardano (except his commentary on the *Tetrabiblos*), Boccaccio, and Dante's *De monarchia* (*On Monarchy*).[98]

In addition to these well-known Paracelsians, we can also use the license data to identify other Italian scholars interested in chemical medicine. For example, while Domenico Mino was a known Paracelsian who twice received licenses to read a list of authors of Galenic and chemical medicine, it is more surprising to learn of the request from his colleague in Monteregali, Perino Condero, who also sought works of chemical medicine in a request from 1624.[99] Only three petitioners for reading licenses requested the works of Heinrich Khunrath, the German physician and alchemist. One of these requests came from Fortunio Liceti, the eminent professor at the University of Bologna (and later Padua) and friend of Galileo, who received a license in 1640 to read a list of prohibited books that was two pages long! This license allowed Liceti to read 120 prohibited works and authors, though the Roman Inquisition meticulously crossed Galileo's name from the list.[100] Despite being denied a license to read Galileo's works, Liceti and Galileo remained in close correspondence throughout 1640 and 1641. The other two physicians who requested to read Khunrath were Venetians. Domenico Tirillo and Michelangelo Rota applied to the Holy Office in Rome together in January 1626 and received permission to read a long list of medical books, including (among others) the most important prohibited names in chemical medicine: Paracelsus, Libavius, and Croll in addition to Khunrath.

Other readers interested in chemical medicine emerge from the license data. Based on his request for a long list of prohibited books of chemical medicine, Alfonso Alettino ("alias Grimaldo"), a physician from Reggio, must have had an interest in the field in 1626 when his license was granted by the Holy Office.[101] So, too, it is safe to assume a similar interest based on the requests of the physicians Giuseppe Trivellino of Venice, Giulio Fererolo and Giovan Battista Soncino of Brescia, and Mario Schipano of Naples. However, physicians were not the only readers of chemical and alchemical texts. The nobleman from Modena Guido Coccapani and his compatriot, the engineer Antonio Guarino, both applied for and received licenses to read long lists of prohibited chemical texts.[102] From both individual and aggregated information it is clear that reading chemical and alchemical texts in early seventeenth-century Italy was a practice that the Roman Inquisition and the Congregation of the Index regularly condoned. In the 1620s, roughly ten to fifteen petitioners requested to read Paracelsus

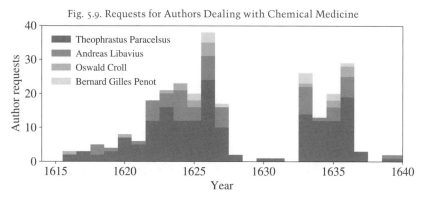

Fig. 5.9. Requests for Authors Dealing with Chemical Medicine

Graph showing license requests (not only by physicians) for authors
who wrote about chemistry and chemical medicine.

each year, while five to nine requested to read Libavius. Croll and Penot
were less popular but consistently requested by a limited number of readers. Figure 5.9 illustrates the variations in demand for Paracelsus, Libavius, Croll, and Penot between 1615 and 1640.

Requests for Robert Fludd's works in the early seventeenth century
present an interesting case of medical readership. Although Fludd built
upon aspects of Paracelsian medicine, it seems clear based on reading licenses requesting his books that the medical content was not the moti-
vating factor in readers' requests. The physicians with permission to read
Fludd included the capacious Catholic consumers of forbidden knowledge
Fortunio Liceti, Francesco Rolando, and Pier Antonio Caballo.[103] Caballo's
close relationship to the Duke of Mantua also sheds some light on other
well-placed readers of this text, including the secretary to the Duke of
Savoy, the great Milanese collector Galeazzo Arconati, and Fabrizio Balneo, the young cardinal nephew to whom Gabriel Naudé dedicated his
Syntagma de studio liberali in 1632.[104] Fludd's work piqued the interest
of the in-house theologian at the Biblioteca Ambrosiana in Milan, Francesco Bernardino Ferrari, and the head of the Jesuit order, Mutio Vitelleschi, who received a license in 1636 not only to read Fludd (and many other
authors) but also received permission from Pope Urban VIII to allow other
Jesuits to read them as well.[105] It seems likely that Vitelleschi added Fludd
to his license based on a recommendation by another Jesuit, Giovanni Battista Zupi, who was also an astronomer and professor of mathematics and
had sought and received a license to read Fludd's works a year earlier.[106]
This range of elite readers licensed to read Fludd reflects the networks of

powerful readers and their diverse interests. Of the eighteen requests to read Fludd by a range of professionals, it is interesting to note that ten of these requests also listed the works of Girolamo Cardano, another eclectic physician and author.

Indeed, among all the requests from medical professionals, the most requested authors were Cardano and Leonhart Fuchs. Cardano, the hometown (heterodox) hero, was named explicitly in a total of 409 license requests and 163 requests by people who were medical professionals. Additionally, requests for Cardano are a consistent proportion of the overall number of requests throughout the period for which we have data. Cardano was no flash in the pan among Italian readers; his impact on Italian culture and readers remained strong a century after his death and regardless of the prohibitions against him. The title page of his *Practica arithmeticae* (1539) included Cardano's portrait encircled by the words of Luke 4:24, "No prophet is accepted in his own country." Yet, based on license requests, this statement seems to have been completely inaccurate. The large number of requests may also be related to the fact that many editions of Cardano's books were published in Italy and were therefore widely available to readers there.

The consistency of interest in Fuchs and Cardano over nearly one hundred years is remarkable. At least two-thirds of requests by medical professionals included works by these two authors. The many licenses granted in the 1650s and 1660s that do not include the names of requested authors obscure the intensity of this interest. If we consider requests for Fuchs without these years, at least 50 percent of the licenses requested by physicians named the Lutheran humanist and botanist. In contrast, Otto Brunfels was requested only a couple of times per year by the seventeenth century. His works had become dated and were less necessary. As an anonymous censor noted in 1590, Fuchs's works had long surpassed those of his predecessor Brunfels.[107] Publications by Cardano and Fuchs both held the attention of Italian readers throughout the seventeenth century.

Girolamo Cardano also had a devoted following among nonmedical readers. Only 40 percent of the requests to read Cardano came from identified medical professionals (163 of 409 total requests; see table 5.1). Of course, we must take into consideration that a large portion of license requests do not provide information about the qualifications of the petitioner. However, we can compare requests for Cardano with requests for Leonhart Fuchs and Amatus Lusitanus, for whom 66 percent and 62 percent of their total requests came from medical professionals. While Fuchs's and Lusitanus's works appealed primarily to medical audiences, the more

TABLE 5.1. TOP TWENTY AUTHORS REQUESTED BY MEDICAL PROFESSIONALS

Author Name	Requests from Medical Professionals	Total Requests
Girolamo Cardano	163	409
Leonhart Fuchs	137	208
Theophrastus Paracelsus	114	190
Arnald of Villanova	89	159
Amatus Lusitanus	78	125
Johannes Jacob Wecker	77	178
Conrad Gessner	73	204
Thomas Erastus	45	80
Antoine Mizauld	42	93
Andreas Libavius	36	63
Theodor Zwinger	35	123
Levinus Lemnius	34	89
Francesco Giuntini	33	99
Heinrich Rantzau	31	64
David Origanus	30	100
Otto Brunfels	27	50
Iovianus Pontanus	26	85
Julius Caesar Scaliger	26	82
Juan Huarte	23	136
Janus Cornarius	23	47

Note: Authors are listed in descending order by number of requests by medical professionals. In the case of an equal number of requests, the order is determined based on total number requests for the author.

varied works of authors such as Cardano, Conrad Gessner, and Theodor Zwinger were widely read beyond the medical community and had strong appeal across Italian society.[108]

Outside of medical readers, the two largest groups requesting Cardano were members of the nobility or secular government and lawyers. The lawyer Alberico Settala was granted a reading license for not only Cardano but also a broad array of medical authors, including Erastus, Mizauld, and Gessner, alongside prohibited jurists including Eberhard Bronchorst and Simon Schard. Settala may have been a lawyer, but his library came from his father Ludovico Settala, a well-known physician, whom Alberico took pains to mention in his application for the license.[109] Annibale Marescotto, a professor of law at the University of Parma (though living at the time in Bologna), wrote to the Holy Office in 1626 with a query about his family's recently inherited collection of prohibited books. Marescotto reported that his nephew had received a license to keep a list of prohibited books primarily comprised of philosophical and literary works. Since Marescotto's nephew's son was showing signs of "the same genius as his father in his studies," Marescotto proposed that he be made responsible for maintaining these prohibited works for his relative.[110] The Holy Office consented to this arrangement, allowing Marescotto to preserve forty-nine books, including quarto copies of Cardano's *De somniis libri decem* (*Ten Books on Dreams*), his *De vita propria liber* (*Book of My Life*), "and others by him" in a locked, separate part of his library.[111] Cardano was an integral part of sixteenth-century libraries and remained important to seventeenth-century collections, even if not always relevant to the professional work of the next generation of readers.

Lawyers who requested to read Cardano tended to do so from a literary or humanist perspective. Giovanni Francesco Scribani, a doctor of law in Genoa, was granted a license to read the works of Cardano alongside a list of other ancient and contemporary authors including Ovid, Apuleius, Lucian, Dante, Enea Silvio Piccolomini, Teofilo Folengo, and Ludovico Ariosto. His requests to read Pietro Aretino's satires, the works (presumably unexpurgated) of Boccaccio, and Petrarch with commentary in Latin and Italian were all denied.[112] Camillo Richelmi, the celebrated jurist and president of the senate of Turin, was granted a license in 1625 to read Cardano's *De subtilitate, De varietate,* "and his other works."[113] The other works requested in his license pertained primarily to his legal work, including Francois Hotman, Valentin Forster, Giacomo Antonio Marta, Agostinho Barbosa, and Matthew Wesenbeck. While legal scholars, like physicians, had a particular and discipline-specific canon of prohibited books that they frequently requested, Cardano's status in the humanist and literary world made his works an important part of legal libraries as well.[114]

For members of the nobility, the other authors requested alongside Cardano indicate that often this author's appeal lay in his approach to astrology. Apelle Lancio of the Tuscan order of the Knights of Saint Stephen received a license to read many prohibited astrology texts. In addition to Ptolemy's *Tetrabiblos* and Cardano's commentary on it, he received permission to read a series of works "On judiciary [astrology]" by Johannes Schöner, Omar Tiberiades, Abraham ibn Ezra, Leopold of Austria, Erasmus Oswald Schreckenfuchs, Haly Abenragel, Albubater, David Origanus, Antonius de Montulmo, Cyprián Karásek Lvovický, and Julius Firmicus Maternus.[115] Similarly, the Venetian government official Luigi

Querini requested to read "all the works of Cardano including those that deal with judicial astrology" and followed the request for Cardano with works by Luca Gaurico, Francesco Giuntini, Messahalla, Omar Tiberiades, Alchabitius, and Sahl ibn Bishr.[116] Ecclesiastical officials also requested Cardano and other astrologers. In 1620, Giovanni Battista Altieri, the canon of St. Peter's Basilica in Rome (but soon elevated to the episcopacy and eventually made a cardinal by Pope Urban VIII), received a license to read his copies of Caradano's *De sapientia, De subtilitate,* and commentary on the *Tetrabiblos,* though Altieri specifically noted that he was not including the horoscope of Christ or the discussion of it in his request. Altieri also requested Haly Abenragel, Albubater, Alchabitius, Giuntini, Origanus, and Iovianus Pontanus. In all three of these lists requested by nobility and secular and ecclesiastical officers, we see the sustained importance of and interest in medieval Persian and Arabic astrologers, a group which included Muslims, Jews, and Syriac Christians.

This interest in Cardano alongside other astrological works does not signify that these licensed readers were necessarily astrologers. The Grand Duchess of Tuscany, Christina of Lorraine, requested a license in 1598 to read Cardano alongside Ptolemy, Alchabitius, and the mathematician Cyprián Karásek Lvovický.[117] By April 1621, Cardinal Mellini had granted Christina an expanded license to read all books that did not deal with religion and "to discuss them with another person of her choosing."[118] The Bolognese noblemen Carlo and Ottavio Ruini received licenses in 1627 to read the works of Cardano, Gessner, and Copernicus.[119] In this case it seems likely that Cardano and Gessner were important humanists to include in their personal library, and the work of Copernicus was of increased interest not for astrological reasons but because of its recent prohibition following Galileo's discoveries. Yet the tendency for the nobility to request Cardano alongside prohibited astrological works is a generalization that only partially captures these political elites' interests. Girolamo Alzano, a "nobleman from Bergamo of a mature age," requested to read the works of Cardano and Paracelsus "in order to study simples and medicine."[120] There is reason to believe that Alzano was serious about this interest; his name appears as one of the citizens locally appointed to establish protections for the poor during the plague outbreak of 1630.[121]

Overall, the list of the top authors requested recalls the sixteenth-century medical republic of letters and the importance of a broad definition of the field of medicine. Many authors in the top twenty listed in table 5.1 reflect both the disputes and the alliances that had defined the sixteenth-century medical community. Conrad Gessner, Thomas Erastus, and Theodor Zwinger's connections with Italian colleagues were likely influential in their continued readership on the Italian peninsula. The polemical disputes of the sixteenth-century medical republic of letters also became canonical aspects of seventeenth-century medical reading. Leonhart Fuchs's enmity with Janus Cornarius would be widely remembered, as would Thomas Erastus's polemic against Paracelsus and Girolamo Cardano's disputes with Julius Caesar Scaliger. Antoine Mizauld, Francesco Giuntini, Heinrich Rantzau, David Origanus, Iovianus Pontanus, and, of course, Girolamo Cardano represent a deep interest in prohibited astrological texts among medical professionals. The increasing interest in chemical medicine is represented on this list by Paracelsus and Andreas Libavius. The popular texts by Arnald of Villanova, Johannes Jacob Wecker, and Levinus Lemnius about regimen, recipes, and medical secrets show the continued interest among elite physicians in popular remedies.[122] While acknowledging that license data are inconsistent and incomplete, these requests nevertheless reveal the long-term importance of the medical republic of letters and offer a window into the ever-shifting allied disciplines that comprised early modern learned medicine.

CONCLUSION

Many physicians sought out prohibited books and applied for licenses to read them, but to what end? Physicians who wanted to read books for illicit purposes, such as embracing religious heresies, would not willingly inform inquisitors of their goals. There were underground religious communities in many Italian cities that furnished books for these purposes. As a vicar of the Venetian Inquisition described in 1559, "hidden, they go lending them [heretical books] from hand to hand."[123] Additionally, among all of the licenses requested, only eighteen requests (about one-third of 1 percent) mentioned prohibited theological texts (and often these were not granted). The vast majority of applicants for reading licenses were not trying to access these works for religious or theological purposes. Instead, physicians framed these requests by emphasizing their professional credentials and in terms of the professional benefit they could obtain by accessing prohibited books.

Physicians who explained their requests cited professional necessity, the importance of healing, and the utility of the texts themselves. The physician Francesco Benucci requested permission to read the works of Girolamo Cardano, Leonhart Fuchs, and Johannes Lange, and the books of secrets of Johannes Jacob Wecker and Albertus Magnus, explaining that he "desired to advance himself in his profession as much as possible and to arrive at an understanding of things that the listed authors dealt with diffusely."[124] The elderly *protomedico* of Lodi, Giulio Inzagi, requested to read Gessner's medical works and Wecker's book of secrets, "with the intent of using them legally and to good ends for the health of his parish."[125] Giacomo Bruni of Colonella, evidently confusing the demands of a reading license request with that of a library catalog, listed more than three hundred texts in his request, most of which were not prohibited. He asked the Master of the Sacred Palace to let him "read and study the books named on this list for the health of the sick."[126] Baldassare Rusca from Como suggested that he needed to read certain prohibited medical books "for the honor of God and the benefit of the public," a sentiment echoed in his townsman and fellow physician Amantio Ripa's explanation of the license as "for the use of his profession and for the service of the public."[127]

This discourse of utility was echoed in license requests by other readers. Alessandro Mazzante, a canon at the Orvieto Duomo, described the decrees from the Council of Trent and the handful of legal books he requested as being "very useful and necessary for his profession."[128] In another instance, Giovan Battista Vertova, a nobleman from Bergamo, submitted his request to read the *Mercurius Gallobelgicus* and works by Gerard Mercator, John Barclay, Sebastian Münster, and Jean Bodin (though this final author was denied) with an explanation that the books were "for his taste and for the utility that comes forth from public service through securing the peace, as he does continuously, not desiring them for any other effect."[129] In a similar vein, legal scholars repeatedly emphasized their professional need for prohibited books by authors including Giacomo Antonio Marta, Tommaso Zerula, and Agostinho Barbosa. The ecclesiastical prohibitions placed on important professional books forced physicians and jurists to appeal to the public utility of their professions. Drawing on the language of utility that justified the correction of these texts, learned readers sought licenses to read books written by heretics that were paradoxically essential to Catholic society.

By the end of the seventeenth century, the importance of prohibited medical books to professional expertise was widely appreciated. In October 1688, when Paolo Bettucci, a physician, astrologer, and poet from Forlì, wrote to the Congregation of the Index for permission to read astrological texts, he couched the request explicitly in terms of lives that could be saved. Referring to himself humbly in the third person as "the orator," Bettucci described how he had argued that God rules the world using the stars as instruments. He continued, "The universities themselves testify that no one medicated by the orator has died, therefore he decided to reveal to the world, by writing a book, the way these things operate for the common good and in order to enrich that book he would need to read books by astrologers who deal with medicine, and to that end he requests a license."[130] To write a book that would save the most lives, the physician, astrologer, and "orator" Bettucci needed to consult prohibited astrologers: he sought a license to read prohibited books in order to save lives.

Reading licenses made a wealth of otherwise prohibited materials legitimately available to many physicians across Italy. The system of licensing and professional justification that became standardized over the course of the seventeenth century lasted well into the nineteenth century. The formula for licensing medical readers in 1850 as overseen by the secretary of the Index, Vincenzo Modena, was distinct from licenses issued for priests, lawyers, and "for those who took a course in literature and philosophy."[131] The form called on the authority of Pope Pius IX and licensed doctors "to read and to keep, in custody, however, such that they come into the hands of no one else, prohibited books about medicine, anatomy, chemistry, and surgery. Also prohibited books of grammar, rhetoric, logic, philosophy, mathematics, astronomy, and secular history."[132] The Catholic understanding of the expertise of the physician as requiring a broad grounding in medical, scientific, and humanist disciplines persisted far beyond the Renaissance. Broad reading was recognized as an essential part of a physician's work, and the system of reading licenses, combined with the 1607 *Index Expurgatorius*, made this reading permissible within the Italian culture of censorship.

CHAPTER SIX

Creating Censored Objects

Individual books allow us to unpack the ways that the censor, the book, and the reader were all involved in creating the censored objects we find in libraries and archives today. I turn now to the material practices of expurgation—how these practices have been studied by scholars, how they were executed by readers and censors, and how they shed light on the individuals involved in expurgation. I conclude by questioning the early modern goal of expurgation in light of the vast range of material interpretations of the rules of the Indexes. If texts were not always censored with the intent of entirely removing forbidden material, what did expurgation accomplish? The process of disciplining books written by Protestant authors, however incomplete and inconsistent, was ultimately a way of reinforcing the boundaries of Catholic community in an age of confessional difference.

In 1991, John Tedeschi described the archives of the Roman Inquisition as a dispersed archive. He traced documents created by the tribunal in Rome on the peregrinations that took them to Paris with Napoleon, then back across the Alps (with substantial losses) to Rome, while still others of these documents found their way to libraries as far afield as Dublin. The archives of the Roman Inquisition, he concluded, have been fundamentally dispersed in ways that shape the record that has come to us.[1] When the Archive of the Congregation for the Doctrine of the Faith (ACDF) opened in 1998, scholars rapidly confirmed Tedeschi's predictions—there were no complete collections of trial records comparable to those that exist for Spain.[2] However, the records of the Congregation of the Index of Prohibited Books were remarkably complete, and much of the administrative business of the body concerned with censorship was now available to scholars. Further research in the ACDF also revealed nearly complete decree registers documenting the official edicts of the Inquisition in Rome.

We are beginning to understand the bureaucratic functioning of the bodies that regulated books and behavior in Italy from the mid-sixteenth through early twentieth centuries, but we still lack a material understanding of Catholic censorship in Italy. In this chapter I turn to the censored books themselves as the most dispersed of all the archives of the Roman Inquisition. Taking these objects together as an archive of practice, I examine censored books as artifacts of how censorship was physically enacted on medical texts in the sixteenth and seventeenth centuries.[3]

Drawing on the methodologies of material history and critical bibliography, I have located and examined copies of expurgated books from libraries across the United States and Italy. Although inquisitors burned some books, initiatives to selectively expurgate texts have left physical evidence of books "corrected" by striking through objectionable words or phrases with a pen, by cutting them out or scraping them away with a knife, or by gluing scraps of paper over controversial sections. Examining these objects reveals how readers and Catholic authorities alike understood the printed book as an intellectual threat and also as a physical object that could be manipulated, regulated, and transformed. Books, like readers, could change, and their opinions could be, at least physically, altered through interactions with ecclesiastical censorship. Whereas the 1607 *Index Expurgatorius* explicitly delineated what bibliographers would call an "ideal copy" of certain prohibited books, my research clearly shows that censors and scholars applied these rules in a variety of ways and with clearly different intentions.[4] From a thin diagonal line across a paragraph to removing pages entirely with a razor, readers participated in the order to "delete and scrape away" (*deletis et abrasis*) in ways that reinterpreted and sometimes even undermined official edicts. Combining historical and bibliographical approaches, I delve into medical books themselves as a lost archive that documents the process of censorship. This archive reveals the varied forms in which readers encountered books and negotiated the unstable relationships between reading, writing, and orthodoxy in the sixteenth and seventeenth centuries. While the processes that readers used to alter their prohibited books differed, the material practices of expurgation were part of a larger project of censorship that sought to transform the ways scholars read and delineated an orthodox Catholic community in the long aftermath of the Reformation.

RUCHESIUS'S FUCHS

This chapter draws on hundreds of copies of censored medical texts that I have examined in dozens of libraries. I begin here with a close reading of a particular artifact: a copy of Leonhart Fuchs's *Libri IIII, difficilium aliquot quaestionum* (*Four Books on Some Difficult Questions*; see figure 6.1). This book currently resides in the Vatican Library in the Raccolta Generale Medicina, one of the subject-themed collections without an easily traceable provenance.[5] The unillustrated work was printed in quarto in Basel in 1540, and it likely came to Italy shortly thereafter—Fuchs's work had been popular on the Italian peninsula since the 1530s. The volume is bound in what is likely an original limp vellum binding, typical for the period, though it probably arrived on the Italian peninsula in loose sheets. The work is also bound with another quarto volume written by Fuchs, the *Apologia Leonharti Fuchsii* (*Defense of Leonhart Fuchs*), printed in Hageneau in 1534. The first text is signed twice in manuscript on the frontispiece: "Bonardo Ruchesio Medico Auctore" between the first and second blocks of text, and then again "Bonardi Ruchesii" between the second block of text and the name of the city where it was printed. Unfortunately, we know nothing about this Ruchesius except that he owned this book and considered himself to be a medical author. He was almost certainly a physician, and if he was indeed a medical author, he published in manuscript rather than in print since there are no books printed by an author with this name.

This artifact reveals that censors, inquisitors, and readers who expurgated books were attentive to the ways in which the materiality of a book impacted how it could be read and censored. This book has been expurgated several times, and possibly by several people: Ruchesius may have been one of the censors but was likely not the only one. On the title page of the book, the author's name, Leonhart Fuchs, has been blacked out in highly acidic ink which is now eating through the paper. The descriptor "medici" has also been blacked out, though the description "ac publici scholae" was not. A censor then pasted over the blacked-out sections with a blank piece of paper, and while "ac publici scholae" was not blacked out, it was covered up. Both acts of expurgation acknowledged that Fuchs's name needed to be obscured, but how that task was accomplished and which descriptors needed to be deleted alongside it were subject to personal interpretation.

It is also worth noting that this book has been "uncensored." After the initial expurgations, a later reader or owner tried to remove the piece of paper covering "Fuchs." This uncensoring was probably a much later intervention from the eighteenth or nineteenth century, and it indicates that the book came into the Vatican Library after that date. Censored books in sixteenth- and seventeenth-century named collections in the Vatican Library

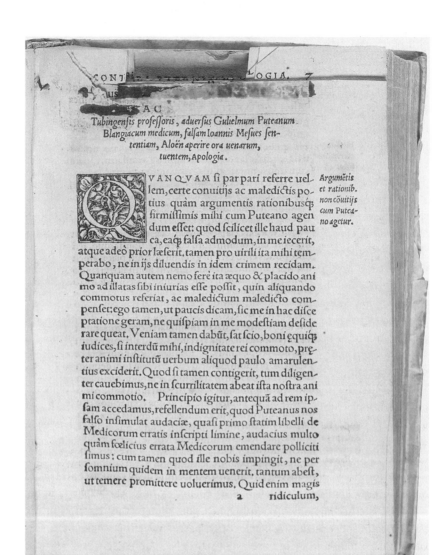

‡ A C
*Tubingenſis profeſſoris, aduerſus Gulielmum Puteanum
Blangiacum medicum, falſam Ioannis Meſues ſen-
tentiam, Aloën aperire ora uenarum,
tuentem, Apologia.*

V A N Q̲ V A M ſi par pari referre uel- *Argumētis*
lem, certe conuitijs ac maledictís po- *et rationib.*
tius quàm argumentis rationibusq̃ *non cōuitijs*
firmiſſimis mihi cum Puteano agen *cum Putea-*
dum eſſet: quod ſcilicet ille haud pau *no agetur.*
ca, eaq̃ falſa admodum, in me iecerit,
atque adeò prior læſerit, tamen pro uirili ita mihi tem-
perabo, ne in ijs diluendis in idem crimen recidam.
Quanquam autem nemo ferè ita æquo & placido ani
mo ad illatas ſibi iniurias eſſe poſſit, quin aliquando
commotus referiat, ac maledictum maledicto com-
penſet: ego tamen, ut paucis dicam, ſic me in hac diſce
ptatione geram, ne quiſpiam in me modeſtiam deſide
rare queat. Veniam tamen dabūt, ſat ſcio, boni equiq̃
iudices, ſi interdū mihi, indignitate rei commoto, præ-
ter animi inſtitutū uerbum aliquod paulo amarulen-
tius exciderit. Quod ſi tamen contigerit, tum diligen-
ter cauebimus, ne in ſcurrilitatem abeat iſta noſtra ani
mi commotio. Principio igitur, antequā ad rem ip-
ſam accedamus, refellendum erit, quod Puteanus nos
falſo inſimulat audaciæ, quaſi primo ſtatim libelli de
Medicorum erratis inſcripti limine, audacius multo
quàm fœlicius errata Medicorum emendare polliciti
ſimus: cum tamen quod ille nobis impingit, ne per
ſomnium quidem in mentem uenerit, tantum abeſt,
ut temere promittere uoluerimus. Quid enim magis
a ridiculum,

Fig. 6.1 Copy of Fuchs's *Libri IIII, difficilium aliquot quaestionum*, expurgated in ways that show the Catholic censor responding to the layout of the printed pages. Leonhart Fuchs, *Libri IIII, difficilium aliquot quaestionum* (Basel, 1540). Biblioteca Apostolica Vaticana, RG Medicina IV.3824 (int.1). © 2020 Biblioteca Apostolica Vaticana.

were never uncensored, though this is a fairly common characteristic of expurgated books that circulated on the antiquarian book market.

The most striking feature of the page shown in figure 6.1 is the large part of it that is missing—part of the top has been cut away, and the pages behind it show similar incisions. Examining the page's verso in an unexpurgated copy immediately reveals the reason for the missing paper: Fuchs's name was printed above the text block on the verso of each page. The missing parts of the page are where Fuchs's name was removed from the book with a blade. In this copy, on the versos of a2 and a3, there is a fringe of pasted-on white paper visible around the edge of where "Leonhart Fuchs" was cut away. These scraps of pasted-on paper reveal that Fuchs's name was originally pasted over also. The censor must have initially intended to paste over the name on the verso of every sheet. However, after doing this labor-intensive process for a few pages, the expurgator of Ruchesius's Fuchs decided that it would be quicker, easier, and arguably more effective to do away with the cut-and-paste project and resort simply to cutting—slicing the name off the page by cutting into the top of the page after the name "Fuchs" and then turning the razor left to slice across to the edge, removing the name entirely. The censor proceeded with this process despite having already pasted over those first few examples, perhaps to maintain a degree of uniformity in how the text looked after expurgation. Readers were attentive to the aesthetic details of how a book was censored. In April 1636, the marquis Vercellino Maria Visconti wrote from Milan to apply for a reading license with the caveat that he did not want "to ruin the books" and that he would "rather not have them than have them mistreated."[6] Meticulously neat or uniform expurgations were often a sign that the person expurgating the volume was not a professional censor (an inquisitor or vicar) but instead an owner who intended to read the book and to use it after it had been expurgated.

The expurgator of Ruchesius's Fuchs used a slicing technique that was efficient, but then he realized after only a couple of pages that this method was flawed. After all, the expurgator was not only living in an age of censorship, he also lived in an era of information overload and information management.[7] It was not enough for an early modern scholar to possess a vast collection of books; it was also important to be able to find the important knowledge within these often hefty volumes. The early modern period saw a proliferation of techniques of information management in both print and manuscript. In this case, the expurgator had realized that removing the top of the page from the inside margin to the outer edge removed Fuchs's name but also had the unfortunate and unintended consequence of removing the page numbers from the text! Fuchs's books regularly advertised their "most copious indices," in multiple languages, but the expurgator's technique effectively eliminated one's ability to use this important text technology. By folio a4, the expurgator realized his mistake and changed his method. Beginning with page 8 (the corresponding verso), he cut down into the page, then across (from left to right when you are facing the verso—he occasionally slipped into the seam margin but never the other way around), and then down again, essentially "biting out" Fuchs's name from the pages.

If you hold Ruchesius's Fuchs closed in front of you and look down at the top, there is simply a hole in the top edge. In fact, there are two holes because the censor applied the same method of removing the name from the verso to the "Apologia adversus Hieremiam Thriverum Brachelium, Medicum Lovaniensem," which begins on page 141 of the text. By the time the expurgator arrived at the second text in the bound volume with the author's name on the verso headers, he had already established a method, and there is no messiness of unnecessary pasting or accidental page number removal. Fuchs's name is carved out in a clean block with no collateral damage to the rest of the book.

The expurgator's reaction to the layout and organizational elements of the printed page was unusually explicit in the case of Ruchesius's book, but this item is not unique. A two-column New Testament with the original Greek printed on the inside column and Erasmus's Latin translation on the outside of the page has been cut in half to completely remove the column that Erasmus translated (see figure 6.2).[8] Like Fuchs, Erasmus was listed on the Index, and thus his "work" on the New Testament (his Latin translation) was removed completely from the book by cutting off the outside half of every page.[9] Around the same time that the book was cut down to only its internal column, it was rebound in its new, narrower state. As in the case of Ruchesius's Fuchs, the page numbers in this volume also fell to the censor's knife. However, page numbers are not the main organizing device in a Bible, so instead a reader of the text added chapters in Arabic numerals to the edges of the Greek text. Readers and censors were aware of both the rules about expurgating books and the exigencies of the book itself, which imposed limitations on, and conversely possibilities for, how it could be censored.

Postpublication censorial interventions were not automated processes; they are the result of a relationship between book and reader, or book and censor, mediated by the materiality of the object. A copy of Thomas Erastus's *Disputationum de noua Philippi Paracelsi medicina pars tertia*

Fig. 6.2. Octavo edition of Erasmus's Latin translation of the Greek New Testament that has been sliced in half vertically to remove Erasmus's translation while leaving the Greek text. It was then later rebound in a narrower format. Uncatalogued. Courtesy of the Biblioteca Comunale Manfrediana in Faenza.

[Third Part of the Disputations Concerning the New Medicine of Paracelsus] has the author's name printed in the header of the verso of every page, as Ruchesius's volume by Fuchs did. Its expurgator censored the text carefully, using pen and ink to cross out "ERASTI" from "ERASTI DISPUTAT." on each folio. On the first verso he censored, he slipped and crossed out both "ERASTI" and part of the title. With the exception of only a couple more slips, he dutifully crossed out "ERASTI" 179 times. When the printers erred and printed "DISPUTAT. DISPUTAT." on page 16 instead of "ERASTI DISPUTAT." the careful expurgator was paying enough attention to not mark the text.[10]

Other expurgators were less patient. In a beautiful 1546 Froben folio edition of the works of Hippocrates edited by Janus Cornarius, the expurgator removed Cornarius's name from the title page by pasting over it with a blank piece of paper.[11] A later reader reinscribed Cornarius's name on

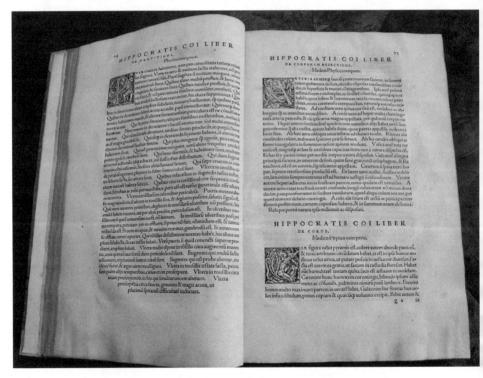

Fig. 6.3. Pages 74 and 75 of *Hippocratis opera*, in which a censor has pasted over the author Janus Cornarius's name with small slips of paper. Janus Cornarius, *Hippocratis opera* (Basel, 1546). BCMF, CINQ.004.002.014. Courtesy of the Biblioteca Comunale Manfrediana in Faenza.

the blank surface. He then cut out the dedicatory epistle and pasted small strips of paper over every heading where Cornarius's name appeared (see figure 6.3). This technique was meant to be unobtrusive and to avoid marring the beauty of this gorgeously crafted edition, but the labor involved in cutting and pasting slips for the whole volume was insurmountable. On page 228, evidently overwhelmed by the magnitude of the task (the book is almost 700 pages long), the expurgator's ambitions flagged and the censorship became haphazard or nonexistent. The material traces of expurgation testify to the human labor of censorship and the intimate work involved in altering texts.

MATERIAL MEANS OF EXPURGATION

Expurgated books are arresting objects, at times strikingly mutilated, at other times subtly transformed. Given their visual impact, expurgated objects have featured regularly in exhibitions, but there have been relatively few scholarly publications that attempt to explain the ways that books were physically expurgated in the early modern period.[12] Rodolfo Savelli's study of the library of the Genoese physician Demetrio Canevari (1559–1625) is the most careful examination of the place of expurgated texts in a private book collection. Savelli recounts the "physical testimony of the volumes" to consider how the censored books speak to the orthodoxy of the collection and reproduces several examples of texts that are expurgated in surprising ways.[13] Silvana Seidel Menchi's article "Seven Ways to Censor Erasmus" laid out an initial morphological classification of types of censorship that she identified among the many copies of works by Erasmus that she examined.[14] Her list includes prepublication censorship and "ritual censorship" alongside a broad category that she calls "material destruction." This chapter considers the ritualized aspects of material destruction to classify the material processes that censors used to expurgate books and to understand how these alterations influenced readers.[15]

The material analysis of expurgated books treats expurgated objects as an archive of practice, combining the bibliographical description of individual books with attention across copies to better understand practices and contexts of expurgation. The following pages are not informed by a complete census of extant works but rather by a strategic sampling of printed works written by the prohibited authors named regularly in requests for reading licenses in Italy. As such, the works of Leonhart Fuchs feature prominently because his works were among the most popular and regularly requested of prohibited medical texts. Additionally, because Italian censors in general determined that Fuchs's books did not deal with religion, readers were often allowed to expurgate copies themselves. Since so many people engaged in this process of expurgation (rather than solely professional censors like inquisitors and their vicars), we are left with a remarkable range of approaches to the material enactment of censorship. The material practices described in the following pages reveal the many ways that readers and censors navigated the blurred line between expurgation as book destruction and expurgation as book preservation. To save these useful books, expurgators stoked fires and sliced pages, altered text with ink and ceruse, obscured passages with paper, and transformed the experience of reading into an act of Catholic piety.

FIRE

Book expurgation took time, patience, and effort to alter texts and render them Catholic. It was an effort that was reserved only for certain kinds of books. Protestant books in Catholic Italy were systematically burned in piazzas across the peninsula, following the biblical precedent set in Acts 19:19 (DV): "And many of them who had followed curious arts, brought together their books, and burnt them before all." Public burning was meant to instruct common people and to be a public spectacle. This method was taken up to great effect by Martin Luther himself in 1518 and by Theophrastus Paracelsus, who had publicly burned the works of Galen to emphasize that his new chemical medicine replaced the written corpus of the ancient author. Local bishops and inquisitors oversaw the public burning of vernacular translations of the Bible and the Talmud.[16] In 1597, the inquisitor of Perugia also observed that the public nature of these acts was essential because otherwise parishioners were loath to turn over their books, suspecting that the inquisitor was keeping them for himself.[17]

Book preservation and book burning were complementary impulses in Counter-Reformation Italy. The frescoes in the Salone Sistino, the main reading room of the recently renovated Vatican Library, depict three historical book burnings, which are part of an iconography meant to show both the evolution of knowledge and the limits that faith, in the Vatican's view rightly, imposes.[18] The first image is of one of the earliest Roman stories about book burning: the conflagration of the Sibylline books. Depicted alongside Augustus's Palatine Library, these two images under the heading "Roman Libraries" point to the long traditions of both book destruction and book preservation in antiquity. The other two images of book burnings appear in a section of the Salone Sistino depicting Catholic church councils. These two images are particularly telling because Pope Sixtus V was personally involved in dictating the iconography in the series on church councils, which highlight the continuity of Catholic dogma and the central role of removing error through repentance and the purging of texts. The first painting in this series depicts the Council of Nicaea (AD 325), which formally condemned Arianism. On the wall to the left of this fresco is another fresco depicting the burning of the Arian books (figure 6.4).[19] The Fourth Council of Constantinople (869–70) is represented by kneeling penitent figures on the right, and the conflagration of Photios's books on the left.[20] The symbolism in both of these images is straightforward. With one hand the Church burns books to remove their sins, and with the other hand converts the heretic. At the end of the sixteenth century when Sixtus V commissioned and helped dictate these messages, he was also preparing his Index of Prohibited Books, which was promulgated in 1590 and led to book burnings across the Italian peninsula.

Although book burning as expurgation may seem counterintuitive since it affected the whole book and not just select portions, sixteenth-century inquisitors lived in a world where images of book burnings decorated the walls of libraries, and they did not see expurgation and book burning as fundamentally opposed. To them, both acts brought about the purification of the text. On December 27, 1603, Arcangelo Calbetti, the inquisitor of Modena, wrote to the Congregation of the Index with a copy of the errors he found in a number of books. After reading the burlesque poetry of Francesco Berni, Calbetti concluded, "Even if a few words are removed, nevertheless the sense of the whole chapters remains obscene, which can easily be seen by whoever reads them. In my opinion, for this reason it cannot be expurgated except by fire."[21] Burning may have expurgated the sin from a book, but in so doing it destroyed the whole object.[22] Sometimes the misdeeds of a book were so great that censors also sought to apply this same total expurgation to the author. As the bishop of Cagli reported, he and his congregation of correctors had decided that there were so many errors, both lascivious and against the faith, in Gian Maria Velmatio's *Christiade* that "we are all agreed in our opinion that not only should this book be burned, but if the author were still living he should be burned also."[23]

However, as we have seen from the expurgations proposed by censors, medical books were prohibited primarily because of the authors' religion and secondarily because of their content. In many cases, the name of a Protestant author was often the only part of the book that needed to be removed. As a result, medical books were burned in their entirety less frequently. One of the few references to the burning of medical texts is a

Fig. 6.4. Fresco depicting the burning of the Arian books in the Salone Sistino of the Vatican Library. The plaque at the bottom reads, "By decree of the Council of Emperor Constantine the books of the Arians were ordered burned." © 2020 Biblioteca Apostolica Vaticana.

list of books burned in Montepulciano in 1598, during the morning mass outside the doors of the Church of Saint Francis. This account mentions works by "several medical commentators of the first class."[24] More often, instead of burning the whole book, readers and religious officials turned to partial expurgation to correct these works. The archbishop of Aquila, Giuseppe De Rubeis, wrote to the Congregation of the Index on April 10, 1602, to recount the diligence he had used in correcting Leonhart Fuchs's *Paradoxa medicinae*. De Rubeis noted that the dedicatory epistle expounded impiously against Catholics and that in book 1, chapter 29, Fuchs used the third chapter of Romans "for similar ends . . . applying to it his perverse Lutheran and heretical sentiment." He continued, "Although I cancelled those sections . . . I believe it would be good to burn it."[25] For some aggressive censors like De Rubeis, selective expurgation was not enough. However, his report is an outlier, and the vast majority of readers preferred to obtain reading licenses and correct their copies rather than consign a whole work to the flames.

BLADE

While books were sometimes burned whole, the fresco of the burning of the Arian books in the Salone Sistino (depicted earlier in figure 6.4) shows a figure in the front right, crouched with muscles tensed from the effort of tearing out pages to feed to the flames. If the intent was to destroy a whole book, ripping pages by hand was an effective way to remove them and ensure that they caught fire more quickly. However, books from which only certain passages, pages, or names were to be expurgated called for a more surgical approach using a blade. Slicing or razing sections of books was an effective way to remove unwanted pages and was typically applied to preliminary materials such as title pages and dedicatory letters. The case of Ruchesius's copy of Fuchs with the author's name cut from the top of the page (as shown earlier in figure 6.1) is a unique example among the censored books I have examined, but it likely happened in other circumstances as well. In his request for a reading license in 1626, Ascanio Bulgarino from Siena requested permission to keep and read a 1541 copy of Julius Pollux's *Onomasticon*, edited by Rudolf Gwalther. However, in place of Gwalther's name Bulgarino wrote instead a series of ellipses and the note: "Here the name has been cut out and blank paper has been put there."[26] In this case we might imagine the work to look today more like the 1536 copy of Otto Brunfels's *Herbarium vivae eicones* (*Living Images of Plants*) shown in figure 6.5, with the author's name cut out of the center

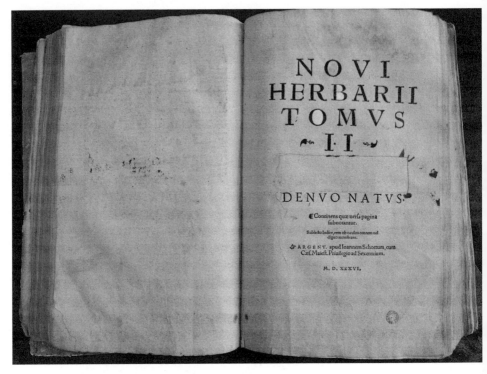

Fig. 6.5. Title page of Otto Brunfels, *Herbarum vivae eicones* (Strasbourg, 1532)
from which a censor has cut Brunfels's name out of the center of the page. The
page was then reinforced by pasting blank paper onto the back to prevent the
rest of the page from ripping when it was turned. BCMF, CINQ.004.003.013.
Courtesy of the Biblioteca Comunale Manfrediana in Faenza.

of the page and then filled in with blank paper, instead of leaving a frag-
ile and unsightly hole.[27] Reinforcing the expurgated page is one of many
measures readers took to protect their books at the same time that they
censored them.

Removing pages from a book with a knife could be executed in a way
that either hid the censorial intervention or drew attention to it. Large
stubs left part of the page and even shortened lines of text waving from
the gutter like flags loudly announcing to readers that there was material that had been removed.[28] In other examples, removed pages were carefully cut close to the gutter. A reader of Thomas Erastus's *Disputationum
de nova Philippi Paracelsi medicina pars tertia* noted in the gutter next
to a short stub from the excised page that only the preface was missing
(see figure 6.6).[29] The ink and handwriting indicate that this note calling

attention to the expurgation was written long after the book's publication, and for many years the missing prefatory letter from Erastus to Crato von Krafftheim must have passed quietly unannounced.

The dull side of a blade could also be used to scrape away portions of the text from the page. Abrasion could be used as a technique to remove a word, or at most two, since the risk of collateral damage by wearing a hole through the paper was very high. A Jesuit-owned copy of Gessner's *Historia plantarum* has Gessner's name, "Conrado," scraped from where it appeared in the middle of a page of text in the dedicatory epistle.[30] A blade could also be used to remove the name of a prohibited author from a leather binding of a book by either scraping it off or cutting out the part of the binding where the name appeared.[31] Figure 6.7 includes two images of bindings that have been altered to remove Leonhart Fuchs's name. In the first image, the censor scraped the name from the spine, leaving the title of the book below. In the second image, the red text of the binding pages is visible, peering out of the hole cut into the middle of the binding to remove the word *Fuchs*. The beginning of Fuchs's name, "Laeon," remains to the left of the hole, as do the first words of the title of his book, *"de*

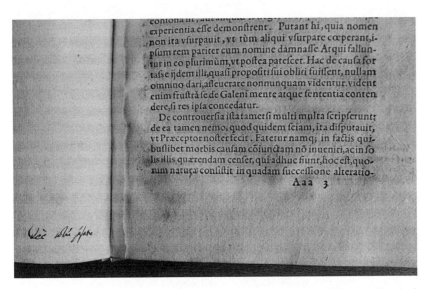

Fig. 6.6. Excerpt from an Erastus volume, with expurgation indicated by a small stub of the excised page. A handwritten note added at the bottom of the facing page explains that only the preface is missing [*Dee[st] solum praefatio*, "it is missing only the preface"]. BSVP, Thomas Erastus, *Disputationum de noua Philippi Paracelsi medicina pars tertia . . .* [Basel, 1572]. Call number 500.ROSSA.SUP.C.6.-43.3. Reproduced with permission from the Biblioteca Antica del Seminario Vescovile di Padova.

Fig. 6.7. Two examples illustrating how a blade was a convenient tool for expurgating authors' names from books' bindings as well as pages. In the top image, prohibited author Leonhart Fuchs's name has been scraped off a binding. In the bottom image, a piece of the binding containing his last name has been cut out, leaving the threads and the rest of the binding intact. Reproduced with permission from the Ministero per i Beni e le Attività Culturali e per il Turismo and the Biblioteca Universitaria di Padova.

medend[is]," to the right. Expurgated books in original bindings often bear signs of censorship even from the exterior.

INK

Using ink to change characters and images was by far the most common way to expurgate a text, and there are endless variations on how this was accomplished. The most common evidence of censorship in medical books is the existence of names or words blacked out with thick lines of iron gall ink. These black lines appear regularly in books written, edited, or printed by Protestants and housed in Italian libraries. Peter Stallybrass has proposed that so-called professional censors—and by this he means not lay readers but instead ecclesiastics charged with correcting texts—primarily censored Petrarch's Babylonian sonnets with a brush rather than a pen.[32] While this may have been the case for Petrarch, it was a rare technique for medical books.[33] This is not to say that the practice did not occur but rather that it was much less prevalent than censoring passages with a pen. Since most medical texts only rarely required that large passages be expurgated, a pen, rather than the broad stroke of a brush, was a more accurate and practical implement. Some pen strokes are regular and similar enough to those that Stallybrass describes to suggest that there are certain marks, such as a thick squiggly line through text, that stand out as being particularly common in book expurgation (see figure 6.8).[34] This kind of pen mark often indicates that the book was censored by professional censors such as local inquisitors and their vicars, whereas irregular or inconsistent marks indicate a censor who was also the owner and reader of the text.

While blacking out text with a dark line or thick squiggle of ink was common, readers could also be more subtle and less visually intrusive with their inked expurgations. At times it seems clear that a censor or reader used thin lines to make sure that passages remained legible. A large cross through whole pages or thin lines through words did little to obscure the text beneath these marks. One of the great paradoxes of censorship is that certain thin lines of expurgation begin to look more like underlining that calls attention to expurgated text rather than obscuring it. Similarly, highly diluted ink could look more like a wash over parts of the text than an impediment to readers. These kinds of expurgations raise the question of the intent of the expurgator since they leave the text completely readable.

Transforming letters into other letters is an especially creative method of expurgation, which I have found in numerous copies of medical books

Fig. 6.8. Regular thick squiggles of ink obscuring objectionable text in a banned
volume by Fuchs. Markings like those shown here were a typical method of
expurgation used by professional censors such as local inquisitors and their
assistants to cover prohibited names and text. BSVP, Leonhart Fuchs, *Paradoxorum
medicinae* (Venice, 1547), call number: 500.ROSSA.SUP.APP.-5.2.-8.a/b. Reproduced
with permission from the Biblioteca Antica del Seminario Vescovile di Padova.

from libraries across Italy and the United States (see, e.g., figure 6.9).[35]
These transformations of letters are not in the same hand, and the volumes containing this kind of expurgation are dispersed across a range
of libraries such that it is highly unlikely that they are the work of only
one "transformer." More intriguingly, transforming letters was a known
practice of expurgation employed by multiple readers and censors in Italy
in this period. These readers, using a method that was consistent but not
identical across volumes, expurgated their texts by changing letters and
rendering prohibited names instead as innocuous gibberish and random
strings of characters. This form of censorship obscured the name of the
author without detracting from the aesthetic of the book.

Whatever the motivation for using this technique, transforming letters
was certainly not the work of a professional censor with limited time and
little interest in preserving the aesthetics of the text. These transforma-

tions are the work of readers and book collectors who, in many cases, were probably also physicians and who altered these important books with care and patience. Transformations of the letters of prohibited names turned these forbidden words into a form of decoration, like a letter-based type ornament framing the text that, with a license and without the author's name legible, could transform the book into licit reading material.

Although in the many copies I examined, the characters of an author's name are not transformed into other words, one book with transformed letters has two locations where words were written on top of Fuchs's name rather than squiggles.[36] This copy of Leonhart Fuchs's *De humani corporis fabrica* (*On the Fabric of the Human Body*) from 1551 is expurgated using several methods. The first method, visible still on the title page, includes writing over Fuchs's name and then pasting a slip of blank paper over the name. The expurgator used this same strategy on Fuchs's name in the dedicatory letter. In the instances of the author's name that followed, the expurgator transformed the letters into nonsense characters. The words under the slips on the title page and dedicatory letter are visible only because someone uncensored the book, and the slips are partially torn off. Even so, the words the reader inscribed over Fuchs's name are impossible to read without a very bright light and manipulations made possible by digital photography.[37] The notes this expurgator left under the slips of paper were intended for his eyes alone.

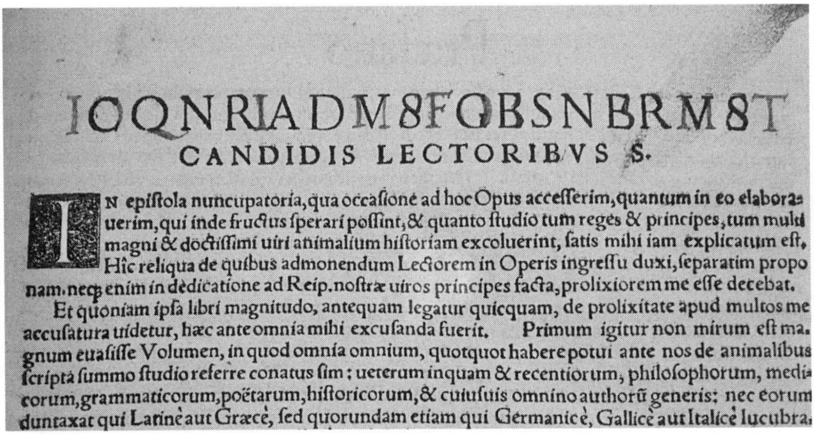
IOQNRIADM8FOBSNBRM8T
CANDIDIS LECTORIBVS S.

IN epiſtola nuncupatoria, qua occaſione ad hoc Opus acceſſerim, quantum in eo elaborauerim, qui inde fructus ſperari poſſint, & quanto ſtudio tum reges & principes, tum multi magni & doctiſſimi uiri animalium hiſtoriam excoluerint, ſatis mihi iam explicatum eſt. Hîc reliqua de quibus admonendum Lectorem in Operis ingreſſu duxi, ſeparatim proponam, neq; enim in dedicatione ad Reip. noſtræ uiros principes facta, prolixiorem me eſſe decebat.
Et quoniam ipſa libri magnitudo, antequam legatur quicquam, de prolixitate apud multos me accuſatura uidetur, hæc ante omnia mihi excuſanda fuerit. Primum igitur non mirum eſt magnum euaſiſſe Volumen, in quod omnia omnium, quotquot habere potui ante nos de animalibus ſcripta ſummo ſtudio referre conatus ſim; ueterum inquam & recentiorum, philoſophorum, medicorum, grammaticorum, poëtarum, hiſtoricorum, & cuiuſuis omnino authorū generis: nec eorum duntaxat qui Latinè aut Græcè, ſed quorundam etiam qui Germanicè, Gallicè aut Italicè lucubra-

Fig. 6.9. Page from a work by prohibited author Conrad Gessner censored by transforming the letters of his name into a nonsense jumble of letters and characters. Conrad Gessner, *Historiae animalium* (Zurich, 1551), call number QL41.G37 1551 F v.1, f. β[1]r. Courtesy of the Department of Special Collections, Stanford University Libraries.

With these help of these technologies, the expurgator's chosen phrase peers through the remains of the pasted slip: "Seben il simular sia le piu volte." The phrase is a near exact quotation from the fourth canzone of Ariosto's *Orlando Furioso*. The one difference is that the expurgator renders "quantunque" with the more colloquial synonym "seben," short for "sebbene."

> *Quantunque il simular sia le più volte*
> *ripreso, e dia di mala mente indici,*
> *si trova pur in molte cose e molte*
> *aver fatti evidenti benefici,*
> *e danni e biasmi e morti aver già tolte;*
> *che non conversiam sempre con gli amici*
> *in questa assai più oscura che serena*
> *vita mortal, tutta d'invidia piena.*[38]

> Although deceit is mostly disapproved,
> Seeming to show a mind malevolent,
> Many a time it brings, as has been proved,
> Advantages that are self-evident,
> And mortal threats and dangers has removed.
> Not all we meet with are benevolent
> In this our life, so full of envious spite,
> And gloomier by far than it is bright.[39]

In this passage Ariosto confronts the question of truth and when it is acceptable to dissimulate, a topic that was clearly on the mind of this expurgator as he censored his medical books and reflected on the necessity of hiding one's true feelings to avoid mortal threats and dangers. The second text written over Fuchs's name reveals still more about the meaning of censorship for this expurgator. This example in the dedicatory letter is easier to read since when the pasted slip was removed, it came off completely and cleanly, revealing text scribbled over Fuchs's name that reads, "tut il di piango e poi la notte piango." Once again, this is a slight misremembering of a famous literary passage, this time, the first line of Petrarch's *Canzoniere* 217, "Tutto 'l dì piango e poi la notte quando."

> *Tutto 'l dì piango, e poi la notte, quando*
> *Prendon riposo i miseri mortali,*
> *Trovomi in pianto, e raddoppiarsi i mali;*

Così spendo 'l mio tempo lagrimando.
In tristo umor vo li occhi consumando,
E 'l cor in doglia; e son fra li animali
L'ultimo, sì che li amorosi strali
Mi tengon ad ogni or di pace in bando.
Lasso, che pur da l'un a l'altro sole
E da l'un' ombra a l'altra, ò già 'l più corso
Di questa morte che si chiama vita.
Più l'altrui fallo che 'l mi' mal mi dole,
Ché Pietà viva, e 'l mio fido soccorso
Vèdem' arder nel foco, e non m'aita.

All day I weep; and then at night, when miserable mortals take rest, I find that I am in tears and that my pains are doubled; thus I spend my time weeping.

With sad moisture I am consuming my eyes and with sorrow my heart; and I am the most wretched of animals, so that the arrows of Love keep me ever banished from peace.

Alas! for from one sun to the next, and from one night to the next, I have already run through most of this death which is called life!

I grieve more for the fault of an another than for my ills; for living pity and the help I have relied on see me burn in the fire and do not aid me.[40]

Given the differences between the expurgator's text and Petrarch's sonnet (orthography, different words that sound similar), it is clear that the expurgator was not copying from Petrarch's text into the pages of his anatomy book. Instead, he was probably remembering, perhaps even humming to himself, one of the late sixteenth- or early seventeenth-century versions of the lachrymose sonnet, which was set to music eighteen times during the period. In a hidden form, covered by a slip of paper that dissimulated both the book's author and the expurgator's sentiments, the censor of this copy of Fuchs memorialized the emotional conflict of censorship. The act expurgation, of preserving a book by destroying parts of it, was a compromise that also entailed loss.

For other expurgators who held mixed feelings about expurgating their books with black ink, applying white gesso to a page to hide prohibited text was an appealing alternative. On January 31, 1572, Bernardino Mazorin, an agent for the Giunti family, testified in a trial before the Venetian Inquisition about books that he sold. The inquisitor asked if they were selling copies of Galen with the letter of a certain Slapner in it. Mazorin responded, "I blotted it out with ceruse so as not to mar the volume, and if it pleases you I will also glue a piece of paper over the letter. Otherwise, they are not sold."[41] Mazorin explicitly articulated that adding a white substance over the text, rather than applying black ink or removing pages with a razor, was a strategic move on the part of booksellers who, like readers, were aware of the aesthetic damage that censorship imposed on books. We might suspect that early modern booksellers similarly used ceruse to remove Erasmus Oswald Schreckenfuchs's name discreetly from his commentary on Georg von Peuerbach's explication of the Ptolemaic system (figure 6.10). However, in the centuries that have followed, the pages of the book have yellowed and darkened while the bright white gesso has not. The white now jumps out at readers as clearly as black would have done. Whether by transforming the letters to disguise them or using white pigment or blank paper rather than black ink, book owners interpreted the laws pertaining to prohibited books in ways that suited their intellectual, commercial, and aesthetic needs in addition to religious law.

Fig. 6.10. Copy of Erasmus Oswald Schreckenfuchs's commentary on Georg von Peuerbach (Basel, 1556) in which Schreckenfuchs's name has been expurgated with white gesso. While this form of censorship may have been visually nonintrusive when it was first applied, as the book's pages have yellowed, the still-white gesso now stands out. Call number 500 Cinq.C.0556. Reproduced with permission from the Biblioteca Civica di Verona.

PAPER

Bernardino Mazorin's offer to paste blank paper over the prohibited dedicatory epistle reflected another common method used to correct and expurgate books. Blank pieces of paper were attached with glue or sometimes wax to the original page. It is no surprise that this is a method a bookseller would adopt since the practice was adapted to censorship from the printer's practice of using blank slips to correct press errors after sheets had been printed.[42] Pasting blank paper over prohibited names was also the style of expurgation to which the inquisitor of Verona referred when he described Girolamo Donzellini's copy of Gessner's *Historiae animalium* as having "the name of the author covered with paper."[43] Cardinal Michele Ghislieri suggested pasting over prohibited pages in a 1555 letter to Giovan Battista Brugnatelli, the auditor to the papal nuncio in Venice. Ghislieri explained that booksellers could sell works that were written by permissible authors but which contained letters (dedicatory epistles) written by heretics. However, he explained, the letter "should be erased, that is to say, a blank sheet of paper should be glued to it so that it cannot be read."[44]

In the vast majority of the books I have examined, the paper used to hide prohibited text has been blank—unlike, for example, pieces of paper used for binding, which were often discarded printed or manuscript pages covered in text. However, expurgation reflected the personal preferences and projects of particular readers, so there were, of course, exceptions. The copy of the Gessner work shown back in chapter 2 as figure 2.1 replaced the author's name above the image of the moose with a piece of paper bearing the printed words PIETATE DOCTRINA (piety learning). Elsewhere in the text, the header of Gessner's letter to readers was replaced with a printed slip with the words "Henric Petrus to the Candid Reader."[45] In a copy of the works of Jerome edited by Erasmus, a reader or censor disguised Erasmus's commentaries by pasting over them with other printed pages (see figure 6.11). These pages are printed in different type that is roughly the same size, but the censor pasted them upside down so that readers would not be confused and accidentally read the text. The exquisite 1565 Froben folio looks uncensored at a quick glance, though the prohibited passages are actually obscured by the addition of pasted upside-down text.[46] Strangely, if readers were to turn the book upside down, they might realize that the flipped text pasted over Erasmus's introductions was actually from Erasmus's preface to the fourth tome of Jerome! Adding the text upside down no doubt slowed readers, causing them to think carefully about whether they should read text that was prohibited. At the same time, the

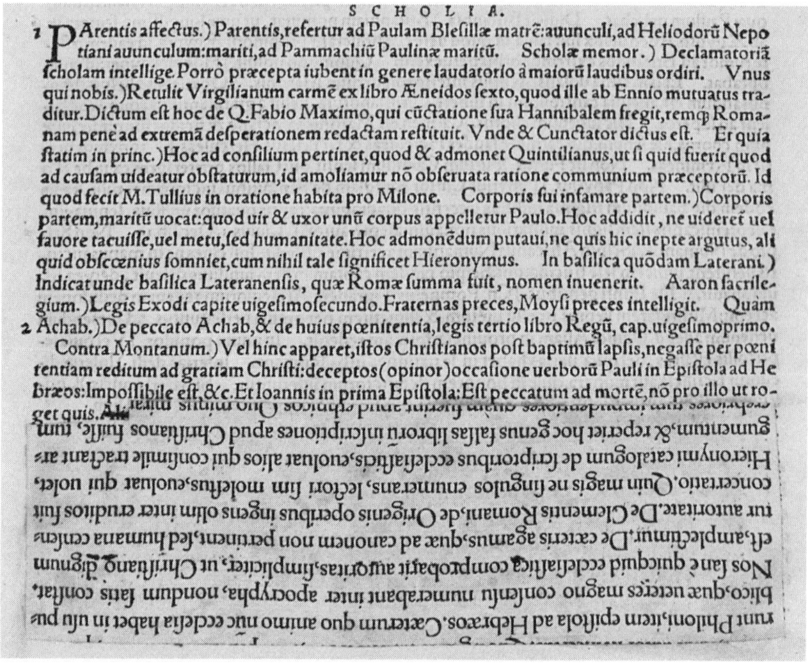

SCHOLIA.

1 PArentis affectus.) Parentis,refertur ad Paulam Blefillæ matrē:auunculi,ad Heliodorū Nepo
tiani auunculum:mariti,ad Pammachiū Paulinæ maritū. Scholæ memor.) Declamatoriā
fcholam intellige.Porrò præcepta iubent in genere laudatorio à maiorū laudibus ordiri. Vnus
qui nobis.)Retulit Virgilianum carmē ex libro Æneidos fexto,quod ille ab Ennio mutuatus tra-
ditur.Dictum eft hoc de Q.Fabio Maximo,qui cūctatione fua Hannibalem fregit,remq́ Roma-
nam penè ad extremā defperationem redactam reftituit. Vnde & Cunctator dictus eft. Er quia
ftatim in princ.)Hoc ad confilium pertinet,quod & admonet Quintilianus,ut fi quid fuerit quod
ad caufam uideatur obftaturum,id amoliamur nō obferuata ratione communium præceptorū. Id
quod fecit M.Tullius in oratione habita pro Milone. Corporis fui infamare partem.)Corporis
partem,maritū uocat:quod uir & uxor unū corpus appelletur Paulo.Hoc addidit, ne uideret uel
fauore tacuiffe,uel metu,fed humanitate.Hoc admonēdum putaui,ne quis hic inepte argutus, ali
quid obfcœnius fomniet,cum nihil tale fignificet Hieronymus. In bafilica quōdam Laterani)
Indicat unde bafilica Lateranenfis, quæ Romæ fumma fuit, nomen inuenerit. Aaron facrile-
gium.)Legis Exodi capite uigefimo fecundo.Fraternas preces,Moyfi preces intelligit. Quàm
2 Achab.)De peccato Achab,& de huius pœnitentia,legis tertio libro Regū, cap.uigefimoprimo.
Contra Montanum.) Vel hinc apparet,iftos Chriftianos poft baptimū lapfis,negaffe per pœni
tentiam reditum ad gratiam Chrifti:deceptos(opinor)occafione uerborū Pauli in Epiftola ad He
bræos:Impoffibile eft,&c.Et Ioannis in prima Epiftola:Eft peccatum ad mortē,nō pro illo ut ro-
get quis.

Fig. 6.11. Part of Erasmus's commentary on the works of Jerome pasted over with printed paper flipped upside down, simultaneously obscuring the prohibited text without detracting from the aesthetics of the page. Desiderius Erasmus, *D. Hieronymi operum* (Basel, 1565), t. 1, p. 201. Call number G-4-IV-12, inventory number 47768. With permission from Biblioteca R. Caracciolo, Lecce. Reproduction prohibited.

use of Erasmian scrap paper suggests that the expurgation was not meant to entirely obliterate Erasmus's work, quite unlike the trimmed copy of the New Testament in Faenza illustrated by figure 6.2 earlier in this chapter.

Covering over prohibited words and passages with blank paper created the materially enticing possibility of a writing surface, which was then available for manuscript additions. Some Venetian printers adopted this blank paper approach in new editions of Petrarch's poetry, which literally left blank spaces on the page where the censored Babylonian sonnets could be filled in by hand. The prohibited sonnets were then bound as extra sheets into the back of the book so that after copying them in pen, the illegal printed pages could be quickly and easily discarded.[47] In the case of prohibited medical books, blank pieces of paper pasted over authors' names were regularly filled in again with the author's name in a later hand.[48] By contrast, the expurgator of a copy in Rome of Fuchs's edition of Hippocrates has written on the blank paper covering Fuchs's name:

"[Ge]rmano quodam" ("a certain German"), thus preserving the original expurgatory intent.[49]

COLLATERAL DAMAGE

Manuscript annotations to expurgated books and the removal of paper pastedowns are usually later attempts to uncensor expurgated texts. Efforts to wash the ink from censored passages or remove the paper pasted over them regularly caused further damage to books. In fact, expurgated texts can sometimes be located in library catalogs when they are described as having suffered from "water damage" because of the staining that the washing process leaves behind. In yet another paradox of censorship, efforts to remove evidence of censorship has left tears and holes in pages that were originally censored in ways that contemporaries had intended to be minimally intrusive.

Expurgating books selectively rather than destroying the whole volume had the advantage of preserving most of the object; however, most methods of expurgation also risked a degree of collateral damage, the unanticipated destruction of licit sections of the book through the material process of expurgation. Cutting pages from a book carried with it the potential loss of material that did not need to be removed. Often, if a dedicatory letter continued onto the recto of a page, the final page of the letter was not removed, and instead the material was cancelled using another means such as ink or paper. Most methods of expurgation attempted to minimize collateral damage, since the purpose of expurgation was to preserve most of a book by removing only certain parts.

Expurgations made with ink have often caused extensive and unintentional collateral damage. When an expurgator took his pen in hand and dipped into his inkwell to remove a name or passage, he likely did so unaware that the highly acidic iron gall ink would eventually burn through the pages, eating holes in the permissible text on the other side of the page. After four hundred years, the effects of expurgation with ink have caused damage that more readily resemble the effects of fire.

NAME EXPURGATION AS *DAMNATIO MEMORIAE*

Many of the examples of expurgation described in this chapter focus on the removal of the names of banned authors from books, rather than the removal of religious or scientific content. The removal of the name of a banned author was a widespread and fundamental part of how early modern Italian readers encountered and engaged with copies of medical books by prohibited authors.[50] I turn now to consider the expurgation of names as a form of *damnatio memoriae* (damnation of memory) that imbued expurgation and reading expurgated books with messages about Catholic community and pious reading that extended beyond the printed page.

For Catholic readers familiar with the Old Testament, expurgation shared certain similarities with the story in Exodus when Moses returned from Mount Sinai with the tablets of the laws to find the Israelites worshipping a golden calf. Enraged, Moses broke the tablets and set the loyal Levites to killing their "brothers, friends, and neighbors" who had betrayed their God. That day the Levites killed three thousand of the Israelites, and Moses made them priests. The next day Moses atoned to God, pleading, "Now if you would only forgive their sin! But if you will not, then blot me out of the book that you have written." And God responded, "Only the one who has sinned against me will I blot out of my book." The Douay-Rheims translation even more clearly captures the Catholic analogy to expurgation, translating "blot me out" instead as "strike me out."[51] Whatever the material method, this story must have resonated for Catholic censors and scholars who quite literally sifted through their libraries to blot, strike out, cover, and cut the names of heretical men from their books. In their view, Christendom had indeed been torn by wrong belief and wrong practice, and the now much-reduced community of believers atoned for their sins and the sins of their community by removing the sinners from their books as they believed God had from his.[52]

The events recounted in Exodus are an example of censorship as a symbolic and material means of delimiting a community. However, the Old Testament is far from the only historical or literary precedent for using the erasure of texts or images to signal removal from society. In a secular context, there was a long tradition in the Roman empire of displaying and then destroying images and inscriptions in order to alter the record of past events. The legal precedent for this practice stretched into the republican era when the images, statues, houses, books, and legal documents of declared enemies of the Roman state were destroyed. In imperial Rome, statues of emperors who were later deposed as tyrants were systematically defaced (literally) and altered into the likenesses of new members of the imperial family.[53]

While scholars agree that that the term *damnatio memoriae* is of modern not classical origin, they have overlooked the fact that versions of the Latin expression became common in the context of ecclesiastical censorship in the sixteenth and seventeenth centuries.[54] In figure 6.12, a

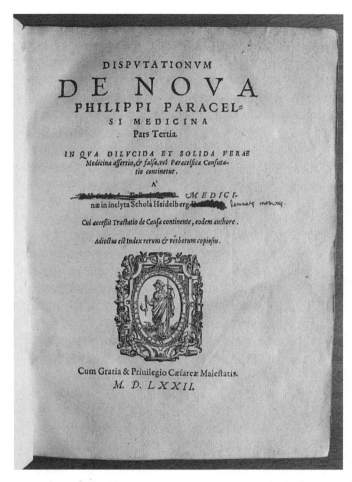

Fig. 6.12. Title page from Thomas Erastus's *Disputationum* in which a censor has blacked Erastus's name from his polemic against Theophrastus Paracelsus and inscribed the reason for the name expurgation beside it: *damnatae memoriae* ("of damned memory"). Thomas Erastus, *Disputationum de noua Philippi Paracelsi medicina pars tertia . . .* (Basel, 1572). BSVP, call number 500.ROSSA.SUP.C.6.-43.3. Reproduced with permission from the Biblioteca Antica del Seminario Vescovile di Padova.

censor removed Thomas Erastus's name and his title (professor) from the title page of his treatise arguing against the works of Paracelsus. The censor then inscribed "damnatae memoriae" ("of damned memory") into the margin. In 1624, the Portuguese inquisitor and bishop Fernando Martins de Mascarenhas published a copy of the Roman Index of Prohibited Books alongside the Portuguese Index and also an Index of Expurgations.[55] The inquisitor's text used the expression "damnatae memoriae" alongside

entries for books by certain heretical authors. In 1632, the new edition of
the Spanish Index of Prohibited Books described all authors prohibited in
the first class as "Auctorum damnatae memoriae opera edita etc. etc." and
included instructions in the expurgations for readers to insert the phrase
damnatae memoriae after the names of certain authors.[56] For example,
the entry with expurgations for Andreas Libavius's works noted that his
Rerum chymicarum epistolica forma (Frankfurt, 1595) included many
references to the physician Ioachim Camerarius the younger, "hominem
damnatae memoriae."[57] By 1643 in Naples, a selection of legal consulta-
tions omitted the names of prohibited authors in the text, replacing them
instead with *"damanatae memoriae authorem."* In this Index, the phrase
was a placeholder or pseudonym for prohibited legal authorities whose professional opinions could be referenced directly, but not their names.[58]

Unlike prohibited content removed from books, the removal of authors'
names and infliction of *damnatio memoriae* was not intended to teach
readers to forget who wrote the books. Indeed, the damnation of memory
(or damnation to memory; the Latin can be translated both ways) was intended to recall the errors of Protestants and to repeatedly and ritually
dishonor them.[59] We saw how Girolamo Rossi listed prohibited authors in
the index of his book *On Distillation* with the adjective "lapsus." Rossi
contentiously included the names of Protestant authors but with the caveat that they should be recalled with their heresies in mind. As censors
like Rossi took up their pens to expurgate names, connecting authors repeatedly to their confessional communities, they might have recalled another passage from the Old Testament. Deuteronomy 25:19 commanded
that the Israelites, once settled in their promised land, "blot out his [Ama-
lek's] name from under heaven," and then implored that they not forget the
name.[60] These Old Testament enemies, the Amalekites, were to be remembered until they could be destroyed completely.

In the case of readers of banned medical books, this passage took on
a more literal and practical meaning. The name of the author was indeed
blotted out, but the order to "not forget" recalled instead the necessity of
retaining knowledge of the author so that the reader could procure the required reading license. Readers were, after all, required to reapply for licenses every three years, and few licenses were requested or granted for
books without bibliographical information. Indeed, when the inquisitor
of Ancona, Fra Arcangelo, inscribed a reading license onto the verso of
the flyleaf of a copy of Fuchs's *Institutiones medicinae* (see figure 6.13), he
quite clearly stated that the licensee could use and read the book as long
as they "delete the name Leonhart Fuchs, a heretic in the first class, and

Fig. 6.13. Reading license recorded in a copy of a work by Leonhart Fuchs that specifically names the author at the same time that it required the author's name to be removed from the volume. Readers were not supposed to forget the names hidden under black ink but instead to remember them in the contexts of punishment and confessional difference. Leonhart Fuchs, *Institutionum medicinae libri quinque* (Basel, 1605). Call number 1P2/7647. Reproduced with permission from the Biblioteca Civica "Romolo Spezioli," Fermo.

others like him."[61] The license itself, recorded on the pages of the prohibited book, announced the name of the author, which was then expurgated from the rest of the volume.

What, then, was the purpose of the expurgation of names in an early modern context? Placing expurgation within the historical practice of *damnatio memoriae* reveals the censorship of names to have been a Catholic practice aimed at delineating a community of the faithful and distinguishing ritualistically those who erred. The Index of Prohibited Books established the connection between an author, his work, and his religion.[62] The expurgation of names from texts meant that the reader, with each

Fig. 6.14. Page from a medical book by Leonhart Fuchs in which the censor inflicted the *damnatio memoriae* (damnation of memory) not only on the letters of the author's name but also on Fuchs's portrait. Leonhart Fuchs, *Plantarum eefigies* [sic] (Lyon, 1551). BAV, call number Stamp.Chigi.VI.1603. © 2020 Biblioteca Apostolica Vaticana.

use of the text, reaffirmed or was reminded of the boundaries between his world and that of the author. Through physically blotting Protestant authors from Catholic books, the community of the faithful constantly redefined the boundaries of their Catholic community. Similarly, each time that a Catholic censor or reader altered his book to remove the name of a heterodox author, he reminded himself of his difference from the Lutheran

Fuchs or the Zwinglian Gessner. Books by heretics became physically
and symbolically different, and by designating them as such, expurgation
promoted a self-defining ritual that materially differentiated the Catholic
reader from his Protestant colleagues. In most cases, expurgation of the
author's name represented this memorialization of damnation, but occasionally readers extended the expurgation to an author's likeness as well.
Some copies of Fuchs's portraits are pasted over with paper, erasing his image from the text as well as his name. The expurgator of the copy shown
here as figure 6.14 took the liberty of using a pen to black out and transform the letters of Fuchs's name and also to draw an X over his face.[63]

CONCLUSION

Instead of damning the memory of Protestant authors to be forgotten, the
defacement of names, images, and works instead reminded readers that
these heretics and their works had been punished. Just as the burning and
destruction of heretical corpses was not intended to obliterate the memory
of the heretic, but instead to commemorate the lives of offenders through
the spectacle of their punishment, so too blacking out names from books
made a spectacle of religious deviance.[64] Even though books by heretical
authors were kept in Catholic libraries, they were retained only in their
altered, punished states. Any later reader who picked up the book would
know immediately that the author was a heretic and would thus approach
the book with the errors of the author in mind. When Cipriano the inquisitor of Rimini wrote to the Congregation of the Index in 1596, he explained
that he kept receiving books from Padua with the author's name crossed
out. He did not know whether he was then allowed to keep these books,
but the expurgation of the names was a signal for him and for other readers to pay attention.[65] The expurgated book embodied the cautionary impulse that Counter-Reformation scholars needed to bring to their reading.

  Although I have discussed four methods of expurgation separately, censored books usually contain multiple forms of expurgations. As I showed
with Ruchesius's copy of Fuchs, certain kinds of materials and methods
were selected depending on the needs of the reader and the physical attri-
butes of the book. More than one material form of expurgation in the same
text could also indicate that a book was censored multiple times and by
several different people. Paying attention to the material methods of expurgation forces us to also keep track of the many individuals who carried
out the physical labor, in addition to the intellectual labor, of censorship.
Silvana Seidel Menchi has pointed to humanist volumes that included

excerpts from Erasmus as the key to measuring "the degree of pervasiveness of censorship."[66] She suggests that by tracking how rigorously these books were censored, we can evaluate how thoroughly the culture of censorship permeated Italian culture. My analysis reveals that close attention to the myriad ways that individual copies were censored demonstrates that expurgation was not only a negotiation between ecclesiastical authorities and readers but also a negotiation between the materiality of the book and its intellectual status as an object between cultures.

Expurgation of medical books ultimately helped define the Catholic medical community on both passive and active levels. Readers who opened censored books knew without further investigation that they were about to confront a work in which the author erred. This book might still be worth reading, but it was also marked as separate from books that were entirely consistent with Catholic orthodoxy. On a more active level, censors, who were often themselves the readers and owners of a given book, took on the work of God, damning the memory and striking out the name of one who had sinned. In both cases, Catholic readers lived in an age in which they were constantly reminded of the divided state of their post-Reformation world and of the important role that books played as intellectual and material objects, simultaneously crossing and helping to define the boundaries of religious conflict. The expurgations of Catholic readers and censors relied on an understanding of a text that both acknowledged its material state and also laboriously attempted to construct medical works that transmitted a single meaning through their altered form.[67] However, as we have seen, the materiality of selective expurgation could also betray its purpose. Over more than four hundred years, the ink of the censors has in some cases faded and in other cases eaten away text that was initially meant to be preserved. Even Catholic censors' "ideal copies" of expurgated books have deteriorated over time, destroying what was meant to be saved and revealing what should have been hidden. The censors' attempts to control both materiality and meaning was ultimately impossible on both levels.

CHAPTER SEVEN

Prohibited Books in Universal Libraries

Expurgated books left tangible reminders about religious divisions throughout the libraries of Europe. After the Englishman Philip Skippon traveled to Italy in 1663, he reminisced about his visit to the Milanese library, the Biblioteca Ambrosiana, where he saw expurgated books. He recalled, "We look'd into *Gesnerus* his works, printed at *Frankfort*, and observed on the top of the title page, *Damnati Authoris, &c.* was written; and all those notes which *Gesner* calls superstitious and magical were blotted out."[1] Expurgated books impressed the English traveler, but objects like the ones Skippon described existed in libraries across Italy (see figure 7.1). Living and working alongside the damned memories of Protestant authors was part of being a Catholic reader in the long aftermath of the Reformation. We turn now to the place of expurgated books in the great Italian libraries at the turn of the seventeenth century to ask: How did prohibited medical texts enter libraries and become integrated into their collections?

A tension arises in this chapter between the late Renaissance scholarly interest in collecting, encyclopedism, and universality, and the cultural imperatives of censorship and prohibition at work in Italy at this time.[2] Adriano Prosperi has framed this question in terms of books by suggesting that the Counter-Reformation in Italy was characterized by two conflicting drives—the one toward building universal libraries and the other toward book burning and censorship. Prosperi suggests that the monument and countermonument of these movements are the Vatican Library in Rome and a mid-eighteenth-century iron woodstove made in Tyrol, cast in the shape of a bookshelf (figure 7.2).[3] The rococo stove is topped by a priest and a performer beating a drum that sits atop an open book bearing the inscription "body of doctrine." The shelves below, covering the oven cavity of the stove, feature an overarching banner proclaiming "Library conse-

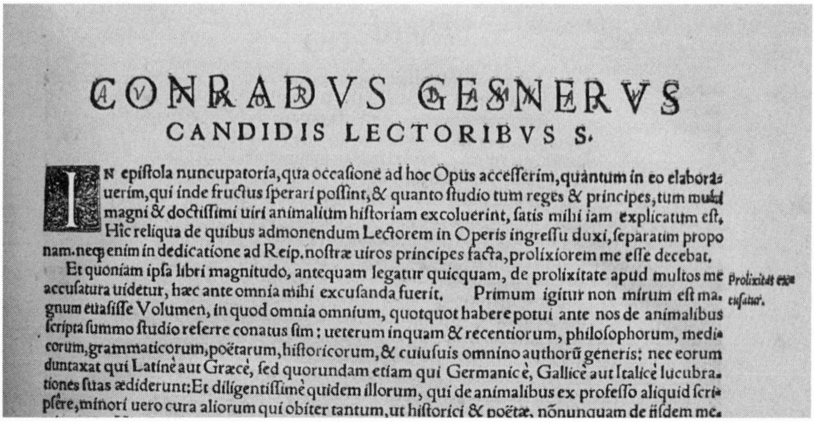

CONRADVS GESNERVS

CANDIDIS LECTORIBVS S.

I N epistola nuncupatoria,qua occasione ad hoc Opus accesserim,quantum in eo elaboraverim,qui inde fructus sperari possint,& quanto studio tum reges & principes,tum multi magni & doctissimi uiri animalium historiam excoluerint,satis mihi iam explicatum est. Hic reliqua de quibus admonendum Lectorem in Operis ingressu duxi,separatim propo nam.neq̃ enim in dedicatione ad Reip.nostræ uiros principes facta,prolixiorem me esse decebat.
Et quoniam ipsa libri magnitudo, antequam legatur quicquam, de prolixitate apud multos me accusatura uidetur, hæc ante omnia mihi excusanda fuerit.   Primum igitur non mirum est ma- gnum etiasisse Volumen, in quod omnia omnium, quotquot habere potui ante nos de animalibus scripta summo studio referre conatus sim : ueterum inquam & recentiorum, philosophorum, medi- corum,grammaticorum,poëtarum,historicorum,& cuiusuis omnino authorũ generis: nec eorum duntaxat qui Latinè aut Græcè, sed quorundam etiam qui Germanicè, Gallicè aut Italicè lucubra- tiones suas ædiderunt:Et diligentissimè quidem illorum, qui de animalibus ex professo aliquid scri- psere,minori uero cura aliorum qui obiter tantum,ut historici & poëtæ. nõnunquam de iisdem me-

Fig. 7.1. Page from Conrad Gessner's *Historiae animalium* in which a censor has inscribed "AUTHORUS DAMNATUS" (damned author) among the letters of Gessner's name in this prefatory letter. Expurgated books were integrated into Catholic libraries in the seventeenth century, and the English traveler Philip Skippon remarked on his surprise at seeing a copy altered like this one on a 1663 trip to the Biblioteca Ambrosiana in Milan. Conrad Gessner, *Historiae animalium* (Zurich, 1551). BNM, call number 50.D.47. With permission from the Ministero per i Beni e le Attività Culturali—Biblioteca Nazionale Marciana. Reproduction prohibited.

crated by fire," and the spines of the books lining the center row of the top shelf proclaim themselves to be the works of Martin Luther, John Calvin, and Huldrych Zwingli. As we have seen repeatedly, book destruction and book preservation went hand in hand in early modern Italy. This chapter confronts the tension between universality and censorship and explores the paradox of preserving prohibited books in Catholic libraries. The chapter also draws a connection between the simultaneous rise of Italy's great libraries and how bibliographical practices for information storage and retrieval were carried out in the face of Catholic censorship. As the figure of the Jesuit Antonio Possevino makes clear, Catholic *bibliothecae* were both collections of books (libraries) and lists of books that Catholics should read. The connection between the physical collection of books that made up a library and the nascent discipline of bibliography was intellec- tually as well as etymologically close.[4]

This chapter examines the treatment of medical books in three early modern Italian libraries: the Biblioteca Ambrosiana in Milan, the Vatican Library in Rome, and the Biblioteca Marciana in Venice. The case study of the Ambrosiana considers the reading licenses that the librarian Antonio

Fig. 7.2. Tyrolean cast-iron woodstove in the shape of a bookshelf, complete with the names of prohibited authors and a banner that reads "Library consecrated by fire." Image copyright by Salzburg Museum, Alpenstrasse 75, 5020 Salzburg.

Olgiati obtained for the newly formed public library and examines an encounter with a reader that played out in the courts of the Roman Inquisition between 1619 and 1621. The story of the Vatican Library recounts the
physical appropriation of one of the great sites of Protestant knowledge,
the Palatine Library at Heidelberg, in order to understand how Protestant
medical knowledge was integrated into the collections of the papacy's library. The example of the Biblioteca Marciana is markedly different from
the other two, and it reveals the relationships between the Venetian printing and publishing industry, the University of Padua, and the fraught relationship between the Catholic Church and the Venetian state.

These libraries were not the "prohibited libraries" that Ugo Rozzo has
revealed in his studies of the Friuli.[5] They were Catholic collections that
actively took steps to integrate Protestant medical books into their hold-
ings. Each of these three libraries negotiated its own relationship between
the concepts of the *Bibliotheca universalis* and the *Bibliotheca selecta*,
and individual scholars straddled these boundaries between the quest for
universal knowledge and their Catholic faith.[6] This chapter follows librarians and administrators of these three libraries to reveal the actors who
negotiated and established the institutional strength that Rozzo praises.
These individuals took on the task of advocating for the importance of prohibited books, and medical books in particular, within their collections.

This chapter also lays the groundwork for the practice of a critical
bibliography that places individual censored books within the context of
early modern collections. Critical bibliography is the emerging discipline
of bibliographical study in which the close examination of books as material objects is placed within an analysis of broader historical and cultural contexts.[7] I present an approach to critical bibliography that models thematic research across the histories of collections in addition to a
methodology for studying individual artifacts. Books are both individual
objects and parts of collections. My research reminds us that the book and
the *bibliotheca* are cultural artifacts that repeatedly changed throughout their histories. The bibliographic study of collections and individual
texts reveals the constellations of actors involved in altering, collecting,
curating, and facilitating access to banned books in early modern Italian
institutions.

This chapter also brings to a conclusion the central story of this book.
Italian intellectuals and ecclesiastical authorities found ways to make
useful, prohibited medical knowledge available to those who needed these
books. My examination of this subject began with prohibitions on books,
authors, and interpersonal relationships in 1559, moved to the expurgation

of medical books by lay and ecclesiastical censors, explained the scientific and religious issues at stake in proposed expurgations, traced licensed readers, and examined how they physically altered books to secure these books positions in collections. This chapter now concludes that narrative by describing the ways that prohibited books that had been deeply contentious in the sixteenth century were integrated into Italy's great Catholic libraries over the course of the seventeenth century.

THE BIBLIOTECA AMBROSIANA

At the turn of the seventeenth century, the place in Italy to find a nearly universal library was in Milan at the Biblioteca Ambrosiana. The Ambrosiana was the first truly public library in Italy and the second in Europe, after the Bodleian. When the Ambrosiana opened in 1609, its collection included about fifteen thousand manuscripts and thirty thousand printed books—for comparison's sake, that was about seven times as many printed books as in the Vatican Library at the same time.[8] In 1627, Gabriel Naudé wrote, "To speak only of the Biblioteca Ambrosiana . . . it surpasses all in grandeur and magnificence. There is nothing more extraordinary than [the fact that] anyone may enter it at nearly any convenient hour and remain there as long as he pleases, to look, to read, and to extract whatever author's works he finds agreeable."[9] The Ambrosiana grew out of the private book collection of Federico Borromeo, the archbishop of Milan. Prohibited books always had a place in the collections that made up the Ambrosiana. The archbishop kept among his papers a long list of books that his father had presented to the inquisitor of Pavia in 1558 and another list from 1588.[10] The list notes that during the inquisitor's first visit in 1558, nine works by Erasmus, Machiavelli, and Cornelius Agrippa von Nettesheim were "taken by the inquisitor." However, the list also includes dozens of books by other prohibited authors, including Conrad Gessner, Sebastian Münster, and Leon Ebreo, which were allowed to remain in the collection.[11] In a candlelit auction in 1608, Borromeo's agents acquired what would become the crown jewel of the Ambrosiana's collections: the Paduan library of the great sixteenth-century collector Gian Vincenzo Pinelli of Padua.[12] Pinelli's library had been a gathering place for intellectual life in late sixteenth-century Padua, and, like Borromeo's collection, it included prohibited books as well as a portrait collection, scientific instruments, and thousands of manuscripts.[13]

The Biblioteca Ambrosiana was founded in 1603. Two years later, its librarian, Antonio Olgiati, wrote to inquire of the vicar general of Milan,

Monsignor Antonio Seneca, whether the library would need a license in order to keep suspended books.[14] The draft of this letter is written in Olgiati's own hand, and the letter reflects the considerable effort the author put into its composition. It contains long, excised passages and blanks left for the names of authorities, which Olgiati later filled in as "Cardinal Arigone or whomever needs it." Finally, Olgiati settled on an introduction:

> In order to calm my conscience, I would like to have one, or really two, briefs from you, the first granting that in the library that is being built we can buy and keep alongside the other books, and not in a distinct location from the others nor locked up, books that are prohibited or that will be prohibited in the Roman Index or in other Indexes with this note and under the condition *donec expurgentur* [until expurgated], and that we be able to do the same with the suspended books that are not outright prohibited.[15]

Olgiati also requested authority to include and read the works of Ramon Lull, whose books had been debated and prohibited for nearly two hundred years and most recently appeared on the 1596 Clementine Index of Prohibited Books.[16] Olgiati wanted the prohibited books to be kept together with the rest of the collection, but he conceded that they could be bound and marked distinctly and that no one would be permitted to read them who did not have a license.[17] The license Olgiati sought would grant the library permission to keep the book, but readers would need individual licenses to access those volumes. Olgiati's request for the Ambrosiana license also clarified that he was not seeking a license for books that dealt directly with heresies. Instead, he sought permission to include classical, medical, legal, and literary works written by Protestants (and errant Catholics) whose works contained offensive passages, but not the pernicious categories of works by such authors as Luther, Calvin, or Machiavelli.

Olgiati's license request highlighted the inherent problem of the Biblioteca Ambrosiana as a universal collection that was also open to the public: it was difficult to control who came in or what they read. Although there is no further documentation about Olgiati's 1605 license request, it seems likely that it was granted. On July 21, 1620, licenses were issued again to Antonio Olgiati, the librarian, and separately to the library itself (see figure 7.3). While it was fairly common for bishops and noblemen to receive licenses for their collections and for those licenses to contain provisions for their librarians, it is notable that the Ambrosiana as an institution received its own license.

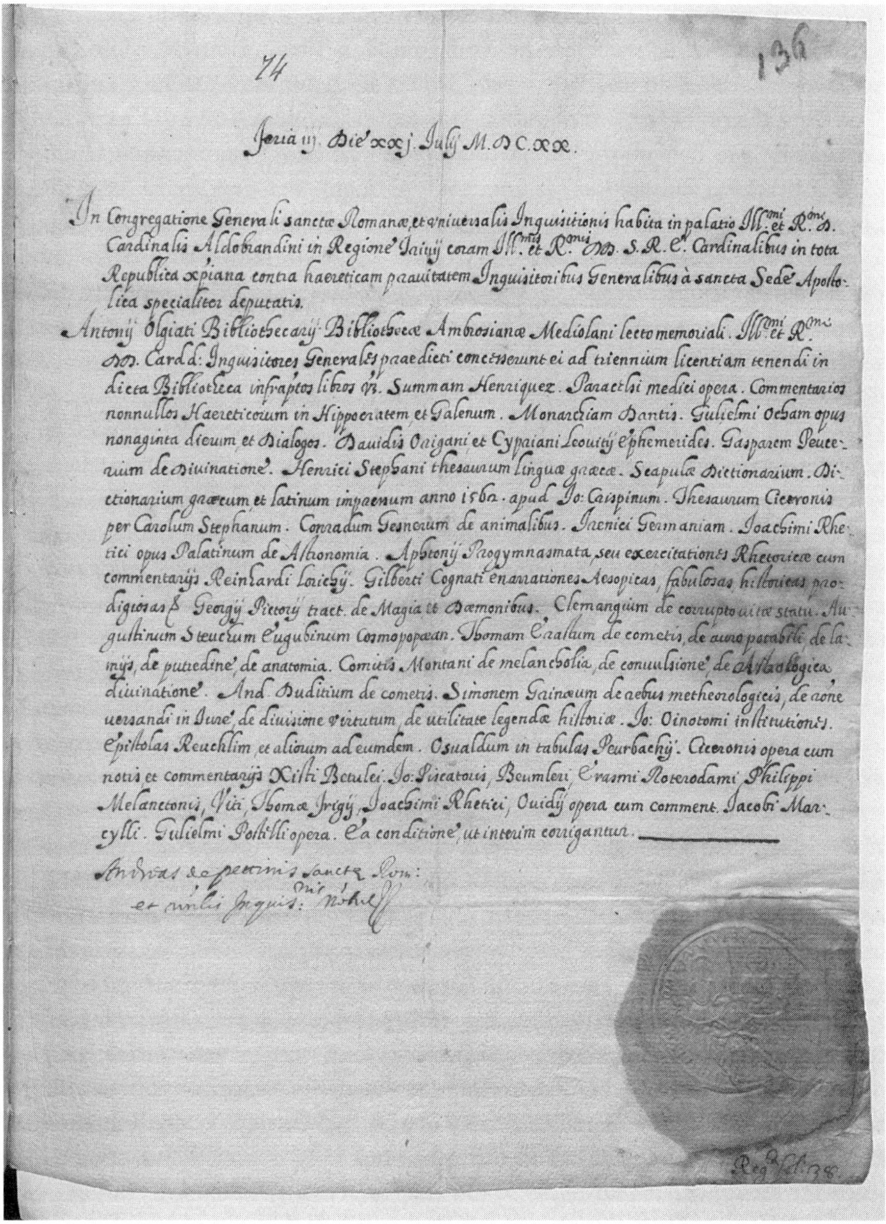

Fig. 7.3. License renewal, dated July 21, 1620, granted to Antonio Olgiati, the librarian of the Biblioteca Ambrosiana, to keep prohibited books in the library's collections. Biblioteca Ambrosiana, G.254 inf., f. 136r.
© Veneranda Biblioteca Ambrosiana. Reproduced by permission.

The renewal of the library's license in 1620 took place amid a conflict that arose when a reader at the Ambrosiana, a certain Carlo Gioseffo Origoni, was arrested and imprisoned by the Inquisition of Milan.[18] The decree registers of the Congregation of the Inquisition in Rome indicate that Origoni was held on the charge of keeping writings by heretics and adducing heretical ideas.[19] On May 20, 1620, Origoni was transferred from the prison in Milan to the prisons of the Holy Office in Rome and assigned defense lawyers. On December 15, 1620, he was tortured for further information about his "use, accomplices, and intention." If the torture did not reveal any further lapses, Origoni would abjure *"de vehemente,"* suffer incarceration for a period to be determined by the cardinals, and be fined 500 scudi to be put toward his debts to the Holy Office in Milan.[20]

The decree registers in which these records appear do not include the trial transcripts, but concern about Origoni's possible heresy must have been brought to the attention of Archbishop Federico Borromeo, who was staying in Rome at the time.[21] On April 28, 1621, Borromeo wrote to the commissary of the Roman Inquisition, Cardinal Desiderio Scaglia, that Antonio Olgiati, as the librarian of the Biblioteca Ambrosiana, would "justify a certain imputation learned from multiple parties that he had allowed this Carlo Gioseffo Origoni to pull out some things that were prohibited from a book in this library."[22] The Ambrosiana, a public library that existed to make useful knowledge accessible, would face repercussions from the Holy Office for the consequences of its public mission and universal collection.

Borromeo's personal archive includes excerpts from Origoni's testimony in which he named Olgiati and the Ambrosiana as the sources of the heretical texts he copied. During his trial, inquisitors confronted Origoni with his volume of heretical writings. In the testimony he declared, "These manuscripts, I wrote them out with my own hand, parts of them I composed myself and parts are taken from others." After naming several texts and sources, he continued, "The oration of Agrippa was in the library of the Signor Cardinal [Borromeo], corrected and in a separate volume that does not deal with Magic . . . that treatise beginning with 'chiromantia satir[. . .]' I think I pulled that from a book by Girolamo Cardano and I don't remember if that book was in the library of the Signor Cardinal or in a different place."[23] Origoni carefully clarified that the work by Agrippa did not deal with magic. Though Agrippa worked as a physician and legal scholar, he was known above all as an occult philosopher. Cardano, too, was a physician, though the inquisitors would have been concerned about the occult and astrological subjects that so fascinated these two authors. Origoni's description of events at the library is an example of the openness at the Ambrosiana that Naudé had extensively praised. Naudé appreciated the cardinal's library as a place where one could at any reasonable time "extract whatever author's works he finds agreeable." Origoni seems to have done just that, but to a heretical end. He visited the cardinal's library, requested and read books, and copied sections that ultimately landed him in inquisitorial prisons and torture chambers.

Borromeo charged the librarian, Olgiati, with the task of submitting a defense of what transpired at the Biblioteca Ambrosiana. Olgiati began by acknowledging the particular responsibility of the library to account for the activities of readers and the movement of books. He explained, "Since I am prefect, I deemed it necessary to show that no guilty fault was committed either on the part of the library or its assistants."[24] Olgiati's defense proceeded in three veins, each supported by direct Latin quotations from the rules of the Index of Prohibited Books.

First, Olgiati confronted the accusation that at the Ambrosiana Origoni read an oration by the Roman emperor Heliogabalus in which the emperor addressed prostitutes. The lewd oration to which Olgiati referred was actually a reimagination of an oration written by the fifteenth-century Tuscan humanist Leonardo Bruni. However, Olgiati's defense suggests that he thought the oration was written by the Roman emperor himself, rather than a humanist imitator.[25] "I confess," wrote Olgiati, "that this oration can be found bound with the lives of the emperors written by Aelius Lampridius and others with notes by Battista Egnazio and printed in Venice by Aldus Manutius."[26] Despite this confession, Olgiati remained unwilling to repent, noting, "You can't find a library, public or private, that does not have [this book]," and it could be kept legally because there was nothing suspect or heretical in it. Similarly, even though the work could be considered obscene, the seventh rule of the Index made an allowance for ancient authors even if they were licentious. Olgiati explained that this was why one could read Catullus, Tibullus, Propertius, Ovid, and Horace, among other ancient authors.[27]

Second, Olgiati confronted Origoni's testimony that at the Ambrosiana Origoni had copied an oration on uncertainty, written by Henricus Cornelius Agrippa von Nettesheim. Olgiati, in what seems the weakest point of his defense, claimed that although the library did have a copy of Agrippa's prohibited book, *De incertitudine et vanitate scientiarum* (*On the Uncertainty and Vanity of the Sciences*), this was a book, not an oration. Moving

on from this semantic quibble, Olgiati added more convincingly that no
one could have copied anything inappropriate from this book because it
had been "diligently and rigorously corrected in the Congregation of the
Index."[28] Olgiati explained that the works Origoni mentioned by Cardano
were allowed by the Index and they had also been corrected in Milan by
the local Congregation of the Index and undersigned by the Inquisitor.[29] As
Olgiati had promised in his original 1605 request for the library's license,
Agrippa's prohibited volume had been dutifully expurgated and the dangerous sections had been crossed off, cut out, or pasted over.[30]

Third, as Olgiati pressed on in his defense, he claimed that Agrippa's
prohibition in the first class did not necessarily imply that all works by
authors in the first class were beyond correction, particularly in subjects
other than religion. Olgiati pointed to other prohibited authors including
Charles Dumoulin, Leonhart Fuchs, Sebastian Münster, Theodor Zwinger,
and Robert Estienne, who "were placed on the Index in the first class and
nevertheless they have been emended and their emendations are registered
in the Roman Expurgatory Index printed in 1607."[31] It was true that certain authors listed in the first class whose works did not deal with religion
were permitted to be read provided that their works were corrected in accordance with the *Index Expurgatorius* of 1607. However, Olgiati's list was
only partially correct. Works by Dumoulin and Fuchs had been corrected,
but Münster's, Zwinger's, and Estienne's works were not included among
the authors expurgated in the *Index Expurgatorius*.

Olgiati acknowledged that there were prohibited books in the Ambrosiana and that Carlo Gioseffo Origoni had, indeed, read those books at the
library. He defended himself on the basis that the library had a license to
possess the books and that they had all been properly expurgated and corrected according to the terms of the license. Origoni's case and Olgiati's
defense reveal a problem with the system of expurgation and licensing,
a flaw that censors had known for many years and that literary scholars
have noticed since: readers use texts to different ends and create meaning
in different ways.[32] Despite attempts to fix texts through expurgation and
the librarian Olgiati's attention to his legal obligations, readers like Origoni might still use corrected texts in unorthodox ways.

From a legal standpoint, Olgiati's argument was lacking. Readers, in
addition to libraries, needed to apply for and receive official licenses to read
prohibited books. Nonetheless, the cardinals of the Inquisition in Rome
were reassured by Olgiati's explanation. On May 15, Cardinal Desiderio
Scaglia wrote to Federico Borromeo confirming that "my Illustrious Cardi-

nal Colleagues are completely satisfied with your library and with the diligence of your assistants, and they do not believe that from it similar lapses could follow."[33] Four days later, Cardinal Giovanni Garzia Mellini, also a member of the Congregation of the Roman Inquisition, wrote to Borromeo that the cardinals were "satisfied with the account of your librarian."[34] In June, Borromeo and Olgiati sent a notarized copy of their previous reading license to Cardinal Arigone. It took nearly a year for Arigone to return it to them with an apology for his delay, which was caused, he claimed, by the inclusion of a book by Charles Dumoulin which, Arigone explained, could not be possessed "because it is exactly that work which principally compelled Clement VIII to prohibit him so severely that not even Cardinals could keep it."[35] While the status of archbishop and cardinal had protected Borromeo and his library, some books were not allowed to any audience. The universal Catholic library could never be truly universal.

The Biblioteca Ambrosiana was one of the seventeenth century's best bids at a universal library. Because of ecclesiastical censorship, it could not achieve the universality described in Conrad Gessner's *Bibliotheca universalis*, but neither did its custodians aspire to the orthodox rigidity of Antonio Possevino's *Bibliotheca selecta*.[36] Catholic book collecting during the Counter-Reformation involved a constant negotiation between collectors and Catholic authorities. The qualifications and mission of each collector and library influenced how catholic a particular Catholic collection could be. Archbishop Federico Borromeo's status in the Catholic Church made it possible for the Biblioteca Ambrosiana to include a vast, though not unregulated, number of prohibited books. The library's public mission meant that someone like Carlo Gioseffo Origoni could enter the Ambrosiana to read and excerpt from the library's materials, potentially putting them to nonorthodox, even heretical, uses. Origoni's case in the Biblioteca Ambrosiana demonstrates the failure of censorship to control interpretation or use even when the books were highly regulated and expurgated. While Cardinal Scaglia was satisfied that Origoni's encounter would not be repeated, there could be no assurance that another reader would not become another Origoni. Origoni's case reminds us that the orthodoxy of a collection relied less on Possevino's approach to assembling the library and more on the intentions of the reader. Licensing a library was important, but it was more essential that readers brought a Counter-Reformation attitude of caution and care to their reading. This essential caution was the orientation that expurgatory censorship had been inculcating in Italian readers.

THE VATICAN LIBRARY

While the Biblioteca Ambrosiana was part of Federico Borromeo's vision of a distinctly seventeenth-century, post-Tridentine Catholic library, the Biblioteca Apostolica Vaticana (the Vatican Library) was a library of a different era.[37] Formally established in 1475 by Pope Sixtus V, the Vatican Library was a Renaissance ambition, though over the course of the Counter-Reformation the place and purpose of prohibited books in the Vatican collections changed drastically.[38] As Anthony Grafton has observed, the Vatican Library was "the chief intellectual arm of the first European state that rested its strength more on learning and art than on dynastic loyalties and military power."[39] Indeed, if the Vatican Library was the central intellectual force of the Catholic Church, the beating heart of the mission of persuasion (*persuasio*), the Roman Inquisition was the muscle charged with enforcement through forcefully monitoring thought and behavior (*coercitio*).[40] In the seventeenth century, the intellectual and inquisitorial tasks went hand in hand, and the Vatican Library held a central role in both. The Vatican Library, as we have already seen, was renovated in the late sixteenth century, and the new reading room of the Salone Sistino included multiple images of book burning alongside book preservation.

The seventeenth century was a period of immense growth for the Vatican Library. The treasure of its recent acquisitions was the seizure of the Palatine Library of Heidelberg in 1622. During the Thirty Years' War, the Catholic League sacked the city of Heidelberg after an almost two-month siege. After much urging, Maximilian of Bavaria gave the library to Pope Gregory XV in what Jill Bepler has described as an act of conspicuous cultural booty.[41] Just over a month later, on October 28, 1622, the Greek scholar Leone Allacci was dispatched to Heidelberg to inventory the captured library and to oversee the removal of the books to Rome, where the vast majority of the seized items still reside to this day.

Allacci's account of the trip, the *Relazione sul trasporto della Biblioteca Palatina*, reads like a swashbuckling adventure. Traveling from Innsbruck to Augsburg to Ingolstadt and then heading west toward Heidelberg, the travelers successfully navigated not only bad weather but also the perils of traversing a war-torn countryside. As Allacci wrote, "We were safe neither in the country nor in villages; in the country [we were in danger] from thieves and soldiers who wander everywhere, and in villages from the countrymen who, when they saw foreigners, if they did not know them, killed everyone without making distinctions between people."[42] When the party finally arrived in Heidelberg at lunchtime on December 13, 1622,

Allacci, weary of travel, decided to stay in town near the library, saving himself further treks up and down the steep approaches to and from the fortified castle.

However, the climax of Allacci's adventure was not his arrival in Heidelberg following a dangerous journey; it was opening the first box of manuscripts to compile an index, which he undertook, "as soon as he was given the keys, without losing any time."[43] Among these boxes, Allacci found the "originals" of works by heretics such as Luther and Philip Melanchthon.[44] He reveled in the collection of printed books, which he described as "infinite," with many volumes in duplicate, triplicate, and "many times centuplicates (*centoplicati*)." Repurposing the wood of the bookshelves to make boxes, Allacci packed up thousands of volumes, selecting among the replicated titles only the copies that were the "most rare" or "old prints" or "in particular those that dealt with the rites and things of the Church." Of the 8,500 volumes that Allacci selected to bring to the Vatican, about 5,000 were printed and 3,500 were manuscript. The Holy Office waylaid approximately 1,000 of these books when the texts arrived in Rome to check their orthodoxy before they were admitted into the Vatican Library.[45]

In addition to Luther's "originals," the Palatine Library held many of the medical texts written by leading members of the medical republic of letters, which Catholic officials had prohibited and then expurgated through the efforts of lay and ecclesiastical censors across Italy. For example, the Vatican Library Palatine Collection contains twenty-two copies of works by Leonhart Fuchs, forty-three by Conrad Gessner, and seventy-seven by Paracelsus. None of the medical books from the Palatine Collection have been expurgated, and it seems likely that none of them passed through the Holy Office before they were shelved in the Vatican Library, as was the procedure for religious texts entering the collection. The books by prohibited Protestant physicians were so desirable that they were carried by mule across the Alps in the middle of winter to their new home in the heart of the Catholic world.

Many of these copies of medical texts were not only prohibited but also prominently signed by their former owner, Achilles Pirmin Gasser, who was himself prohibited in the first class on the Pauline Index of 1559. Gasser was a German physician and astrologer, a supporter of Copernicus and Rheticus, a correspondent of Gessner, and an avid book collector. He purchased works for himself, for his friends, and for his patron Ulrich Fugger, who after Gasser's death acquired his library thereby joining its fate to that of the Palatine Library.[46] The books in Gasser's library document the rich scholarly networks of Protestant Germany, since Gasser assiduously inscribed his volumes with his name and details about each volume's acquisition, including dates, names, and prices. The addition of the Palatine Library to the Vatican Library amid the Thirty Years' War betrayed a changed strategy toward policing the boundaries of confessional communities, ironically archiving a collection of medical books that through Gasser's marginalia documented the thriving Protestant and interconfessional medical community that existed beyond the control of the Catholic Church.

The paradox of censorship revealed in this account merits further consideration relative to medical books in particular. Although Gasser did not write extensive marginalia in his books beyond the initial inscriptions, he paused in his reading of Fuchs's 1566 *Institutiones medicinae* to underline a particular passage: "Medicine emends, corrects, and restores the weakened condition of the human body and thus creates health, and preserves it once it is present."[47] We might consider the parallels here between the work of medicine, textual criticism, and the expurgation of Protestant medical books in order to keep them in Catholic libraries. Through expurgation, Catholic readers could model themselves on the good physician, correcting the vices of texts and preserving that which was healthy. However, as we know from both the history of disease and the history of religion, the categories of vice and health are always shifting. What the Catholic Church perceived as infected with heresy in the middle of the sixteenth century was permissible within the realm of a healthy Catholic society in the middle of the seventeenth century.

In addition to the Palatine collection, the library of the Roman Inquisition (the Holy Office collection) is another of the largest collections of prohibited books in the Vatican Library.[48] It will come as no surprise that the Roman Inquisition's rich archive also contained a library, nor that the library later became part of the Vatican Library. It is more revealing, however, that in the 1630s, someone, perhaps Francesco or Antonio Barberini, sorted the nearly three thousand volumes catalogued in the Holy Office library into four categories meant to attenuate the prohibitions leveled against the texts.[49] The four categories show the author to have been rethinking the classification of prohibited books shortly after the huge influx of prohibited material into the Vatican Library through the Palatine Library. Rather than a complete prohibition or expurgation, this system allowed for a more nuanced articulation of what kind of reading seventeenth-century censorship should prevent and what kind of reading the Catholic Church should conversely facilitate.

The four categories read as follows:

1. Books that have nothing heretical and at this time should be sold without punishment by all of the Roman booksellers.[50]
2. Writings by the Holy Fathers edited by heretics, which contain nothing bad except the name of the interpreter or editor.[51]
3. Ancient and profane authors, or writings by recent authors about things and matters that have nothing pertinent to the faith.[52]
4. Heretical or suspended books that are nevertheless useful for history or erudition or another [field of] knowledge.[53]

Each category was accompanied by a sign which was then added to the bibliographic entry in the library catalog. This document is curious on many levels. Pico della Mirandola's letters, for example, were sorted into the category of works by church fathers edited by heretics, which, of course, they were not. Books such as Conrad Gessner's book on fossils and Marsilio Ficino's writings on mystical theology were suggested to be, if not strictly orthodox, at least explicitly nonheretical, and the author of the list advocated that these works should be sold freely in Rome. (In the case of Gessner, the volume was also listed again in the fourth category of useful books.) Works by Arnald of Villanova, Gessner's *Bibliotheca universalis*, Agrippa von Nettesheim's *De incertitudine et vanitate scientiarum*, and Leonhart Fuchs's and Thomas Erastus's medical works, which had a century earlier prompted huge campaigns for correction and resulted in careful lists of expurgations, were now considered to be in a category that contained things entirely irrelevant to the faith. To further emphasize this point, we need only compare to some Italian literary works. The author of this list placed Giovanni Boccaccio's *Decameron* and Baldassare Castiglione's *Il cortegiano* (*The Courtier*), both of which were extensively expurgated in the sixteenth century, in the same third category of works with "nothing pertinent to the faith." Finally, and most important for this study, medical works by Amatus Lusitanus, Leonhart Fuchs, Janus Cornarius, and Otto Brunfels were classified in the fourth category as "Heretical or suspected books that are nevertheless useful for history or erudition or another [field of] knowledge." The repeated description of medical books as useful had, by the 1630s, come to be understood as its own categorical justification for allowing access to otherwise prohibited texts. The Barberini manuscript places this discourse of professional utility among many disciplines, and medicine is not specifically named. While physicians and medical texts had prompted a widely documented discourse about utility in the sixteenth century, the justification of knowledge through reference to utility was applied broadly beyond this professional category in the seventeenth century.

The Barberini catalog classification confirms that the heightened sensitivity of church officials in the years following the Reformation had lessened, although it also reveals continued antipathy toward certain medical authors. The list also included a group of works that did not deserve to have their prohibition attenuated or revoked. Among the medical authors in the catalog, the Italian Protestant physician Guglielmo Grataroli remained resolutely prohibited with no attenuation.[54] The confusion that often permeated book prohibitions was also present in this list. For example, the *Praxis canonica* by Giovanni Michele Savonarola, the celebrated fifteenth-century physician, appeared on the list and was not mitigated by any marginal notations.[55] Although Giovanni Michele Savonarola was the grandfather of the Dominican preacher and self-proclaimed prophet Girolamo Savonarola, the physician's works had never been prohibited. While Fuchs's works were exonerated as either "useful" or "unrelated to the faith," no works by Paracelsus, the so-called Luther of medicine (though not a Lutheran), were reclassified and were thus judged by the compiler to be worthy of continued prohibition.[56]

Medicine and medical authors who had since the sixteenth century been at the center of learned conversations across Europe were by the midseventeenth century gradually accepted in Italian libraries. At the Vatican Library and the Holy Office of the Inquisition, the embrace of Protestant medical learning was sometimes explicit, as in the case of the Barberini manuscript, and at other times implicit, as in the case of the medical books in the Palatine Library entering unexpurgated into the stacks. Nevertheless, both cases reveal the ways that the prohibited knowledge of the sixteenth century had, by the seventeenth century, come to be seen as desirable, "useful," and at the very least presenting "nothing pertinent to the faith."

THE BIBLIOTECA MARCIANA

The Venetian state's library, the Biblioteca Marciana, grappled with the same challenges of universality, utility, and censorship as the Ambrosiana and the Vatican Library, although Venice's distinct economic and religious context inflected these local debates in the opening decades of the seventeenth century. Like the Vatican Library, the Marciana was an important library prior to the invention of moveable type.[57] One hundred and thirty years before Aldus Manutius and his colleagues would fundamentally alter Venice's economy and culture, transforming it into Europe's leading city for printed books, the Florentine humanist and poet laureate Francesco Petrarca (Petrarch) donated his manuscripts to found a public library at San Marco.[58] In 1468, Cardinal Bessarion followed Petrarch's example and bequeathed his collection to the Marciana, establishing it as a European center for humanistic studies.[59]

Over the course of the sixteenth century, the Marciana was also an important intellectual center for medical scholarship. Venice had not yet established a library in its university town of Padua, and the Marciana served as the region's primary public resource for scholars. The manuscript collections of the Marciana played an important role in the rise of medical humanism. As Giacopo Morelli, the eighteenth-century librarian of the Marciana, gloated in his history of the library, John Caius, the prominent English physician and correspondent of Gessner, traveled to Venice to consult the Marciana's copy of a manuscript of Galen for his *De medendi methodo libri duo* (*Two Books on the Method of Healing*).[60] Johann Rhodius, the Danish physician and prefect of the botanic garden in Padua, had consulted a manuscript of Cornelius Celsus at the Marciana that was essential to his contributions to the field of medical humanism. In his preparations for an expanded edition of the library's history, Morelli added a manuscript supplement to this passage expanding still further on the medical lineage of the Marciana. "Galen's codices were examined by the Englishman Henry Knoll," Morelli noted, and further, the first printed edition in 1538 of the full Greek text of Ptolemy's *Almagest* was compiled "from one of Bessarion's codices."[61] The wide-ranging physician and scholar Hadrianus Junius also spent time in the library at Venice, as did Conrad Gessner in 1543 while compiling his *Bibliotheca universalis*.[62] Between the booming print industry, the humanist lineage and manuscripts of the Biblioteca Marciana, and the "learned men" at the University of Padua, the Republic of Venice was an important hub for both the *peregrinatio medica* and the medical republic of letters.

While the Marciana had flourished for many years, especially under the direction of Cardinal Pietro Bembo between 1530 and 1543, by the end of the sixteenth century this once-great collection had fallen into a state of ruin. Picked over by scholars who failed to return their books and rotting from lack of care in the damp lagoon, the Marciana was in desperate need of investment and renovation. An unsigned document from 1611 or 1612 reported to the Riformatori dello Studio on the state of the library in no uncertain terms: "I attest with the present letter to have seen that everything was disorganized and many of the books of the most illustrious Cardinal Bessarion have been removed and taken by thieves from the shelves where they are attached with only a thin iron chain for security.

Many of the books of Secretary [Francesco] Vianello are wet and ruined, and others are spoiled and terribly treated by the rains."[63] At the beginning of the seventeenth century, at the same time that Antonio Olgiati and Federico Borromeo were building the Biblioteca Ambrosiana in Milan, the Venetian state was taking a number of steps to attempt to restore Italy's first public library to its previous glory and centrality. On May 21, 1603, the Senate of the Republic decreed that any and all who printed books in Venice or within the Venetian state were obliged to "consign the first of any kind of book that they print, bound in parchment, to our library of San Marco. They cannot begin to sell that book until they have a certification (fede) from the librarian of the said library."[64] This law streamlined book production in Venice while simultaneously expanding the prestige of the Marciana. Enhancing the library in this manner also served the civic purpose of documenting and preserving the commercial and entrepreneurial patrimony of the printing industry in Venice.

The impetus to reform and restore the once-great Marciana derived from events both abroad and at home in the Veneto. The growing prestige of libraries in Milan, Florence, and Rome must have seemed incongruous to Venetians who knew their own city to be one of the great book centers in Europe. Another motivation for reform arrived at the Marciana in twenty-three locked chests. The chests were full of books that had previously been the personal library of Melchior Wieland, the Prussian botanist, naturalist, and caretaker of the botanical garden in Padua. The chests were Wieland's final intellectual contribution to his adopted state. Wieland, who died in 1589, had been a fastidious book collector and a close friend of Benedetto Giorgio, who was elected librarian of the Marciana in 1588. In addition to donating his collection of volumes, Wieland also left the library a sum of money to inventory the books, to copy the inventory in triplicate, and to pay for the cost of labor and boats to transfer the chests from Padua to Venice in January 1590.[65]

Melchior Wieland's donation to the Marciana included approximately 2,200 printed books, roughly 450 of which he classified as works of medicine or natural philosophy. Like most of the learned physicians of the sixteenth century, Wieland collected books that included many works by prohibited authors. Folio editions by Leonhart Fuchs, Conrad Gessner, Girolamo Cardano, Amatus Lusitanus, and Thomas Erastus shared space on his shelves with quartos by Guglielmo Grataroli, Johannes Crato von Krafftheim, and Otto Brunfels, among others.[66] Wieland also owned an extensive collection of mathematics books that showcased his interest in astrology and included a copy of the not-yet-prohibited *De revolutio-*

nibus by Copernicus.[67] While the prohibited medical, philosophical, botanical, and natural historical texts in Wieland's collection were part of,
and perhaps even necessary for, his professional work as a physician and
director of the botanical garden, Wieland's inventory shows that he read
and collected prohibited books well beyond the requirements of his profession. Works on dueling had been banned since the Council of Trent, yet
Wieland had multiple copies in this genre written by Andrea Alciati and
Girolamo Muzio. He owned Jean Bodin's *De republica*, Baldassare Castiglione's *Il cortegiano*, Petrus Ramus's *De militia*, and a copy of a work
listed as "Senesi dell'Aretino etc.," which must have been a volume of Pietro Aretino's erotic sonnets.[68] These volumes all entered the collection of
the Marciana in 1590, where they are still housed today. The experience
of the Wieland donation to the Marciana was unique in scale but not in
concept. Through their wills, Ulisse Aldrovandi donated his library to the
University of Bologna, Pompeo Caimo left his books to the University of
Padua, and Romolo Spezioli bequeathed his library to the city of Fermo.
All of these medical libraries contained copies of books by prohibited authors, expurgated to varying degrees, and possessed with and without reading licenses.[69] Prohibited books were an essential part of private, medical
collections and through donations and acquisitions entered en masse into
public libraries across Italy at the turn of the seventeenth century.

At the public library of San Marco, the position of the custodian of the
library further consolidated Venice as a center of print culture and established the importance of censorship for regulating book circulation and
readership. Since 1609, the Cypriot Giovanni Sozomeno had been in charge
of reorganizing the library while simultaneously serving as the head of the
prepublication censorship apparatus in Venice, a position known as the
"overseer of books" (*revisore de' libri*). Until 1617, Sozomeno also oversaw
a crucial aspect of book censorship in the city as the appointed "customs
overseer" (*revisore alla dogana*)—the official who inspected packages of
imported books at the customhouse.[70] In 1626, a decree by the Senate officially installed Sozomeno as the first custodian (*custode*) of the Marciana.
His position included assisting students using the library, and he was selected for the post both because of his familiarity with the collections and
also because of his reputation as a scholar and teacher of Greek.[71] In a testament to the continued legacy of medical humanism in Venice, an elegy by one of his pupils praised Sozomeno in particular for his translation
from Greek to Latin of Galen's two commentaries on the second book of
common illnesses by Hippocrates.[72]

Giovanni Sozomeno was also charged with creating the first library catalog of manuscripts and printed books at the Marciana. His 239-page quarto manuscript contains a 51-page inventory of the library's manuscripts and a 170-page list containing the titles of printed books at the library.[73] Sozomeno divided the books, much as Wieland had, by subject and then by format. Distinctive titles owned by Wieland appear in the catalog, such as his 1543 edition of Copernicus's *De revolutionibus*, although the inventory did not distinguish between books by provenance. The catalog also listed, unselfconsciously and without differentiation, the medical works of many prohibited authors, including Leonhart Fuchs, Conrad Gessner, Girolamo Cardano, Amatus Lusitanus, Levinus Lemnius, Thomas Erastus, Guglielmo Grataroli, and even Girolamo Donzellini, the infamous Venetian physician and book importer who had been drowned in the lagoon on the orders of the Venetian Inquisition only a few decades earlier.[74] These texts, though prohibited, were books that were primarily and explicitly relevant to medicine. Works on plague, treating illness, the natural history of plants, anatomy, and translations and editions of scientific texts by classical authors also fell under the heading "Medicine" in Sozomeno's 1624 catalog.

In 1624, medicine and medical learning had long encompassed a broad range of disciplines, often represented in the works of these same prohibited authors. But where were the books of secrets, the natural histories of minerals and animals, and the works of astrology, alchemy, and chemistry that featured regularly in seventeenth-century physicians' requests for reading licenses? Sozomeno, the diligent sifter and sorter of bibliographical information, did not drop these prohibited titles from his catalog; instead, he listed the prohibited books from these allied and controversial disciplines separately. Nestled in his inventory between books of jurisprudence and texts in Greek, Sozomeno inserted a cryptic subject heading titled "Various (*Diversi*) books in folio, quarto, octavo, duodecimo, and sixteenmo."[75] This catchall category is a who's who of the controversial books and authors that physicians had been requesting to read and censors had been expurgating throughout the sixteenth and early seventeenth centuries. The first entries in the "Diversi" section were Conrad Gessner's natural history books on quadrupeds, birds, fish, animals, and snakes; his lexicographical *Onomastikon* and *Phisicarum meditationum*; and his *Bibliotheca universalis*.[76]

Sozomeno's separate listing of prohibited works might best be understood as an example of dissimulation through which he disguised controversial works and authors within the chaos and disorganization of a miscellaneous heading. This strategy was not exclusive to Catholics. Johann

Heinrich Alsted put his discussion of Copernicus's moving earth in a miscellaneous section of his encyclopedia but gave readers the tools to find it in the index by listing it as "the earth, does it move?"[77] In Sozomeno's catalog, his treatment of the works of Girolamo Cardano is particularly illuminating. The Marciana owned several works written by Cardano, and the entries for his books that appeared under Sozomeno's heading "Medicine" included his *De sanitate tuenda* (*On the Preservation of Health*, 1580 and 1582 editions), *Opuscula medica* (*Little Works of Medicine*, 1559), *De prognost[icis]* (*On Prognostics*, 1568), *De aere* (*On Air*, 1570), and *De rerum varietate* (*On a Variety of Matters*, 1557). These works were among Cardano's least controversial and most obviously medical publications. By contrast, the books by Cardano listed only as "various" (*diversi*) included again his *De rerum varietate* (1557) and the uncontroversial *De proportionibus* (*On Proportions*, 1570), but also the more problematic *De subtilitate* (*On Subtlety*, 1582), two editions of the *In Cl. Ptolemaei Pelusiensis IIII de astrorum iudiciis . . . libroros commentaria* (*Commentary on Ptolemy's Astrology*, 1554 and 1578), two editions of his *Opuscula varia* (*Various Works*, 1547 and 1562), his philosophical *De Consolatione* (*On Consolation*, 1542), and his autobibliographical *De libris propriis* (*On My Own Books*, 1557).

While the bulk of Cardano's eclectic intellectual interests does not necessarily fall within a strictly defined realm of medicine, this treatment in Sozomeno's catalog is suspicious because of the company that these works keep under the heading "Diversi." The "Diversi" category also included the vast majority of Paracelsus's works, books of astrology by Antoine Mizauld, the chemical and alchemical works of Andreas Libavius and Geber (Jabir ibn Hayyan), the *Secretum secretorum* by pseudo-Aristotle, the works of Marsilio Ficino, and an assortment of works by his student Giovanni Pico della Mirandola. The close categorical proximity between these "miscellaneous" medical texts and works by the heretical monks Giordano Bruno and Tommaso Campanella reveals more blatantly the intellectual rationale linking many authors in the "Diversi" category.[78] Sozomeno's catalog discreetly offered a solution to navigating the fraught territory between censorship and universality by scattering problematic texts within a long list of miscellany. Only those scholars who diligently searched for prohibited and controversial medical texts would find them. Perhaps when they found the listings for these prohibited volumes, they paused to consider the Latin irony of the antonyms *diversus* and *universus* that facilitated the location of prohibited knowledge in Catholic libraries.

Sozomeno's catalog is another example of the intrinsic place of prohibited books in Italian libraries at the turn of the seventeenth century. His career as censor, scholar, and public servant exemplifies the uniquely Venetian version of seventeenth-century book collecting in which learning and intellectual control worked together to further the civic goals of the Venetian state. In seventeenth-century Venice—a city bent on restoring and enshrining its own cultural patrimony as a center of print culture—prohibited medical books were both openly and stealthily advertised as having an essential place on the shelves of the Biblioteca Marciana.

STRATEGIC MANAGEMENT AND DESIGN

While the late sixteenth- and early seventeenth-century histories of the Biblioteca Ambrosiana, the Vatican Library, and the Biblioteca Marciana each have characteristics particular to their local contexts, it is no coincidence that the themes of universality, utility, and the role of censorship are relevant to all three. These libraries were undergoing significant transformations during the same period, and the organization of these institutions was self-consciously modeled on contemporary Italian libraries.[79] The librarian of the Ambrosiana, Antonio Olgiati, regularly referred to himself as the "prefect" of the library (though this was not his official title) in imitation of the heads of the Vatican Library, the Biblioteca Casanatense, and the Biblioteca Angelica.[80] Similarly, Giovanni Sozomeno's newly founded position as *custode* of the Marciana was modeled on similar positions at the Vatican Library and the Laurenziana in Florence.

Venice, in particular, set out in search of information about how other states' libraries functioned. The Riformatori dello Studio wrote to their diplomatic official in Florence on December 6, 1622, requesting information about "the method and order in which the books are kept in the library of the Grand Duke." They urged him to be as "copious, distinctive, and particular" as possible in his description, so as to be of greatest service.[81] The Riformatori also sent similar letters to Milan and Rome, soliciting details about the administration of the collections at the Biblioteca Ambrosiana and Vatican Library. A few weeks later, Valentino Antelmi responded from Florence, reporting that he had been to the library of the Grand Duke in San Lorenzo (the Laurenziana) and had even consulted the index, "a book that lists all the volumes that exist in the library, making distinctions about their content and language."[82] This letter represents the possible impetus that led to Sozomeno's catalog less than two years later. Antelmi went on to describe how the books were chained to the desks in a long room full of beautiful, expensive works and overseen not by a *custode* but by one of the canons of San Lorenzo who supervised the care and conservation of the collection.[83]

The unsigned response to the Riformatori from Rome was still more "copious, distinctive, and particular" regarding the operations of the Vatican Library: "All of the administration of the Vatican Library is overseen by a Cardinal of the Holy Church, who is called the Librarian, though he is served by assistants and officials."[84] The report specified that there were two *custodi*, two revisors, four writers (who were in essence copyists), a bookman, and a sweeper, all of whom spent three hours every morning conducting their business. The report laid out the duties of each staff member. Where Antelmi had been impressed by the architecture and adornment of the Laurenziana, this Venetian dignitary was struck by the personnel and the infrastructure of knowledge production and maintenance at the Vatican Library. Building the great Catholic libraries of seventeenth-century Italy was a project that turned the Italian states reflexively toward examples close at hand. While local exigencies and personalities were crucial to how these collections developed, neither librarians nor their patrons operated in a vacuum. The cultures of collecting and of censorship were persuasive and pervasive, and the libraries in Milan, Rome, and Venice developed and grew in ways that accommodated the ideal of universality and mitigated prohibitions.

The three libraries discussed in this chapter—the Ambrosiana, the Vatican, and the Marciana—all had extensive collections of prohibited medical books. Catholic, institutional libraries at the beginning of the seventeenth century were repositories of both licit and prohibited learning.[85] However, accommodating prohibited and Catholic books in the same collection required certain concessions and strategic designs.[86] After all, prohibited books were censored as both physical and intellectual objects. I turn here to the library of Romolo Spezioli, a physician from the late seventeenth and early eighteenth century who donated what may be the most complete surviving early modern medical library to his hometown of Fermo.

Romolo Spezioli received his medical degree from the university in his native city of Fermo in 1664, but he spent most of his medical career in Rome, where he taught at La Sapienza and served as personal physician to eminent patrons including Cardinal Decio Azzolini (junior), Queen Christina of Sweden, and Pope Alexander VIII.[87] In 1703, Spezioli decided to donate his books to the local library in Fermo, which had been opened to the public in 1688.[88] At his death in 1723, Spezioli transferred the whole of his nearly twelve thousand-book collection to the Biblioteca Civica di Fermo.

Unsurprisingly given his position and connections, Spezioli's collection included many medical books that had been listed on the Index of Prohibited Books.[89] Since the library and its records remain intact, we can examine precisely how a late-seventeenth-century physician materially and spatially integrated these prohibited books into his Catholic collection.

In Milan, when Antonio Olgiati, the librarian at the Biblioteca Ambrosiana, asked for a license for the library, he did so specifying that he would concede to binding prohibited books differently, but he wanted them shelved within the collection, not separately. The prohibited volumes in Romolo Spezioli's library also indicated on their bindings that they were prohibited. Each cover was marked with the Latin symbol for the abbreviation "Pro" (for *prohibitus* in Latin or *proibito* in Italian; see figure 7.4). Potential readers were warned about the author or content by the binding. No reader would be at risk of absentmindedly selecting a title that the Catholic Church believed had the potential to endanger their souls without advance warning.

Spezioli's labeling system drew readers' attention to the prohibited status of the book after it had been removed from the shelf. But how did readers find these prohibited volumes among the many thousands of books in Spezioli's collection? Spezioli's manuscript catalog, delightfully titled *Bibliotheca universalis index auctorum et voluminum (Universal Library, Index of Authors and Volumes)*, arranged his books by subject and size— and sometimes denoted with a stylized asterisk that a particular volume was prohibited.[90] In an early iteration of the catalog, the volumes were listed with a shelf mark in a particular format that included a number, a Greek letter, and the mark "oo" all separated by dashes.[91] In later catalogs, the full bibliographical information about the prohibited book was provided (title, author, date), but the shelf mark was often omitted.[92] The change in the method of cataloging prohibited books over time suggests that the volumes were originally shelved separately, but over time the books were integrated onto the regular shelves of the library. However, by including the entry but removing the shelf number from his *Bibliotheca universalis*, Spezioli's library took an approach analogous to Sozomeno's catalog of the Biblioteca Marciana: a reader needed to know how to find a book in order to read it.

Although Antonio Olgiati declined to shelve the Ambrosiana's books separately, this solution was widely accepted in the eighteenth century. Initially, Spezioli separated his prohibited volumes from the rest of his collection. Lorenzo Corsini, the future Pope Clement XII, opened his library, the Corsiniana, to the public on May 1, 1754. The library was outfitted

Fig. 7.4. Inscription "Pro" added to the cover of prohibited books from the
collection of the physician Romolo Spezioli, who donated his extensive medical
library to his hometown of Fermo. Although the prohibited books were labeled
this way, the library catalog obscured their location on the shelves. Reproduced
with permission from the Biblioteca Civica "Romolo Spezioli," Fermo.

with specially labeled shelves built to contain the *libri heterodoxi* (heterodox books) in the collection (see figure 7.5).[93] Today, these shelves still
contain astrological and theological texts. The books are labeled and on
display, not hidden away, designated as prohibited only by the sign on the
otherwise uniform shelving. Over the course of the seventeenth century,
prohibited books had come to have a place that was, by the eighteenth
century, physically and intellectually integrated into the space of Catholic
libraries.[94]

Fig. 7.5. Shelf of "Heterodox Books" at the Biblioteca Corsiniana in Rome, which
still contains books that were banned in the early modern period. Reproduced with
permission from the Biblioteca dell'Accademia Nazionale dei Lincei e Corsiniana.

CONCLUSION

Gabriel Naudé is the best known early seventeenth-century scholar (trained
as a physician) who reflected extensively on the place, role, and importance
of libraries.[95] Naudé considered the place of prohibited books in Catholic collections and was famously open to having the works of "the most
learned and famous heretics" in a library, which he described as "neither
an absurdity nor a danger." However, he did go on to stipulate that readers needed to be licensed by responsible authorities in order to read these
works.[96] For Naudé, authors who studied and excelled in the same subjects
should be grouped together regardless of their religious leanings: "Such
authors as have best handled the parts of any science or field of learning, whatever it be: as Bellarmine for controversial theology; Toledo and
Navarre, cases of conscience; Vesalius, anatomy; Mattioli, the history of
plants; Gessner and Aldrovandi, that of animals; Rondelet and Salviani,
that of fishes; Vicomercatus, that of meteors; and the like."[97] When Naudé
published his *Advis pour dresser une bibliothèque* (*Advice on Establish-*

ing a Library) in 1627, he (perhaps unwittingly) reunited the medical republic of letters that had been torn asunder by personality and religious difference in the middle of the sixteenth century. Here we read, side by side, the names of the great observers of fish, Guillaume Rondelet and Ippolito Salviani, who had been first on Gabriele Falloppio's list of "traitors" to the medical republic of letters for "continuously provoking each other, the one writing against the other."[98] Naudé also philosophically reunited the works of Ulisse Aldrovandi and Conrad Gessner, the devoted colleagues who, despite the religious and intellectual climate, persisted in their exchange of ideas and plant specimens. Naudé's *Advis*, which replaced the censorship of books in libraries with a system of licensing and regulating readers, mirrors the strategies that the Catholic Church took toward regulating medical knowledge in the seventeenth century.

We should not read Naudé's libertinism into the policies of the Catholic Church, but there is no doubt that both Naudé's position and the lived reality of nearly all Catholic readers lay somewhere between that of true universality and the boundaries officially prescribed by Catholic censorship. Indeed, Giovanni Sozomeno's 1624 catalog of the Biblioteca Marciana unabashedly began its list of books in the humanities with Antonio Possevino's *Bibliotheca selecta*, followed immediately on the next line by Gessner's *Bibliotheca universalis*.[99] From the medical republic of letters to the Index of Prohibited Books, the transnational and multiconfessional community of medical learning was, by the mid-seventeenth century, rubbing spines (if not shoulders) on the shelves of Italy's great Catholic libraries.

Epilogue

This book has followed the fortunes of medical books as they moved from the republic of letters, onto the Italian Indexes of Prohibited Books, and then back again onto the shelves of Catholic Italy's great libraries. Throughout this fraught process, we have encountered physicians in their roles as humanists, healers, university professors, historians, book smugglers, and even censors. We have also observed the many physical ramifications of censorship on medical books. From thin lines through words to missing pages and effaced passages, these censored books are archives of their own histories as much as the letters, reports, and treatises written about them. The Catholic censorship of medical books was a project that encroached on the lives of physicians and scholars and altered methods of scholarship in many communities across Counter-Reformation Italy.

In each of these contexts, I have also traced the intensification and maturation of a discourse of utility surrounding medical works. Claims about the utility of medicine were repurposed and confessionalized in response to the Catholic Church's prohibitions on certain medical books through the protests of physicians, the ruminations of censors, the petitions for licenses to keep these books, and the material ways in which they were changed to eventually find their permanent homes in Catholic libraries. Throughout the sixteenth and seventeenth centuries, physicians and ecclesiastical officials alike appealed to the utility of prohibited texts to support the project of expurgation and the licensing of texts to qualified professionals.

This epilogue shifts our focus from the subject of the censorship of medical texts to reconsider the most famous censorship controversy of the early modern period: the Catholic Church's condemnation of Copernicanism in 1616 and of Galileo's *Dialogue* in 1632. The prohibition of medical books and the systems of censorship that mitigated those prohibitions also laid the groundwork for the development of utility and professional expertise as central aspects of early modern medicine. Examining the paradoxes of censorship and the discourses of scientific expertise and utility reveals the ubiquity of the censor as an interlocutor throughout the Galileo affair.

As we saw in the preceding pages, sixteenth-century censors came to the conclusion that the utility of medical knowledge did not need to be sacrificed wholesale to comply with Catholic faith. Even before the period discussed in this book, there were persuasive precedents from late antiquity and the Middle Ages for compromise between Catholic theology and useful knowledge. Augustine, the fifth-century bishop of Hippo, suggested that Greek and Roman philosophy should not be discarded in favor of theology but rather that philosophy should serve theology, especially to avoid embarrassment on the part of the Church.[1] As Augustine reasoned, even non-Christians knew something about "the earth, the heavens, and the other elements of this world, about the motion and orbit of the stars and even their size and relative positions." It would reflect poorly on Christianity to ignore this body of knowledge: "Now it is a disgraceful and dangerous thing for an infidel to hear a Christian, presumably giving the meaning of Holy Scripture, talking nonsense on these topics; and we should take all means to prevent such an embarrassing situation, in which people show up vast ignorance in a Christian and laugh it to scorn."[2] Philosophy could therefore serve Christianity and should not be dismissed out of hand. This hierarchical relationship, which bolstered the Catholic Church's patronage of scientific endeavors, has come to be known as the *handmaiden formulation*: theology was the queen and philosophy her handmaiden. In the seventeenth century, Galileo Galilei, among others, challenged this formulation.

Let us revisit this seminal drama in the history of science with an eye toward how it played out in light of what we have learned about the prohibition of medical texts. The reception history of Copernicanism starts in 1543, when Nicolaus Copernicus published his *De revolutionibus orbium coelestium* (*On the Revolutions of the Heavenly Spheres*).[3] In one of history's great coincidences connecting the history of astronomy and that of medicine, 1543 was also the year that Vesalius published his *De humani corporis fabrica* (*On the Fabric of the Human Body*).[4] These two remarkable books are often cited together to mark the beginning of a so-called Scientific Revolution. We have examined the complex, iterative process by which medical books in Catholic Europe were subsequently prohibited,

expurgated, and licensed to readers. The reception of Copernican astronomy, however, unfolded differently. Unlike many medical books, Copernicus's work was not banned in any of the sixteenth-century Indexes of
Prohibited Books.[5] Whereas most prohibited physicians were Protestants,
Copernicus was Catholic (even a church canon), and he dedicated his
book to Pope Paul III. Although there was an early attack on Copernicus's
work by the Dominican Giovanni Maria Tolosani, it had few immediate
consequences due to the death of the Master of the Sacred Palace and the
more pressing concerns of the Council of Trent.[6] Church officials were not
widely concerned about the possible theological implications of Copernicus's theory, and the assertions of *De revolutionibus* were blunted by the
Lutheran Andreas Osiander's unsigned preface, which declared the central
heliocentric thesis of a moving Earth to be a mere hypothesis. Moreover,
the new calculations and tables in Copernicus's dense book were useful,
with applications for both astronomy and the Catholic Church.[7] The reception and censorship of *De revolutionibus* was also distinct from that of
the medical texts discussed in this book because Copernicus's work was
not as widely read in the sixteenth century. Some scholars have gone so far
as to say, with more than a hint of irony, that *De revolutionibus* was the
"book that nobody read."[8] While banned medical books were rapidly and
extensively debated because of their widespread readership in the personal
and intellectual networks of the republic of letters, Copernicus's book did
not enjoy such immediate fame or notoriety, or the censorial gaze that
came with it. That would all change in the early years of the seventeenth
century, due in large part to Galileo.

In the wake of Galileo's discoveries with his telescope and his highly
public arguments in favor of a moving Earth, Copernicus's work emerged
as a central part of these cosmological debates.[9] The well-known chain
of events might be reconstructed roughly as follows, with this condensed
account giving particular weight to the similarities and differences between the prohibition and correction of *De revolutionibus* and the medical works described throughout this book.[10] First, Galileo, tinkering with
lenses, created a telescope that could be turned to the skies, thus revealing
the moon's textured surface, satellites around Jupiter, a seemingly infinite
number of stars, spots on the sun, the phases of Venus, and protuberances
around Saturn.[11] With this new evidence in hand, Galileo argued publicly
and polemically in favor of the Copernican (heliocentric) model of the solar
system. Galileo then argued in a semipublic letter to his friend Benedetto
Castelli in 1613, and then in a longer public letter to his patron Christina of Lorraine in 1615, that the book of nature and the book of scripture

could not be in conflict.[12] In particular, Galileo believed his astronomical discoveries indicated that the earth rotated on its axis and around the stationary sun, and these empirical truths should therefore inform how one interpreted an oft-cited passage in Joshua about the movement of the sun. In his letter to Christina, Galileo outlined a new relationship between the queen and the handmaiden of the Middle Ages.[13] According to Galileo, what one learns about nature must inform how scripture is read and interpreted. In the wake of nearly sixty years of Roman censorship, Galileo's reformulation of the handmaiden metaphor should be understood as a continuation of the discourse about expurgation that had, paradoxically, promoted the expertise of lay professionals as distinct from that of theologians. Indeed, Robert Bellarmine—the theologian, member of the Congregation of the Index, and future saint charged with managing the Copernican controversy—understood this connection and sought to control the situation by minimizing the separation between these realms of authority, consolidating power within the hands of the papacy.[14]

At the same time that Galileo's letter to Castelli was circulating, another letter in favor of heliocentrism was printed in Naples and swiftly made its way to Rome. Paolo Antonio Foscarini, an otherwise unknown Carmelite priest, had joined Galileo in his public advocacy of Copernicanism and in interpreting the Bible in ways that corroborated this cosmology. Bellarmine admonished both men that Copernicanism was a theory, not truth, and that it had no bearing on how the Bible should be interpreted.[15] Foscarini received this rebuke in a letter, and Galileo in a murky and poorly documented meeting. Bellarmine formalized his pronouncement in a meeting of the Congregation of the Index on March 1, 1616, proposing that Foscarini's work be banned and Copernicus's "suspended until corrected."[16] This verdict left open two important questions: Who would correct Copernicus's book? And what made *De revolutionibus* worthy of correction, rather than outright prohibition?

The answer to both questions can be found in the person and work of Francesco Ingoli, a lawyer and ecclesiastic from Ravenna. Ingoli had earned his degree from the University of Padua in 1601, and it is possible that he knew Galileo during their shared time there.[17] Ingoli had debated Galileo publicly in Rome and had regularly joined in philosophical discussions at the house of Prince Cesi in 1612–13.[18] In late 1615 or 1616, Ingoli circulated a manuscript treatise on the location and immobility of the earth and against the Copernican system, refuting Galileo's arguments posed in the letter to Castelli.[19] Following the death in 1617 of Cardinal Caetani, for whom Ingoli served as secretary, Ingoli inherited the cardi-

nal's role as a censor and began to compose a series of expurgations of Copernicus's De revolutionibus.[20] On April 2, 1618, Ingoli submitted his first
report, which concluded that "the already-prohibited book by Copernicus
(which falsely asserts a mobile earth and a stationary sky) is considered
useful and necessary to astronomers and it was desired by all."[21] He proposed that Copernicus's De revolutionibus be corrected and emended, and
he forwarded his corrections to the Congregation of the Index.

Ingoli's expurgations focused on emphasizing the hypothetical nature
of the Copernican thesis, removing praise of heretics, and deleting parts of
the text related directly to biblical exegesis.[22] From the preface, Ingoli took
pains to remove one of Galileo's favorite Copernican statements, which he
had quoted at the beginning of the Letter to Christina and which emphasized the expertise of mathematicians over that of "triflers who though
wholly ignorant of mathematics nevertheless abrogate the right to make
judgements about it because of some passage in Scripture wrongly twisted
to their purpose."[23] After reviewing Ingoli's expurgations, the Congregation of the Index delayed ruling on the proposed corrections and instead
submitted them to "others who would consider [the matter] better," for-
warding Ingoli's report to the Jesuit fathers who were lecturers in mathematics at the Collegio Romano.[24] In this respect, the examination of Copernicus's De revolutionibus followed a similar procedure to that of the
medical texts that we have examined in this study. For both, expurgations
were composed by theologians with input from content area experts.

This process would have been familiar to Galileo as well. He was living in Padua and teaching at the university while the expurgation of medical books was being foisted upon his colleagues and rivals in the schools
of philosophy and medicine. In a series of letters to his friend, the well-
connected ecclesiastical official Piero Dini, Galileo suggested that ecclesiastical experts trained in mathematics should examine Copernicus's
theory. Galileo explicitly requested that Dini discuss the validity of Copernicanism with the Jesuit Father Grienberger, "a famous mathematician
and my great friend and patron." In 1615, Galileo wrote to Dini:

> I merely ask that its [Copernicus's book's] teachings be examined and
> its arguments be evaluated by the most Catholic and most expert per-
> sons, that its propositions be confronted with sense experience. . . .
> Since there is no shortage in Christendom of men who are most expert
> in this profession, it would seem that the truth or falsity of this doc-
> trine ought not to be deferred to the judgment of those who are unin-
> formed and who are clearly known to be affected by feelings of bias.[25]

Galileo explained further in his postscript, "I believe that the quickest remedy would be to call upon the Jesuit Fathers, who are so much more educated than these [Dominican] Fathers."[26] Galileo wanted any decision about the validity of Copernicanism to be determined by scholars and experts, not just by theologians. In Galileo's view, the Jesuits, whose support Galileo had courted during his time in Rome, were a point of intersection between theological and natural philosophical expertise.[27] Galileo believed that under close inspection by those attentive to sense experience and properly educated in mathematics, his telescopic evidence supporting the truth of the Copernican system would speak for itself.

In 1618, the Congregation of the Index followed Ingoli's careful, though amateur, expurgations with input from the Jesuit fathers, as Galileo had suggested in his letter to Dini. To Galileo's dismay, Christopher Grienberger, along with fellow Jesuit Orazio Grassi, did not respond as the professors at the University of Padua had when faced with the task of expurgation. Rome was not Padua, and the Jesuits, though cosmopolitan, did not resist the Congregation of the Index's request for cooperation and corroboration of the expurgations.

Early in May 1620, the Congregation of the Index issued its formal decree regarding the correction of *De revolutionibus*. The decree read, "Since Copernicus's work contains many things that are very useful to the commonwealth, by unanimous consent they came to that decision: . . . that it should be permitted . . . provided that those passages in which he argues about the movement and location of the earth as fact and not hypothetically are corrected according to the following correction."[28] The rationale and wording that Ingoli and eventually the Congregation of the Index used to justify the expurgation of Copernicus's work centered on utility and expertise. Copernicus's book, as Ingoli reported, was considered "most useful and necessary to astronomy." The Congregation of the Index similarly described it as "very useful to the commonwealth." Additionally, as Ingoli's report explained, the book was "desired by all." These are the same motivations and justifications that were established in the sixteenth century to support expurgation and licensing of medical texts, rather than banning and burning them. Copernicus's work was useful, everyone wanted to read it, and expurgations were composed with the support of learned professionals—in this case Jesuit fathers, rather than lay practitioners.

Meanwhile, the prohibition of *De revolutionibus* was changing the ways that people in Italy read astronomical texts, linking the fates of Galileo and Copernicus in the minds of readers. In January 1625, Bartolomeo Balbi, a Genoese merchant and enthusiastic correspondent of Galileo, requested a license from the Roman Inquisition to read a list of sixty-three prohibited books and authors.[29] Alongside his request for the works of Giro_lamo Cardano, Julius Firmicus Maternus, Francesco Giuntini, and Copernicus, Balbi requested the "works of Galileo." Balbi's request presumably referred to the *Starry Messenger* and the *Assayer*. However, neither of these works had been prohibited. Instead, we see how Balbi's engagement with Galileo and his ideas led him to apply the 1616 and 1620 condemnations of Copernicanism and books that argued for a moving Earth to Galileo's published, though not yet prohibited, works. Censorship, understood as the complex processes of "prohibition, permission, and correction" of books, changed the ways that people read and approached texts, in this case linking Galileo's works to Copernicus's prohibition.[30]

At its most basic level, the process that the Catholic Church took toward banning Copernicus was parallel to that of the expurgation of medical books. The work was suspended, widely requested by professionals, justified as useful, and expurgated as a form of compromise with the input of experts. Just as expurgation was not a welcome outcome for some readers of prohibited medical books, Galileo polemically argued that an expurgatory compromise that denied heliocentrism would be "the worst judgment that could fall upon Copernicus's book."[31] In the wake of that unwanted outcome, Galileo, for a time, stopped publicly advocating for Copernicanism. However, in 1624 the intellectual tides had changed, and Galileo's patron and friend Maffeo Barberini had been elevated to the papacy as Urban VIII. Galileo took the opportunity to write and circulate his next great defense of Copernicanism. The first iteration of what would become his masterpiece—the *Dialogo sopra i due massimi sistemi del mondo* (*Dialogue Concerning the Two Chief World Systems*), which was eventually published in Florence in 1632—began as a treatise in the form of lengthy letter addressed to Franceso Ingoli.[32] Blaming him for spurring the Congregation of the Index to reject the "Copernican opinion," Galileo repudiated Ingoli's analysis at length. Galileo resorted repeatedly to ad hominem attacks on Ingoli's expertise and his incapacity to judge the issue, accusing the ecclesiastical censor of Copernicus of having spent fewer days studying these complicated matters than Copernicus spent years.[33] Galileo's final refutation of Ingoli turned once again on his lack of professional credentials: "I want to bring forth a certain fitness which I once was in the custom of pointing out to nonprofessionals who were incapable of esoteric demonstrations."[34] Ingoli himself never got to read the "letter to Ingoli" in its original form because Galileo's supporters withdrew it from circulation when circumstances in Rome became suddenly less favorable.

However, we must remember that one initial context for the composition of Galileo's *Dialogue* was an encounter with Ingoli, the censor of Copernicus who was a priest, and not a professional astronomer.

Foregrounding the context, processes, and people involved in expurgation brings new issues to light in Galileo's framing of his *Dialogue*. The prefatory material of the *Dialogue* capitalized on Galileo's readers' fluency in the themes of professional expertise and the utility of knowledge. In his note "To the Discerning Reader," Galileo's opening sentences revealed his impetus for writing the text in light of the bans of 1616. The passage reads: "There were those who impudently asserted that this decree had its origin not in judicious inquiry, but in passion none too well informed. Complaints were to be heard that advisers who were totally unskilled at astronomical observations ought not to clip the wings of speculative intellects by means of rash prohibitions."[35] The *Dialogue* drew on the same themes and language that Galileo laid out in his letter to Dini nearly twenty years earlier. In both, Galileo noted the rashness of the prohibitions and the personal biases that entered into these decisions. Additionally, he set scholars with astronomical experience against those without it. His choice of language parallel to that in the letter to Dini was no accident on the part of an author as rhetorically skilled as Galileo.[36]

Galileo's target in his introduction to readers—the unnamed adviser who was "totally unskilled at astronomical observations"—was surely Francesco Ingoli.[37] Early modern readers would have noticed that Galileo specifically chose the word *consultor* to refer to this adviser, referring not to advisers in general, but in particular to the title for people who provided censorship opinions to the Congregation of the Index.[38] Readers would also have understood Galileo's phrase about "clipped wings" to be a description of book expurgation, referring to the blade used to slice pages out of a volume. The metaphor alluded to the dual nature of expurgation as a physical as well as intellectual reality of scholarship in the seventeenth century.

The passive-aggressive sabotage of the expurgation of medical books by professors at Padua is also important background for the authority Galileo repeatedly asserted over Ingoli. Galileo's triumph in publishing his *Dialogue* with its attack on Ingoli clearly visible in the opening pages reenacted the Paduan professors' acts of disobedience in the face of the *honorata impresa*—the "honorable enterprise" of censoring books. Despite the Catholic Church's insistence on linking the fates of the two men, Galileo did not take the same approach as Cesare Cremonini, who opportunistically joined forces with censors and inquisitors. Instead, Galileo doubled down on his expertise as separate from and superior to that of the theologian-censor Ingoli and, by extension, the Catholic Church.

The final page of Galileo's letter of dedication to Ferdinando II, the Grand Duke of Tuscany, lies across the book's gutter from Galileo's letter to readers, and it emphasizes the utility of his work. Galileo asked the grand duke to accept the dedication of the *Dialogue*, hoping that from the work "lovers of truth can draw the fruit of greater knowledge and utility."[39] In so doing, readers would acknowledge that all such rewards stemmed from their positions as happy subjects of the grand duke. Galileo's framing of his long-pondered defense of Copernicanism as ultimately "the fruit of utility" deriving from the protection and sponsorship of the grand duke sought to position the *Dialogue* as both essential knowledge and fundamentally Florentine. Knowledge on its own was important, but Galileo's description of his work as useful knowledge was a further justification, honed by a long experience of living in a culture of censorship. The positioning of his work emphasized its controversial nature and ultimate potential for orthodoxy under the protection of the right patron.

In the end, neither Galileo's expertise nor the utility of the work nor even the protection of the grand duke could save the *Dialogue* from prohibition.[40] The *Dialogue* was prohibited without mitigation, and Galileo observed that licenses to read it had been so restricted that only the pope could grant them.[41] His friend Fortunio Liceti was granted a license to read 120 prohibited books and authors, but the Holy Office explicitly denied his request to read Galileo's *Dialogue*.[42] Carlo Emanuele Vizzani, the Bolognese nobleman and professor of logic at the University of Padua, reflected on Galileo's position among the most notorious prohibited authors in his 1639 request for a reading license:

> He [Carlo Emanuele Vizzani] requests any and all books by damned authors and other books in any way prohibited dealing with literature and philosophy contained on the Roman Index of Prohibited Books, except those books that deal *ex professo* with religion and judicial astrology, speak negatively in any way of the jurisdiction of the papacy, or works by authors of damned memory (*damnate memoriae*) Charles Dumoulin, Niccolò Machiavelli, and the book by Galileo Galilei on the motion of the earth and the stability of the sky.[43]

In the eyes of this learned reader, Galileo had joined the ranks of the most infamous and dangerous political enemies of the Catholic Church. He was "of damned memory," not to be forgotten, but to be always set apart.

After Galileo's death, his disciple Vincenzo Viviani, together with Galileo's grandson, the priest Cosimo Galilei, set about on a mission to restore Galileo in the memory of the Catholic faithful. Cosimo, ever attentive to pious details, applied conscientiously for a license to open the trunk that contained copies of his grandfather's papers and to read his copy of the *Dialogue* as part of this project, which also included burning incriminating manuscripts.[44] It ultimately took a hundred years after Galileo's death for Viviani and Cosimo's project to bear fruit and for Galileo's philosophical utility to overcome the damnation of his memory and justify an expurgated edition of the *Dialogue*. This "corrected" edition of Galileo's (almost) complete works was sanctioned by Pope Benedict XIV and finally published by Giuseppe Toaldo in 1744.[45]

We have often treated the famous story of Galileo and Copernicanism in a kind of hermetic isolation, focused on narratives of authority, truth, and epistemology. In so doing, we have underestimated the many ways that censorship and expurgation were deeply embedded in all stages of Galileo's defense of Copernicanism. Considering the prohibition and expurgation of medical texts has illuminated a set of historical mechanisms and discursive techniques that can help us better understand early modern conflicts between religion and knowledge. The utility of knowledge and the unique expertise of professionals are omnipresent in both the Catholic Church's response to Copernicanism and Galileo's defense of it. These highly developed discourses emerged from conversations about the censorship of medicine and came to be applied to scientific knowledge more broadly in the seventeenth century.

When we consider Galileo's approach to defending Copernicanism, we must remember that the astronomer became a scientific reader over the course of the sixteenth century.[46] In fact, Galileo was a professor in Padua for nearly twenty years (1592–1610) during which he rubbed shoulders with, sharpened elbows against, and read in Pinelli's library alongside the faculty members there who were appointed by the Catholic Church to expurgate medical texts. We must understand Galileo as a sixteenth-century reader whose intellectual development took place in a community that advocated for the utility of scientific works in the face of Catholic prohibitions. Further, the advocacy for utility by lay readers was encouraged by the Catholic Church's acknowledgement of the expertise of medical professionals as separate from that of ecclesiastics. In his *Dialogue*, Galileo was speaking to a system of censorship established through the prohibition and expurgation of medicine in the sixteenth and early seventeenth centuries.

When considering these discourses of utility and expertise, it is somewhat misleading to dwell exclusively on Galileo's condemnation by the
Roman Inquisition. Although Galileo was condemned, the justifications of
utility and professional expertise that he advocated were the winning discourses of scientific rhetoric in the seventeenth and eighteenth centuries.
Studying the effects of censorship on science requires attention to how
the important debates about what constituted Catholic knowledge recognized the role of experts outside the ecclesiastical hierarchy. Expurgation
and licensing activated and amplified a discourse of utility that became
one of the defining features of scientific culture and was, fundamentally,
intertwined with Catholic piety. The complex processes of ecclesiastical
censorship ultimately acknowledged and helped to establish the role of
professional expertise and to define the utility of prohibited knowledge for
Catholic society.

ACKNOWLEDGMENTS

Researching and writing this book has required countless hours spent reading and writing quietly in libraries and archives. It has also led to countless more hours spent talking animatedly with friends and colleagues about what I was reading and writing. I am deeply grateful to many people and organizations for helping facilitate these two scholarly traditions. It is my hope that this research will continue to inspire conversation and debate in the years to come.

My early work on this book was supported by research grants and fellowships from the Lane History of Science Fund at Stanford University, the Council on Library and Information Resources and Mellon Foundation, the Rare Book School and Andrew W. Mellon Fellowship of Scholars in Critical Bibliography, the Renaissance Society of America, the Gladys Krieble Delmas Foundation, and the Mellon-ACLS Dissertation Completion Fellowship. Since joining the faculty at Harvard University, I have been grateful for exceptional support from colleagues, administrators, and staff, and for funding for research trips, image permissions, childcare grants, and especially a manuscript workshop that provided me with essential feedback at a critical stage.

Most of chapter 3 of this book was published as "The Mind of the Censor: Girolamo Rossi, a Physician and Censor for the Congregation of the Index" in the journal *Early Science and Medicine*. I am grateful to Brill and to *Early Science and Medicine* for allowing me to reproduce that material here.

I have visited many libraries and archives to conduct the research for this project. I have cited these collections with gratitude in the notes and bibliography of this volume. Additionally, the staff at these institutions have facilitated my work through direct advice and their deep knowledge of their collections. I am especially grateful to Floriana Amicucci, Giovanna Bergantina, Mons. Alejandro Cifres, Pietro Gnan, John Mustain, John Overholt, Daniel Ponziani, Shannon Supple, and Fabiano Zambelli for their help.

It is with deep gratitude that I acknowledge Enrico Ramirez-Ruiz and Abraham Loeb—my partner's mentors at UC Santa Cruz and Harvard— for their flexibility and support, which have made it possible for our family to travel together on my many research trips to Italy.

This project has benefited from numerous careful and generous readers. I would like to offer my sincere thanks to Paula Findlen—for reading countless messy drafts, for having the patience to let me find my own way to better analysis, and for modeling collegiality and intellectual generosity at every turn. I also received extensive support during my time at Stanford from Jessica Riskin and Laura Stokes, and from Anthony Grafton at Princeton. In May 2018, Mordechai Feingold, Paula Findlen, Sachiko Kusukawa, and Katharine Park joined me at Harvard for a daylong workshop on my manuscript. Their comments have made this book richer, bolder, and much longer, and I am deeply grateful for the time and care they put into reading it.

Additionally, Federico Barbierato, Ann Blair, Anthony Grafton, and Corey Tazzara provided essential comments on full drafts of the book at various stages. Two anonymous peer reviewers offered readings of my manuscript that helped me see my own arguments more clearly. Thomas Wilson has diligently tamed my more creative Latin translations, and Anne Harrington gets full credit for the title. Lori Meek Schuldt provided careful copyediting advice, and Madeline McMahon, Thomas Wilson, and Valentina Frasisti helped check the proofs. I have also been lucky to have two dedicated editors: Karen Merikangas Darling at the University of Chicago Press, and Benjamin Marcus (my dad), who has brought his lawyer's eye to my academic prose.

It is a pleasure to thank a long list of friends, colleagues, and mentors who have supported me and this project with research and editorial assistance, careful readings of chapters, translation suggestions, sharp questions, obscure citations, delicious food, honest advice, and heartfelt celebrations of the many book stages and life events along the way. To Claire Rydell Arcenas, Andreea Badea, Rhae Lynn Barnes, Chris Basich, Ian Beacock, Nicola Bell, Richard Bell, Alex Bevilacqua, Brad Bouley, Bruno Boute, Allan Brandt, Brian Brege, Janet Browne, Giorgio Caravale, Marco Cavarzere, Alessandra Celati, Roger Chartier, Andrew Chittick, Cindy Chittick, Catherine Chou, Jon Connolly, Mackenzie Cooley, Maria Pia Donato,

Caroline Duroselle-Melish, Stephanie Frampton, Peter Galison, Roger Gaskell, Susan Glazer, Ashley Gonik, Mira Goral, Emily Hainze, Crystal Hall, Earle Havens, Benjamin Holtzman, Jordan Howell, Karl Johnson, Alan Charles Kors, Francisco Malta Romeiras, Angélica Márquez-Osuna, Maureen Miller, Yair Mintzker, Ann Moyer, Nicki Olivier Hellenkamp, Ada Palmer, John Pollack, Meredith Quinn, Ahmed Ragab, Jonathan Regier, Trish Ross, Sarah Gwyneth Ross, Paolo Savoia, Cotton Seed, Susannah Shoemaker, Gabriela Soto Laveaga, Peter Stallybrass, David Stern, Lynne Stillings, Daniel Stolzenberg, Michael Suarez, Chris Suh, Molly Taylor-Poleskey, Michael Tworek, Xenia von Tippelskirch, Stefano Villani, Yvon Wang, Nick Wilding, Caroline Winterer, Leah Whittington, Benjamin Wilson, Joan Yospin, Matthew Yospin, Richard Yospin, and Fabiola Zurlini, thank you!

Over the course of this project, my family has provided me with constant love, sympathetic phone calls, appropriate doses of skepticism, and last-minute child care. I have been sustained by trips home to Maine for gardening and boat trips in the summer and reading by the fire in both summer and winter. Thank you to my parents, Anita Bernhardt and Benjamin Marcus; my parents-in-law, Bruce and Arlene MacLeod; my brother, Nathan Marcus; and my grandmother, Angeline Bernhardt.

As I write these acknowledgments from the living room couch, I am listening to my partner, Morgan MacLeod, discuss trucks, books, trains, and Italian translations of words with our child, Simon. I want to thank Simon for his heartwarming enthusiasm about someday reading this book and Morgan for his unending support of my career and our family. I dedicate this book to Morgan—it has been a labor of love on his part as much as mine.

This appendix includes a list of physicians and people who studied medicine who were prohibited on the 1559 Pauline Index of Prohibited Books, presented in alphabetical order by first name. The information is compiled from the names and dates in Jesús Martínez de Bujanda, *Index des Livres Interdits* (Sherbrooke, QC: Centre d'études de la Renaissance, Editions de l'Université de Sherbrooke, 1984–96), vol. 8.

1. Arnald of Villanova (1238–1311)
2. Achilles Pirmin Gasser (1505–77)
3. Bernardino Tomitano (1517–76)
4. Bernardo Ochino (1487–1564)
5. Bruno Seidel (1530–93)
6. Conrad Gessner (1516–65)
7. Euricius Cordus (Heinrich Ritze Solden) (1484–1535)
8. Francesco Stancaro (1501–74)
9. François Rabelais (1483–1553)
10. Gaudentio di Treviso (Merula?) (1500–1555)
11. Georg Aemilius (Oemler) (1517–69)
12. Georg Agricola (Bauer) (1494–1555)
13. Gerardus Listrius (1470–1546)
14. Hadrianus Junius (1511–75)
15. Heinrich Cornelius Agrippa von Nettesheim (1486–1535)
16. Heinrich Pantaleon (1522–95)
17. Hieronymus Schurff (1481–1554)
18. Jakob Milich (Mylichius) (1501–59)
19. Jakob Schegk (Degen) (1511–87)
20. Janus Cornarius (1500–1558)

21. Jodocus Willich (1501–52)

22. Johannes Carion (1499–1538)

23. Johannes Dryander (Eichmann) (1500–1560)

24. Johannes Guinterius (Winther von Andernach) (1497–1574)

25. Johannes Philonius Dugo (d. 1553)

26. Johannes Postellus (Posthius) (1537–97)

27. Joseph Grunpeck (1473–1532)

28. Justus Velsius (Welsens) (1502–82)

29. Kaspar Peucer (1525–1602)

30. Leo Jud (1482–1542)

31. Leonhard [Leonhart] Fuchs (1501–66)

32. Luca Gaurico (1475–1558)

33. Marsilius of Padua (1290–1343)

34. Michael Servetus (1509–53)

35. Michael Toxites (Schutz) (1514–81)

36. Ortensio Lando (1512–55)

37. Otto Brunfels (1488–1534)

38. Paulus Ricius (1480–1541)

39. Pietro d'Abano (1246–1320)

40. Pompeo della Barba (1521–82)

41. Raymond of Sabunde (1385–1436)

42. Robert Constantin (1530–1605)

43. Sebald Havenreuter (1508–89)

44. Serafino da Fermo (Aceti de' Porti) (1496–1540)

45. Valerius Anshelm (Ryd) (1475–1546)

46. Veit Winsheim (Oertel, Ortelius) (1501–70)

47. Wolfgang Fabricius Capito (1478–1541)

NOTES

INTRODUCTION

1. Many thanks to Paula Findlen for sharing this fantastic quote. Francesco Redi to Leopoldo de' Medici, Florence, 13 May 1670, Biblioteca Marucelliana, Florence (hereafter cited as BMF), Ms. Redi, 7 (Lettere della Corte ed alla Corte di Toscana), f. 140r: ". . . credo che l'Anima mia anderà di sicuro in perdizione a conto di libri proibiti. Se Iddio invece di creare Adamo avesse creato me nel Paradiso Terrestre ed in vece di vietarmi quel Fico, e quella mela mi avesse vietato il leggerci i libri, io son così debole, che di sicuro averei fatto peggio di Adamo." Unless otherwise noted, all translations are my own.

2. Lorella Mangani and Giuseppe Martini, *La Biblioteca di Francesco Redi e della sua famiglia, Catalogo* (Arezzo: Accademia Petrarca, 2006), 60.

3. Redi owned Israel Spach's *Nomenclator scriptorum medicorum* (Frankfurt, 1591). Mangani and Martini, *La Biblioteca di Francesco Redi*, 627.

4. Paschal Le Coq (Gallus), *Bibliotheca medica* (Basel: Conradum Waldkirch, 1590), 425: "Leonhartus Fuchsius primum in paradoxis suis lib. I. multa utiliter super quibusdam simplicibus medicamentis, & circa eadem erroribus scripsit." The instructions for this expurgation can be found in *Index librorum prohibitorum et expurgatorum* (Madrid: Ludovicum Sanchez Typographum Regium, 1612), 641, identified in the list of Index editions as *Spain 1612.*

5. Neil James Tarrant, "Censoring Science in Sixteenth-Century Italy: Recent (And Not-So-Recent) Research," *History of Science* 52, no. 1 (2014): 1–27.

6. Mario Infelise, *I libri proibiti: Da Gutenberg all'encyclopédie* (Rome: GLF editori Laterza, 1999).

7. On the hermeneutics of reading in an age of censorship, see Annabel Patterson, *Censorship and Interpretation: The Conditions of Writing and Reading in Early Modern England* (Madison: University of Wisconsin Press, 1984); and Marco Cavarzere, *La prassi della censura nell'Italia del seicento: Tra repressione e mediazione* (Rome: Edizioni di storia e letteratura, 2011).

8. See, for example, Peter Galison, "Removing Knowledge: The Logic of Modern

Censorship," in *Agnotology: The Making and Unmaking of Ignorance*, ed. Robert Proctor and Londa Schiebinger (Stanford, CA: Stanford University Press, 2008), 37–54. For a comparative approach to censorship, see Robert Darnton, *Censors at Work: How States Shaped Literature* (New York: W.W. Norton, 2014).

9. On rethinking the Counter-Reformation Church, see John W. O'Malley, *Trent and All That: Renaming Catholicism in the Early Modern Era* (Cambridge, MA: Harvard University Press, 2000); Francisco Bethancourt, *The Inquisition: A Global History, 1478–1834*, trans. Jean Birrel (Cambridge: Cambridge University Press, 2009); and Alexandra Bamji, Geert H. Janssen, and Mary Laven, eds., *The Ashgate Research Companion to the Counter-Reformation* (Farnham, UK: Ashgate, 2013).

10. For this prosopographical approach, see in particular Ugo Baldini and Leen Spruit, eds. *Catholic Church and Modern Science: Documents from the Archives of the Roman Congregations of the Holy Office and the Index* (Rome: Libreria editrice vaticana, 2009), hereafter cited as *CCMS*; Martin Austin Nesvig, *Ideology and Inquisition: The World of the Censors in Early Mexico* (New Haven, CT: Yale University Press, 2009); Kimberly Lynn, *Between Court and Confessional: The Politics of Spanish Inquisitors* (Cambridge: Cambridge University Press, 2013); and Thomas F. Mayer, *The Roman Inquisition: A Papal Bureaucracy and Its Laws in the Age of Galileo* (Philadelphia: University of Pennsylvania Press, 2013). There is also a multiyear prosopographical project about Roman inquisitors and censors. The editions on the sixteenth and seventeenth centuries are forthcoming, but the eighteenth-century edition is Herman H. Schwedt et al., eds., *Prosopographie von Römischer Inquisition und Indexkongregation 1701–1813* (Paderborn, Ger.: Ferdinand Schöningh, 2010). Schwedt also wrote a prosopographical introduction to Oscar di Simplicio's edited volume; see Herman H. Schwedt, "Gli inquisitori generali di Siena, 1560–1782," in *Le lettere della Congregazione del Sant'Ufficio all'inquisitore di Siena, 1581–1721*, ed. Oscar di Simplicio (Trieste, It.: Università di Trieste, 2009).

11. Most scholarship on reading prohibited books in early modern Italy has examined people who read them illegally, despite the prohibitions. This literature has become extensive, but for a good starting place, see the following: Paul Grendler, *The Roman Inquisition and the Venetian Press* (Princeton, NJ: Princeton University Press, 1977); Federico Barbierato, *The Inquisitor in the Hat Shop: Inquisition, Forbidden Books and Unbelief in Early Modern Venice* (Burlington, VT: Ashgate, 2012); Federico Barbierato, *La rovina di Venetia in materia de' libri prohibiti: Il libraio Salvatore de' Negri e l'Inquisizione veneziana (1628–1661)* (Venice: Marsilio Editori, 2007).

12. "Essendo grandissima et importantissimo consideratione per salute dell'anime il negotio di prohibire, permettere, correger, et stampar li libri . . ." Archive of the Congregation for the Doctrine of the Faith, Vatican City (hereafter cited as ACDF), Index V, f. 24r.

13. Adriano Prosperi, *Tribunali della coscienza: Inquisitori, confessori, missionari* (Turin: Einaudi, 1996); Carlo Ginzburg, *The Cheese and the Worms: The Cosmos of a Sixteenth-Century Miller*, trans. Anne and John Tedeschi (Baltimore: Johns Hopkins University Press, 1980); Gigliola Fragnito, *La Bibbia al rogo: La censura ecclesiastica e i volgarizzamenti della Scrittura: 1471–1605* (Bologna: Il Mulino, 1997).

14. P. Grendler, *Roman Inquisition and Venetian Press.*

15. Anne Jacobson Schutte, "Palazzo del Sant'Uffizio: The Opening of the Roman Inquisition's Central Archive," *Perspectives on History* 37, no. 5 (May 1999): 25–28; *L'Apertura degli archivi del Sant'Uffizio romano: Giornata di studio* (Rome: Accademia nazionale dei Lincei, 1998); Peter Godman, *The Saint as Censor: Robert Bellarmine between Inquisition and Index* (Leiden, Neth.: Brill, 2000); Cavarzere, *La prassi della censura*; Gigliola Fragnito, ed., *Church, Censorship, and Culture in Early Modern Italy*, trans. Adrian Belton (Cambridge: Cambridge University Press, 2001).

16. The literature here is extensive. For a few examples, see Gigliola Fragnito, "Girolamo Savonarola e la censura ecclesiastica," *Rivista di storia e letteratura religiosa*, 35 (1999), 501–29; Silvana Seidel Menchi, *Erasmo in Italia, 1520–1580* (Turin: Bollati Boringhieri, 1987); Giorgio Caravale, *Forbidden Prayer: Church Censorship and Devotional Literature in Renaissance Italy*, trans. Peter Dawson (Burlington, VT: Ashgate, 2011); and Stefan Bauer, *The Censorship and Fortuna of Platina's Lives of the Popes in the Sixteenth Century* (Turnhout, Belg.: Brepols, 2006).

17. Fragnito, *La Bibbia al rogo*; and Gigliola Fragnito, *Proibito capire: La Chiesa e il volgare nella prima età moderna* (Bologna: Il Mulino, 2005).

18. On Italian medical practitioners beyond the elite and university-trained men discussed here, see Katharine Park, *Doctors and Medicine in Early Renaissance Florence* (Princeton, NJ: Princeton University Press, 1985); Katharine Park, *Secrets of Women: Gender, Generation, and the Origins of Human Dissection* (New York: Zone Books, 2006); Sharon T. Strocchia, *Forgotten Healers: Women and the Pursuit of Health in Late Renaissance Italy* (Cambridge, MA: Harvard University Press, 2019); David Gentilcore, *Healers and Healing in Early Modern Italy* (Manchester: Manchester University Press, 1998); David Gentilcore, *Medical Charlatanism in Early Modern Italy* (Oxford: Oxford University Press, 2006); Margaret Pelling, *The Common Lot: Sickness, Medical Occupations, and the Urban Poor in Early Modern England: Essays* (London: Longman, 1998); and James E. Shaw and Evelyn S. Welch, *Making and Marketing Medicine in Renaissance Florence* (Amsterdam: Rodopi, 2011).

19. Bouley demonstrates that "emerging scientific practice" was "supported and defined by religion." Bradford A. Bouley, *Pious Postmortems: Anatomy, Sanctity, and the Catholic Church in Early Modern Europe* (Philadelphia: University of Pennsylvania Press, 2017), 6.

20. Frederick Schauer, "The Ontology of Censorship," in *Censorship and Silencing: Practices of Cultural Regulation*, ed. Robert C. Post (Los Angeles: Getty Research Institute for the History of Art and the Humanities, 1998), 162.

21. On medical travel, see Ole Peter Grell, Andrew Cunningham, and Jon Arrizabalaga, *Centres of Medical Excellence? Medical Travel and Education in Europe, 1500–1789* (Farnham, UK: Ashgate, 2010); and Thomas Bartholin, *Thomas Bartholin on the Burning of His Library and on Medical Travel*, trans. Charles Donald O'Malley (Lawrence: University of Kansas Libraries, 1961).

22. Ian Maclean, "The Medical Republic of Letters before the Thirty Years War," *Intellectual History Review* 18, no. 1 (2008): 15–30. On the republic of letters writ large, see Hans Bots and Françoise Waquet, *La République des lettres* (Paris: Belin, 1997);

and Anthony Grafton, "A Sketch Map of a Lost Continent: The Republic of Letters," in *Worlds Made by Words: Scholarship and Community in the Modern West* (Cambridge, MA: Harvard University Press, 2009).

23. Very few Jewish and Muslim scholars were prohibited on the Index. The Roman Index of 1559 focused on heresy and in particular the threat of Protestantism.

24. On the status of the Galenic and Hippocratic corpora, see Ian Maclean, *Logic, Signs, and the Order of Nature*. On rhetoric in early modern medicine, see Stephen Pender and Nancy S. Struever, *Rhetoric and Medicine in Early Modern Europe* (Burlington, VT: Ashgate, 2012). On early modern medicine in general, see Harold J. Cook, "Medicine," in *Early Modern Science*, ed. Katharine Park and Lorraine Daston, vol. 3 of *The Cambridge History of Science* (Cambridge: Cambridge University Press, 2006), 407–34; Nancy G. Siraisi, *Medieval and Early Renaissance Medicine: An Introduction to Knowledge and Practice* (Chicago: University of Chicago Press, 1990), 187–93; and A. Wear, R. K. French, and Iain M. Lonie, *The Medical Renaissance of the Sixteenth Century* (Cambridge: Cambridge University Press, 1985). On humanism in the lives of medical practitioners in Venice, see Sarah Gwyneth Ross, *Everyday Renaissances: The Quest for Cultural Legitimacy in Venice* (Cambridge, MA: Harvard University Press, 2016).

25. On the Catholic side, see Maria Pia Donato, "Scienza e teologia nelle congregazioni romane: La questione atomista, 1626–1727," in *Rome et la science moderne*, ed. Antonella Romano (Rome: École française de Rome, 2009), 595–634; Bouley, *Pious Postmortems*. For Lutherans, see Tricia M. Ross, "Care of Bodies, Cure of Souls: Religion and Medicine in Early Modern Germany" (PhD diss., Duke University, 2017). On Jewish physicians, see Andrew D. Berns, *The Bible and Natural Philosophy in Renaissance Italy: Jewish and Christian Physicians in Search of Truth* (New York: Cambridge University Press, 2015).

26. John Brooke and Ian Maclean, eds., *Heterodoxy in Early Modern Science and Religion* (Oxford: Oxford University Press, 2005); Ole Peter Grell and Andrew Cunningham, eds., *Medicine and the Reformation* (New York: Routledge, 2013); Alessandra Celati, *Medici ed eresie nel Cinquecento italiano* (PhD diss., Università degli studi di Pisa, 2016); Riccarda Suitner, "Radical Reformation and Medicine in the Late Renaissance: The Case of the University of Padua," *Nuncius* 31, no. 1 (January 2016): 11–31; Richard Palmer, "Physicians and the Inquisition in Sixteenth-Century Venice," in Grell and Cunningham, *Medicine and the Reformation*, 118–33; Alessandra Celati, "Heresy, Medicine and Paracelsianism in Sixteenth-Century Italy," *Gesnerus* 71, no. 1 (2014): 5–37. There is also a vast literature on the intersection between healing and witchcraft, which appeared regularly in Inquisition trials. See, for example, Domizia Weber, *Sanare e maleficiare: Guaritrici, streghe e medicina a Modena nel XVI secolo* (Rome: Carocci, 2011).

27. On this overlap, see Maria Pia Donato and Jill Kraye eds., *Conflicting Duties: Science, Medicine, and Religion in Rome, 1550–1750* (London: Warburg Institute, 2009); Maria Pia Donato, "Les doutes de l'inquisiteur: Philosophie naturelle, censure et théologie à l'époque modern," *Annales HSS* 64, no. 1 (2009): 15–43; Rodolfo Savelli, "La biblioteca disciplinata: Una 'libraria' cinque-seicentesca tra censura e dissimulazione," in *Tra diritto e storia: Studi in onore di Luigi Berlinguer promossi dalle Università di Siena e di Sassari*, vol. 2, ed. Luigi Berlinguer (Soveria Mannelli, It.: Rubbettino, 2008) 856–944; Saverio Ricci, *Inquisitori, censori, filosofi sullo scenario della Contrariforma*

(Rome: Salerno Editrice, 2008), 363–76; Giuseppe Gianluca Cicco, "La Censura e le opere di argomento medico-scientifico," in *Dal torchio alle fiamme: Inquisizione e censura; Nuovi contributi della più antica biblioteca provinciale d'Italia*, ed. Vittoria Bonani (Salerno: Biblioteca Provinciale, 2005); and Hervé Baudry, *Livro medico e censura na primeira modernidade em Portugal* (Lisbon: Cham Ebooks, 2017).

28. Francesco Petrarca and David Marsh, *Invectives*, I Tatti Renaissance Library 11. (Cambridge, MA: Harvard University Press, 2003), 127.

29. Park, *Doctors and Medicine*, 120–21, 124.

30. On the utility of ancient mathematics, see Serafina Cuomo, *Pappus of Alexandria and the Mathematics of Late Antiquity* (Cambridge: Cambridge University Press, 2000), esp. 30–55.

31. On the utility of astronomy in the Renaissance, see Pietro D. Omodeo, "*Utilitas astronomiae* in the Renaissance: The Rhetoric and Epistemology of Astronomy," in *The Structures of Practical Knowledge*, ed. Matteo Valleriani (Cham, Switz.: Springer, 2017), 307–31.

32. This literature has become extensive. The place to start is with the following: Pamela Smith, *The Body of the Artisan: Art and Experience in the Scientific Revolution* (Chicago: University of Chicago Press, 2004); Pamela Long, *Artisan/Practitioners and the Rise of the New Sciences, 1400–1600* (Corvallis: Oregon State University Press, 2011).

33. Juan Caramuel y Lobkowitz, *Syntagma de arte typographica*, ed. and trans. Pablo Andrés Escapa (Salamanca: Instituto de Historia del Libro y de la Lectura, 2004), 70–71. See also Horace, *Satires; Epistles; The Art of Poetry*, trans. H. Rushton Fairclough (Cambridge, MA: Harvard University Press, 1926), 478–79.

34. Galen, *On the Usefulness of the Parts of the Body*, trans. Margaret Tallmadge May (Ithaca, NY: Cornell University Press, 1968), 67n3–4. Here Galen was following Aristotle's understanding of the soul as the efficient, formal, and final causes of the body.

35. Galen, *On the Usefulness of the Parts*, 731. Galen's final lines of the work explain this explicitly: "By 'epode' I do not mean the magician who uses enchantments; for we know that the melic poets, called lyric by some, have not only a strophe and an antistrophe but a third song as well, an epode which they used to chant standing before the altars and, as they say, singing hymns of praise to the gods. And so, likening this book to such an epode, I have given it that name" (733).

36. I see this discourse of utility as an extension of the phenomenon of rationalization explained by Mordechai Feingold in "Jesuits: Savants," in *Jesuit Science and the Republic of Letters*, ed. Mordechai Feingold (Cambridge, MA: MIT Press, 2003), esp. 6–15; and "Science as a Calling? The Early Modern Dilemma," *Science in Context* 15, no. 1 (March 2002): 79–119. For a similar discourse in the Académie Royale, see Robin Briggs, "The Académie Royale des Sciences and the Pursuit of Utility," *Past & Present* 131, no. 1 (May 1991): 38–88.

37. On the connection between utility and Catholic censorship, see Daniel Stolzenberg's study of the Jesuit censorship of Athanasius Kircher: Daniel Stolzenberg, "Utility, Edification, and Superstition: Jesuit Censorship and Athanasius Kircher's *Oedipus Aegyptiacus*," in *The Jesuits II: Cultures, Sciences, and the Arts 1540–1773*, ed. John W. O'Malley, Gauvin Alexander Bailey, Steven J. Harris, and T. Frank Ken-

nedy (Toronto: University of Toronto Press, 2006), 336–54. On early modern utilitarian thinking in the context of Bacon, see especially Vera Keller, *Knowledge and the Public Interest, 1575–1725* (Cambridge: Cambridge University Press, 2015); Stephen Gaukroger, *Francis Bacon and the Transformation of Early-Modern Philosophy* (Cambridge: Cambridge University Press, 2001); and Charles Webster, *The Great Instauration: Science, Medicine, and Reform, 1626–1660* (London: Duckworth, 1975).

38. Elisa Andretta describes the deployment of the concept of utility in the dedications of medical books in the sixteenth century. She points out that appeals to utility increased in the 1570s. I would argue this is a direct result of the amplification of this discourse through the hurdles imposed by ecclesiastical censorship. Elisa Andretta, "Dedicare libri di medicina: Medici e potenti nella Roma del XVI secolo," in Romano, *Rome et la science moderne*, 207–55.

39. For recent scholarship in this vein, see Evan R. Ragland, "'Making Trials' in Sixteenth-and Early Seventeenth-Century European Academic Medicine," *Isis* 108, no. 3 (2017): 503–28; and Nancy G. Siraisi, "Medicine, 1450–1620, and the History of Science," *Isis* 103, no. 3 (2012): 491–514. On the history of this division, see Harold Cook, "The History of Medicine and the Scientific Revolution," *Isis* 102, no. 1 (2011): 102–8.

40. For Cardano's tallies, see Girolamo Cardano, *The Book of My Life (De vita propria liber)*, trans. Jean Stoner (New York: New York Review Books, 2002), 191. His account of Alciati's praise can be found on 221–22.

41. Peter Murray Jones, "Medical Libraries and Medical Latin, 1400–1700," in *Medical Latin from the Late Middle Ages to the Eighteenth Century*, ed. Wouter Bracke and Herwig Deumens (Brussels: Koninklijke Academie Voor Geneeskunde Van Belgie 1999), 131.

42. Nancy G. Siraisi, *History, Medicine, and the Traditions of Renaissance Learning* (Ann Arbor: University of Michigan Press, 2007), 19.

43. Siraisi, *History, Medicine*, 19; Ian Maclean, "The Diffusion of Learned Medicine in the Sixteenth Century through the Printed Book," in *Learning and the Market Place: Essays in the History of the Early Modern Book* (Leiden, Neth.: Brill, 2009), 59–86.

44. I find myself tempted by the solution of Ottoman librarian Atufi Hayreddin Hizir, who, faced with these difficult decisions, decided in favor of classifying subjects that fit between categories as "toward medicine" or "toward philosophy." With thanks to Cemal Kafadar for sharing his work in progress on Hizir.

45. Paula Findlen, *Possessing Nature: Museums, Collecting, and Scientific Culture in Early Modern Italy* (Berkeley: University of California Press, 1994); Ann Blair, *Too Much to Know: Managing Scholarly Information Before the Modern Age* (New Haven, CT: Yale University Press, 2010).

46. On Aldrovandi's library, see Caroline Duroselle-Melish and David A. Lines, "Editor's Choice: The Library of Ulisse Aldrovandi (†1605): Acquiring and Organizing Books in Sixteenth-Century Bologna," *Library* 16, no. 2 (2015): 133–61; and Maria Cristina Bacchi, "Ulisse Aldrovandi e i suoi libri," *L'Archiginnasio: Bollettino della Biblioteca Comunale di Bologna* 100 (2005): 255–366.

47. Karl Heinz Burmeister, *Achilles Pirmin Gasser, 1505–1577: Arzt und Naturforscher, Historiker und Humanist* (Wiesbaden, Ger.: G. Pressler, 1970).

48. Frederick G. Meyer, John Lewis Heller, and Emily W. Emmart, eds., *The Great*

Herbal of Leonhart Fuchs: De historia stirpium commentarii insignes, 1542 (Notable Commentaries on the History of Plants) (Stanford, CA: Stanford University Press, 1999), 37.

49. Girolamo Mercuriale to Giulio Santori, Card. of S. Severina, in Rome (Pisa, 21 May 1595), *CCMS*, vol. 1, t. 3, pp. 2668–69: "Perche facendo la professione che io faccio di leggere mi conviene haver molti libri, et particularmente quelli, ne quali ho studiato." Volume 1 of Baldini and Spruit's *Catholic Church and Modern Science* was presented in four physical books known as *tomes*; abbreviated citations for *CCMS* here and throughout refer to volume, tome, and page numbers.

50. On Mercuriale, see the collection of essays in Alessandro Arcangeli and Vivian Nutton, eds., *Girolamo Mercuriale: Medicina e cultura nell'Europa del Cinquecento (Forlì, 8–11 novembre 2006)* (Florence: L. S. Olschki, 2008). On Mercuriale's library, see J. M. Agasse, "La bibliothèque d'un médecin humaniste: L'*Index librorum* de Girolamo Mercuriale," *Cahiers d'humanisme*, 3–4 (2002–3): 201–53.

51. On libraries as archives of physicians' practice, see Hannah Murphy, "Common Places and Private Spaces: Libraries, Record-Keeping and Orders of Information in Sixteenth-Century Medicine," *Past & Present* 230 (November 2016): 253–68. On the paperwork of physicians, see also Andrew Mendelsohn and Hess Volker, "Case and Series: Medical Knowledge and Paper Technology, 1600–1900," *History of Science* 48, no. 3–4 (2010): 287–314. On Galen's use of libraries for a classical precedent, see Matthew C. Nicholls, "Galen and Libraries in the Peri Alupias," *Journal of Roman Studies* 101 (2011): 123–42.

52. Richard J. Durling, "Girolamo Mercuriale's *De modo studendi*," *Osiris* 6 (1990): 181–95, on classical literature, see esp. 188–91.

53. On note-taking, see Ann Blair, "Textbooks and Methods of Note-Taking in Early Modern Europe," in *Scholarly Knowledge: Textbooks in Early Modern Europe*, ed. Emidio Campi et al. (Geneva: Droz, 2008), 39–73; Ann Blair, "The Rise of Note-Taking in Early Modern Europe," *Intellectual History Review* 20, no. 3 (September 2010): 303–16; Blair, *Too Much to Know*, esp. 62–116.

54. Cardano, *Book of My Life*, 65.

55. However, this penchant was not exclusive to physicians. We must recall that nonphysicians also read and owned many medical books. Peter Murray Jones, "Book Ownership and the Lay Culture of Medicine in Tudor Cambridge," in *The Task of Healing: Medicine, Religion and Gender in England and the Netherlands 1450–1800*, ed. H. Marland and M. Pelling (Rotterdam: Erasmus, 1996), 49–68. On book collecting by Venetian physicians and apothecaries, see S. Ross, *Everyday Renaissances*.

56. Vivian Nutton, "Greek Science in the Sixteenth-Century Renaissance," in *Renaissance and Revolution: Humanists, Scholars, Craftsmen, and Natural Philosophers in Early Modern Europe*, ed. J.V. Field and Frank A. J. L. James (Cambridge: Cambridge University Press, 1993), 15–28; Vivian Nutton, "Hellenism Postponed: Some Aspects of Renaissance Medicine, 1490–1530," *Sudhoffs Archiv* 81, no. 2 (1997): 158–70.

57. On medical anti-Arabism in this period, see Peter E. Pormann, "La querelle des médecins arabistes et hellénistes et l'héritage oublié," in *Lire les médecins grecs à la Renaissance: Aux origines de l'édition médicale*, ed. Véronique Boudon Millot and Guy Cobolet (Paris: De Boccard Edition-Diffusion, 2004), 113–41.

58. On Leoniceno, see Daniela Mugnai Carrara, *La biblioteca di Nicolò Leoniceno:*

Tra Aristotele e Galeno (Florence: Leo S. Olschki Editore, 1991); and "Profilo di Nicolò Leoniceno 1428–1524," *Interpres* 2 (1979): 169–212.

59. Conrad Gessner, *Thesaurus Euonymi Philiatri, De remediis secretis, liber physicus, medicus, et partim etiam chymicus . . .* (Zurich, 1552); Levinus Lemnius, *Occulta naturae miracula* (Antwerp, 1559).

60. On the history of anatomy and print, see Sachiko Kusukawa, *Picturing the Book of Nature: Image, Text, and Argument in Sixteenth-Century Human Anatomy and Medical Botany* (Chicago: University of Chicago Press, 2012); and Andrea Carlino, *Books of the Body: Anatomical Ritual and Renaissance Learning*, trans. John Tedeschi and Anne C. Tedeschi (Chicago: University of Chicago Press, 1999).

61. Ian Maclean, *Scholarship, Commerce, Religion: The Learned Book in the Age of Confessions, 1560–1630* (Cambridge, MA: Harvard University Press, 1998) 68; Maclean, "Diffusion of Learned Medicine," 72.

62. Maclean, "Diffusion of Learned Medicine," 64.

63. Maclean, "Diffusion of Learned Medicine," 73.

64. Maclean, *Scholarship, Commerce, Religion*, 63. Maclean also notes that physicians published more extensively than legal scholars, who were a comparable professional group.

65. On Erastus, see the recent biography by Charles D. Gunnoe, *Thomas Erastus and the Palatinate: A Renaissance Physician in the Second Reformation* (Leiden, Neth.: Brill, 2011).

66. On Junius, see Dirk Van Miert, ed., *The Kaleidoscopic Scholarship of Hadrianus Junius (1511–1575): Northern Humanism at the Dawn of the Dutch Golden Age* (Leiden, Neth.: Brill, 2011).

67. François Rabelais and Andrew Brown, *Pantagruel: King of the Dipsodes Restored to His Natural State with His Dreadful Deeds and Exploits Written by the Late M. Alcofribas, Abstractor of the Quintessence* (London: Hesperus, 2003), 3.

68. On the relationship between medical and historical writing, see Siraisi, *History, Medicine*. See also Gianna Pomata and Nancy Siraisi, eds., *Historia: Empiricism and Erudition in Early Modern Europe* (Cambridge, MA: MIT Press, 2005).

69. Girolamo Fracastoro, *Syphilidis, sive Morbi Gallici*, 1530.

70. On medical lives and life writing, see Siraisi, *History, Medicine*, 106–33; and Cardano, *Book of My Life*.

71. On Cardano's astrology and medicine, see Nancy G. Siraisi, *The Clock and the Mirror: Girolamo Cardano and Renaissance Medicine* (Princeton, NJ: Princeton University Press, 1997); Anthony Grafton, *Cardano's Cosmos: The Worlds and Works of a Renaissance Astrologer* (Cambridge, MA: Harvard University Press, 1999); and Anthony Grafton and Nancy Siraisi, "Between the Election and My Hopes: Girolamo Cardano and Medical Astrology," in *Secrets of Nature: Astrology and Alchemy in Early Modern Europe*, ed. William R. Newman and Anthony Grafton (Cambridge, MA: MIT Press, 2001), 69–131.

72. Rheticus also studied with Conrad Gessner. On Rheticus, see Dennis Richard Danielson, *The First Copernican: Georg Joachim Rheticus and the Rise of the Copernican Revolution* (New York: Walker, 2006). The literature on Copernicus is vast. For a starting point, see Robert S. Westman, *The Copernican Question: Prognostica-*

tion, Skepticism, and Celestial Order (Berkeley: University of California Press, 2011); André Goddu, *Copernicus and the Aristotelian Tradition: Education, Reading, and Philosophy in Copernicus's Path to Heliocentrism* (Leiden, Neth.: Brill, 2010); Michael H. Shank, "Setting up Copernicus? Astronomy and Natural Philosophy in Giambattista Capuano da Manfredonia's Expositio on the Sphere," *Early Science and Medicine* 14 (2009): 290–315; Martin Clutton-Brock, "Copernicus's Path to His Cosmology: An Attempted Reconstruction," *Journal for the History of Astronomy* 36, no. 2 (May 2005): 197–216; Edward Rosen, *Copernicus and His Successors* (London: Hambledon Press, 1995); and Noel M. Swerdlow and Otto Neugebauer, *Mathematical Astronomy in Copernicus's De Revolutionibus* (New York: Springer-Verlag, 1984).

73. Tricia M. Ross, "Anthropologia: An (Almost) Forgotten Early Modern History," *Journal of the History of Ideas* 79, no. 1 (Janueary 2018): 13.

74. On early modern information management, see Blair, *Too Much to Know.*

75. On Gessner's life, see Alfredo Serrai, *Conrad Gessner* (Rome: Bulzoni, 1990); and Urs B. Leu, *Conrad Gessner (1516–1565): Universalgelehrter und Naturforscher der Renaissance* (Zurich: Verlag Neue Zürcher Zeitung, 2016).

76. Girolamo Cardano, "Books Written by Me: When, Why, and What Became of Them," in Cardano, *Book of My Life*, 192–206. On Cardano's autobibliography, see Grafton, *Cardano's Cosmos*, 182.

77. Conrad Gessner, *Bibliotheca Uniuersalis, siue Catalogus omnium scriptorum locupletissimus . . .* (Tiguri: Apud Christophorum Froschouerum, 1545), ff. 169v–170r. Zentralbibliothek Zürich, DrM 3, https://doi.org/10.3931/e-rara-16206.

78. As cited in Blair, *Too Much to Know*, 56.

79. Grafton and Siraisi have warned against overstating this point, in Grafton and Siraisi, "Between the Election," 69–131. See also Monica Azzolini, *The Duke and the Stars: Astrology and Politics in Renaissance Milan* (Cambridge, MA: Harvard University Press, 2013), 135–66.

80. Conrad Gessner, *Pandectarum sive partitionum universalium libri XXI* 1548, f. 165r.

81. This was an ongoing source of debate from classical antiquity through the early modern period. See Maclean, *Logic, Signs, and the Order of Nature*, 70–72.

82. On Gessner's thanking practices, see Ann Blair, "The Dedication Strategies of Conrad Gessner," in *Professors, Physicians and Practices in the History of Medicine: Essays in Honor of Nancy Siraisi*, ed. Gideon Manning and Cynthia Klestinec (Cham, Switz.: Springer, 2017), 169–209; Ann Blair, "Conrad Gessner et la publicité: Un humaniste au carrefour des voies de circulation du savoir," in *L'Annonce faite au lecteur*, ed. Annie Charon, Sabine Juratic, and Isabelle Pantin (Louvain, Belg.: Presses universitaires de Louvain, 2017), 21–55; and Ann Blair, "Conrad Gessner's Paratexts," *Gesnerus* 73, no. 1 (2016): 73–123.

83. Most of these lists are divided by subject, though Aldus Manutius's is divided by language (Greek, Latin, and Italian). In the *Pandectae* these lists can be found on Gessner, *Pandectarum*, ff. *6r/v (Christoph Froschauer); ff. 107v–109r (Aldus Mantius); ff. 117r–119v (Sebastian Gryphius); ff. 165r–166v (Christian Wechel); ff. 237r–238r (Ioan. Gymnicus); ff. 261r/v (Ioannes Frellonius); and f. 303r (Vincenzo Valgrisi); and in the separately published final book on theology, 1559 a2r–a3r (Hieronymus Froben).

84. Gessner, *Pandectarum*, f. 166v.

85. These small formats were normal for medical textbooks. See Maclean, "Diffusion of Learned Medicine," 63.

86. Le Coq, *Bibliotheca medica*, 72.

87. Le Coq added a division between the sections on herbals and pharmacopoeias, but the text is otherwise the same. Le Coq, *Bibliotheca medica*, 370–449. The edition of Bock is Hieronymus Bock, *De stirpium, maxime earum, quae in Germania nostra nascuntur* (Argentorati: excudebat Wendelinus Rihelius, 1552), ff. a8v–d1r.

88. Spach addresses Le Coq's publication in the dedicatory letter. Spach, *Nomenclator scriptorum medicorum*, f.)(3r.

89. Ian Maclean notes that by the first decade of the seventeenth century, chemical medicine was so widespread that the heading for "medicine" in the catalogs of the Frankfurt Book Fair was temporarily changed to "books of medicine and chemistry." Maclean, "Diffusion of Learned Medicine," 78.

90. On the lack of fixity in early print culture, see Adrian Johns, *The Nature of the Book: Print and Knowledge in the Making* (Chicago: University of Chicago Press) 19, 31–37.

91. Findlen, *Possessing Nature*; Nancy G. Siraisi, *Communities of Learned Experience: Epistolary Medicine in the Renaissance* (Baltimore: Johns Hopkins University Press, 2013).

92. Janus Cornarius, *Hippocratis coi medicorum longe principis, opera quae ad nos extant omnia* [. . .] (Basel: Froben, 1546), ff. †7r–v.

1. Ippolito Salviani to Ulisse Aldrovandi, 15 October 1558, Biblioteca Universitaria di Bologna (hereafter cited as BUB), Fondo Aldrovandi, Ms. 38 v. II, f. 1r. Salviani's original text, written from Rome, reads, "ancora che non si potrano legere senza licentia delli inquisitori, per esser reprobate tutte le sue opere . . . pure io so che mi darano licentia." Also transcribed in Laurent Pinon, "La culture scientifique à Rome à la Renaissance—Clématite bleue contre poissons séchés: Sept lettres inédites d'Ippolito Salviani à Ulisse Aldrovandi," *MEFRIM: Mélanges de l'École française de Rome; Italie et mediterranée*, 114, no. 2 (2002): 490. Pinon has transcribed this letter as "per esser ex-probate," but the manuscript reads "reprobate." For "*De piscibus*," see Conrad Gessner, *Historiae animalium* (Zurich: apud Chist. Froschouerum, 1551), v. 4.

2. Salviani to Aldrovandi, 15 October 1558.

3. On early modern postal systems, see Jay Caplan, *Postal Culture in Europe: 1500–1800* (Oxford: Voltaire Foundation, 2016); Sergio Chieppi, *I servizi postali dei Medici dal 1500 al 1737* (San Giovanni Valdarno, It.: Servizio editoriale fiesolano, 1997); Clemente Fedele, *Per servizio di Nostro Signore: Strade corrieri e poste dei papi dal medioevo al 1870* (Modena, It.: E. Mucchi editore, 1988).

4. Brian W. Ogilvie, *The Science of Describing: Natural History in Renaissance Europe* (Chicago: University of Chicago Press, 2006), 38. On the *peregrinatio medica* and the ties it created between the medical centers of Europe and the periphery, see Grell,

Cunningham, and Arrizabalaga, *Centres of Medical Excellence*. For an example of medical peregrinations, see Conrad Gessner's *Liber amicorum*, discussed in Richard J. Durling, "Conrad Gesner's Liber amicorum: 1555–1565," *Gesnerus* 22, no. 3–4 (1965): 134–59.

5. Fiammetta Sabba, *La Bibliotheca universalis di Conrad Gesner: Monumento della cultura europea* (Roma: Bulzoni, 2012), 20–25.

6. Ogilvie, *Science of Describing*, 80.

7. Florike Egmond, "Clusius and Friends: Cultures of Exchange in the Circles of European Naturalists," in *Carolus Clusius: Towards a Cultural History of a Renaissance Naturalist*, ed. Florike Egmond, Paul Hoftijzer, and Robert Visser (Amsterdam: Koninklijke Nederlandse Akademie van Wetenscappen, 2007), 47.

8. On Mainardi, see Daniela Mugnai Carrara, "Giovanni Mainardi," in *Dizionario Biografico degli Italiani*, 94 vols. (Rome: Istituto della Encyclopedia Italiana, [1960–]; hereafter cited as *DBI* with volume and page numbers), 67:561–64. On the genre of medical epistolary in the sixteenth century, see Siraisi, *Communities of Learned Experience*; and MacLean, "Medical Republic of Letters."

9. For other popular medical genres, see Siraisi, *History, Medicine*, 63–79; Maclean, "Diffusion of Learned Medicine," 59–86; Maclean, *Logic, Signs, and the Order of Nature*, 55–63; and Gianna Pomata, "Sharing Cases: The Observationes in Early Modern Medicine," *Early Science and Medicine* 15, no. 3 (2010): 193–236.

10. *Epistolae medicinales diversorum authorum* (Lugduni, Apud haeredes Jacobi Juntae, 1557). On Lange's *Epistolae*, see Vivian Nutton, "The Reception of Fracastroro's Theory of Contagio: The Seed That Fell among Thorns?" *Osiris* 2, vol. 6 (1990): 215; and Siraisi, *Communities of Learned Experience*.

11. These were published together because the authors had individually made reputations with their published correspondence. For a list of editions of *Epistolae medicinales*, see Ian MacLean's appendix in MacLean, "Medical Republic of Letters," 29–30.

12. Maclean, "Diffusion of Learned Medicine," 79.

13. Giovanni Odorico Melchiori to Ulisse Aldrovandi, 18 August 1554, BUB, Fondo Aldrovandi, Ms. 38² v. I, f. 111r. Melchiori's original text, written from Venice, reads, "Et rari sono quelli che stampino un'opera nuova, se prima non ne hanno il saggio della riuscita in qualche altra cosa." Ian Maclean's research corroborates Melchiori's account (Maclean, "Diffusion of Learned Medicine," 68).

14. Daniel Stolzenberg reminds us of the importance of the role of printed books in facilitating this manuscript community; see Daniel Stolzenberg, "A Spanner and His Works: Books, Letters, and Scholarly Communication Networks in Early Modern Europe," in *For the Sake of Learning: Essays in Honor of Anthony Grafton*, ed. Ann Blair and Anja-Silvia Goeing (Leiden, Neth.: Brill, 2016), 157–72. For later examples of scientific authors, correspondence, and early printing, see Elizabeth Yale, *Sociable Knowledge: Natural History and the Nation in Early Modern Britain* (Philadelphia: University of Pennsylvania Press, 2016), 168–204; and Iain Watts, "Philosophical Intelligence: Letters, Print and Experiment during Napoleon's Continental Blockade," *Isis* 106, no. 4 (December 2015): 749–70.

15. See Egmond, "Clusius and Friends," 18. For the public and private natures of

manuscript correspondence, see Candice Delisle, "The Letter: Private Text or Public Place? The Mattioli-Gesner Controversy about the *aconitum primum*," *Gesnerus* 61, no. 3–4 (2004): 161–76.

16. Paula Findlen, "The Death of a Naturalist: Knowledge and Community in Late Renaissance Italy," in Manning and Klestinec, *Professors, Physicians and Practices*, esp. 153–56.

17. BUB, Fondo Aldrovandi, Ms. 38. v. I–IV.

18. Bots and Waquet, *La république des lettres*, 40–41.

19. Girolamo Rossi to Ulisse Aldrovandi, 1 April 1581, BUB, Fondo Aldrovandi, Ms. 136 Tomo VIII, f. 186r. "Ancor ch'io non la conoschi per presenza, il suo nome è pero si chiaro et celebre, che si fa conoscere et veder per tutto, et da tutti, massime da quelli che si dilettano di leggere et saper le cose naturali, di cui essendo VS intelligentissima."

20. As quoted in Anna Pavord, *The Naming of Names: The Search for Order in the World of Plants* (London: Bloomsbury, 2005), 198.

21. On Luca Ghini, who established a great reputation without publishing for himself, see Findlen, "Death of a Naturalist."

22. Maclean, "Diffusion of Learned Medicine," 79.

23. On polemic surrounding the naming of plants, see Marie-Élisabeth Boutroue, "'Ne dites plus qu'elle est amarante': Les problèmes de l'identification des plantes et de leurs noms dans la botanique de la Renaissance," *Nouvelle revue du XVIe siècle* 20, no. 1 (2002): 56.

24. I would like to thank Dániel Margócsy for sharing his unpublished paper on images of war and battle in the republic of letters.

25. On Cornarius's translations, see Frank Hieronymus, "Physicians and Publishers: The Translation of Medical Works in Sixteenth-Century Basle," in *The German Book 1450–1750: Studies Presented to David L. Paisey in His Retirement*, ed. John L. Flood and William A. Kelly (London: British Library, 1995), 97–101.

26. On Fuchs and Cornarius, see Meyer, Trueblood, and Heller, *Great Herbal*, 804–15, and Kusukawa, *Picturing the Book of Nature*, 125–131.

27. As cited in Nancy G. Siraisi, "Giovanni Argenterio: Medical Innovation, Princely Patronage and Academic Controversy," in *Medicine and the Italian Universities: 1250–1600* (Leiden, Neth.: Brill, 2001), 348, 350.

28. Paula Findlen, "The Formation of a Scientific Community: Natural History in Sixteenth-Century Italy," in *Natural Particulars: Nature and the Disciplines in Renaissance Europe*, ed. Anthony Grafton and Nancy Siraisi (Cambridge, MA: MIT Press, 1999), 369–400.

29. For the life of Lusitanus, see Eleazar Gutwirth, "Amatus Lusitanus and the Location of Sixteenth-Century Cultures," in *Cultural Intermediaries: Jewish Intellectuals in Early Modern Italy*, ed. David B. Ruderman and Giuseppe Veltri (Philadelphia: University of Pennsylvania Press, 2004), 216–34; and Andrew Berns, "The Crypto-Judaism of Amatus Lusitanus" (paper presented at meeting of the Society of Crypto-Judaic Studies, Phoenix, AZ, August 4, 2008). On the prohibition of Lusitanus, see Jesús Martinez de Bujanda, *Index des Livres Interdits*, 12 vols.(Sherbrooke, QC: Centre d'études de la Renaissance, Editions de l'Université Sherbrooke, 1984–96),

9:457–58 (hereafter cited as *ILI* with volume and page numbers); and *CCMS*, vol. 1, t. 1, pp. 744–48.

30. Findlen, "Formation of a Scientific Community," 387.

31. For Wieland's marginalia, see Pietro Andrea Mattioli, *Epistolarum medicinalium libri quinque* (Prague, 1561), 163, in Biblioteca Nazionale Marciana, Venice, 18.D.20.

32. BUB, Fondo Aldrovandi, Ms. 38 v. I, f. 53v: "Con tradittori, che di continuo si pugnono l'un l'altro si scrivono contro. Ne de il Rondellesio se il Salviano. Vedete il Matthiolo, L'Amato, et Melchioro. Vedete il Fuchsio. Vedete il Gesnero. Non vi e se non odii dove dev' essere amore."

33. Gessner, *Historiae animalium*, v. 1 β6v, γ1r: "Catalogus doctorum virorum, qui ut opus hoc nostrum & rempub. literariam illustrarent, vel aliunde imagines animalium, aut nomina & descriptiones miserunt: vel praesentes communicarunt. Horum nonnulli superius quoq[ue] nominati sunt, quod insuper scriptis eorum publicatis adiutus sim."

34. Findlen, "Formation of Scientific Community," 373.

35. On the relationship between personal visits, correspondence, and print into the seventeenth century, see Vivian Nutton, "Dr. James's Legacy: Dutch Printing and the History of Medicine," in *The Bookshop of the World: The Role of the Low Countries in the Book Trade, 1473–1941*, ed. Lotte Hellinga et al., (Goy-Houten: Hes & De Graaf, 2001), 210–11.

36. For a recent survey of the Italian Inquisitions, see Adriano Prosperi, Vincenzo Lavenia, and John A. Tedeschi, *Dizionario storico dell'Inquisizione* (Pisa: Scuola Normale Superiore, 2010). For a comparative study of the European Inquisitions, see Edward Peters, *Inquisition* (New York: Free Press, 1988).

37. *ILI*, 8:28–30.

38. *ILI*, 8:33.

39. *ILI*, 8:37, 753.

40. *ILI*, 8:49–50.

41. Paul Grendler, "The Conditions of Enquiry: Printing and Censorship," in *The Cambridge History of Renaissance Philosophy*, ed. Charles B. Schmitt et al. (Cambridge: Cambridge University Press, 1988), 46.

42. *ILI*, 8:139.

43. As quoted in Siraisi, *Clock and Mirror*, 22. On Fuchs, see Kusukawa, *Picturing the Book of Nature*; Meyer, Trueblood, and Heller, *Great Herbal*; Gerd Brinkhus and Claudine Pachnicke, *Leonhart Fuchs (1501–1566): Mediziner und Botaniker* (Tübingen, Ger.: Kulturamt, 2001); Deutsches Medizinhistorisches Museum, *Leonhart Fuchs zum 500. Geburtstag: Philologe, Mediziner, Botaniker—Ein Universalgelehrter an der Universität Ingolstadt: Ausstellung vom 13. Juli 2001 bis 7. Oktober 2001 im Deutschen Medizinhistorischen Museum Ingolstadt* (Ingolstadt, Ger.: Deutsches Medizinhistorisches Museum, 2001).

44. Nancy G. Siraisi, "Oratory and Rhetoric in Renaissance Medicine," *Journal of the History of Ideas* 65, no. 2 (April 2004): 197.

45. The literature on Servetus is vast. For biographies in English, start with Jerome

Friedman, *Michael Servetus: A Case Study in Total Heresy* (Geneva: Droz, 1978); and Marian Hillar, *The Case of Michael Servetus (1511–1553): The Turning Point in the Struggle for Freedom of Conscience* (Lewiston, NY: Edwin Mellen Press, 1997). Francisco Javier B. González Echeverría's recent biography pays particular attention to Servetus's medical career; see Francisco Javier González Echeverría, *El amor a la verdad: Vida y obra de Miguel Servet* (Tudela, Sp.: Gobierno de Navarra, 2011).

46. Velsius had a fascinating career that brought him into contact and conflict with nearly all of the great intellectual and religious leaders of his generation. See Hans de Waardt, "Justus Velsius Haganus: An Erudite but Rambling Prophet," in *Exile and Religious Identity, 1500–1800*, ed. Jesse Spohnholz and Gary K. Waite (London: Pickering and Chatto, 2014), 97–110.

47. On Ochino, see Roland Bainton's classic *Bernardino Ochino, esule e riformatore senese del Cinquecento, 1487–1563* (Florence: Sansoni, 1941), and the recent dissertation by Andrea Beth Wenz, "Bernardino Ochino of Siena: The Composition of the Italian Reform at Home and Abroad" (PhD diss., Boston College, 2017). On Stancaro, see Joanna Kostyło, "Commonwealth of All Faiths: Republican Myth and the Italian Diaspora in Sixteenth-Century Poland-Lithuania," in *Citizenship and Identity in a Multinational Commonwealth: Poland-Lithuania in Context, 1550–1772*, ed. Karin Friedrich and Barbara M. Pendzich (Leiden, Neth.: Brill, 2008), 179–83.

48. On Capito, see Erika Rummel, ed., *The Correspondence of Wolfgang Capito*, 3 vols. (Toronto: University of Toronto Press, 2005–9). On Jud, see Leo Weisz, *Leo Jud: Ulrich Zwinglis Kampfgenosse, 1482–1542* (Zurich: Zwingli-verlag, 1942).

49. Alessandra Celati, "*Contra medicos*: Physicians Facing the Inquisition in 16th-Century Venice," *Early Science and Medicine*, forthcoming; Suitner, "Radical Reformation," 11–31; E. Feist-Hirsch, "The Strange Career of a Humanist: The Intellectual Development of Justus Velsius (1502–1582)," in *Aspects de la propagande religieuse*, ed. G. Berthoud (Geneva: Droz, 1957), 308–24. John Tedeschi asks this question in relation to his discussion of Simone Simoni in "Italian Reformers and the Diffusion of Renaissance Culture," *Sixteenth Century Journal* 5, no. 2 (October 1974): 85. In fiction, see Sinclair Lewis, *Arrowsmith* (New York: Harcourt, Brace, 1925).

50. The classic account of Brunfel's biography is F. W. E. Roth, "Otto Brunfels, 1489–1534: Ein deutscher Botaniker," *Botanische Zeitung* 58 (1900): 191–232. Carlo Ginzburg has credited Brunfels with founding the influential phenomenon of Nicodemism; see Carlo Ginzburg, *Il nicodemismo: Simulazione e dissimulazione religiosa nell'Europa del '500* (Turin: Einaudi, 1970), 3–28. And for a critique, consider Carlos M. N. Eire, "Calvin and Nicodemism: A Reappraisal," *Sixteenth Century Journal* 10, no. 1 (Spring 1979): 44–69.

51. Kenneth F. Thibodeau, "Science and the Reformation: The Case of Strasbourg," *Sixteenth Century Journal* 7, no. 1 (April 1976): 35–50, esp. 40.

52. Niklas Holzberg, "Ein vergessener Schüler Philipp Melanchthons: Georg Aemilius (1517–1569)," *Archiv für Reformationsgeschichte—Archive for Reformation History* 73 (1982): 94–122.

53. On Gessner, see Serrai, *Conrad Gesner*; and Sabba, *La Bibliotheca universalis*.

54. On Peucer, see Martin Roebel, *Humanistische Medizin und Kryptocalvinismus: Leben und medizinisches Werk des Wittenberger Medizinprofessors Caspar*

Peucer (1525–1602) (Herbolzheim, Ger.: Centaurus Verlag & Media, 2012); and the collection of essays Hans-Peter Hasse and Günther Wartenberg, eds., *Caspar Peucer (1525–1602): Wissenschaft, Glaube und Politik im konfessionellen Zeitalter* (Leipzig: Evangelische Verlagsanstalt, 2004).

55. Melanchthon also had relationships with Jakob Milich (Mylichius), Johannes Carion, and Michael Toxites (Schutz), who were also included on the Pauline Index.

56. *ILI*, 8:515.

57. Archivio di Stato di Venezia (hereafter cited as ASV), Savi all'Eresia, b. 156. Also published in P. Grendler, *Roman Inquisition and Venetian Press*, appendix 1, document 2, pp. 296–301.

58. On Robert Constantin, see Michel Magnien, "Le 'Nomenclator' de Robert Constantin (1555): Première bibliographie française?" *Renaissance and Reformation / Renaissance et Réforme* 34, no. 3 (2011): 65–89. On Johann Günther von Andernach, see Melchior Adam, *Vitae Germanorum medicorum* (Frankfurt, 1620), 223–27.

59. *ILI*, 8:347–48.

60. Nicolaus Copernicus, *De revolutionibus orbium cœlestium libri VI* [. . .] (Basel: Ex Officina Henricpetrina, 1566). The standard reference on Gasser is Karl Heinz Burmeister, *Achilles Pirmin Gasser*.

61. See Franco Bacchelli, "Luca Gaurico," in *DBI*, 52:697–705; and Grafton, *Cardano's Cosmos*, 96–106.

62. On Agrippa, see M. van der Poel, *Cornelius Agrippa, the Humanist Theologian and his Declamations* (Leiden, Neth.: Brill, 1997); and Charles G. Nauert Jr., *Agrippa and the Crisis of Renaissance Thought* (Urbana: University of Illinois Press, 1995).

63. On Schürpf, see Wiebke Schaich-Klose, *D. Hieronymus Schürpf: Leben und Werk des Wittenberger Reformationsjuristen, 1481–1554* (Trogen, Ger.: Buchdruckerei Meili, 1967).

64. *ILI*, 8:274–75.

65. *CCMS*, vol. 1, t. 1, p. 76.

66. C. L. Heesakkers and Dirk Van Miert, "An Inventory of the Correspondence of Hadrianus Junius (1511–1575)," *Lias: Journal of Early Modern Intellectual Culture and Its Sources* 37, no. 2 (2010): 181; and Dirk Van Miert, "The Religious Beliefs of Hadrianus Junius (1511–1575)," in *Acta Conventus Neo-Latini Cantabrigiensis: Proceedings of the Eleventh International Congress of Neo-Latin Studies, Cambridge 30 July–5 August 2000*, edited by Rhoda Schnur et al. (Tempe: Arizona Center for Medieval and Renaissance Studies, 2003), 586–88.

67. *ILI*, 7:138. The entry reads, "Hadriani Iunii nempe medici titulus praefationi praefixus in nonnullis exemplaribus Lexici graecolatini: eo titulo excepto, caetera eius opera legi possunt, quod nihil contra sanam doctrinam habeant, et auctorem ipsum constet catholicam fidem profiteri."

68. As quoted in Antonio Rotondò, "La censura ecclesiastica e la cultura," in *Storia d'Italia*, vol. 5, *I documenti*, ed. Ruggero Romano and Corrado Vivanti (Turin: 1973), pt. 2, pp. 1451–52.

69. Paul Grendler, *The Universities of the Italian Renaissance* (Baltimore: Johns Hopkins University Press, 2002), 19, 21.

70. Archivio di Stato di Bologna (hereafter cited as ASB), Senato, Lettere, s. II. vol. 6, Lettere del Senato all'Ambasciatore 1559, unnumbered folios. The original letter, dated 25 January 1559, reads, "L'essersi publicato in questa città il catalogo dei libri prohibiti ha generato molto dispiacere, et tanto più sendo questa, Terra di studio, dove concorrono ordinariamente genti di diverse bande, le quali quando intieramente s'havesse da osservare tal prohibitione, siamo certi non ci stariano, et già per tal rispetto alcuni se ne sono andati."

71. On the intellectual climate at the University of Padua, see Aldo Stella, *Anabattismo e antitrinitarismo in Italia nel XVI secolo* (Padua: Liviani, 1969), 39–61.

72. ASB, Senato, Lettere, s. II. vol. 6, Lettere del Senato all'Ambasciatore 1559, unnumbered folios. The text of the original letter, dated 25 January 1559, corresponding to the translated quotation reads, "ma quelli anchora delle lettere humane, et di tutte le altre scientie, quali non parlano in modo alcuno di Religione, ne di fede, et in ciò non perdete tempo, sendo che per Editto publico si è dato tempo d'un mese a denonciare tutti li libri, et già ne sono scorsi alcuni, si è havuto ricorso a Mons. Vicel[egat] o a Mons. Vesc[ov]o n[ost]ro, et all'Inquisitione, informandoli di questo negocio, quali benignamente si sono eshibiti di scrivere costi à favore del desidersio n[ost]ro, et delli Dottori."

73. The lists of books have been detached from the letters and are missing. For the correspondence between Aldrovandi and the Senate of Bologna about the Index of Prohibited Books, see ASB, Senato Lettere s. VII v. 34, and ASB, Senato Lettere, s. II. vol. 6, unnumbered folios.

74. ASB, Studio, busta 353a. I am grateful to Paolo Savoia for sharing this document with me.

75. Guido Dall'Olio, *Eretici e inquisitori nella Bologna del Cinquecento* (Bologna: Istituto per la Storia di Bologna, 1999), 243: "Avertendo però detti Collegii di honestarsi talmente nella descrittione delle loro liste che non si facciano danno, et che volendo troppo, non ottenghino niente."

76. Dall'Olio, *Eretici e inquisitori*, 243.

77. P. Grendler, *Roman Inquisition and Venetian Press*, 118–19.

78. P. Grendler, *Roman Inquisition and Venetian Press*, 121.

79. P. Grendler, *Roman Inquisition and Venetian Press*, 120.

80. Kusukawa, *Picturing the Book of Nature*, 56. On Aldrovandi's descent into insolvency while publishing his natural history volumes, see A. Baldacci, "Contributo alla bibliografia delle opere di Ulisse Aldrovandi," in *Intorno alla vita e alle opere di Ulisse Aldrovandi*, ed. Ludovico Frati (Bologna: Beltrami, 1907), 69–139. On the costs of printing a book in early modern Italy, see Conor Fahy, *Printing a Book at Verona in 1622: The Account Book of Francesco Calzolari Junior* (Paris: Foundation Custodia, 1993); and Ian Maclean, *Scholarship, Commerce, Religion*, 211–34.

81. Kusukawa, *Picturing the Book of Nature*, 55.

82. For reference books and information overload in early modern Europe, see Blair, *Too Much to Know*.

83. Kusukawa, *Picturing the Book of Nature*, 60.

84. Chapter 6 will describe ways of including and disguising prohibited books in library catalogs.

85. *CCMS*, vol. 1, t. 3, p. 2603. See also, *ILI*, 8:42n39.

86. This particular copy of Zwinger's *Theatrum vitae humanae* (Basel, 1586 v. 1) is housed in BUB, A.M.RR.III.23. More discussion on reading licenses will be provided in chapter 5.

87. As cited in Paolo Prodi, *Il cardinale Gabriele Paleotti (1522–1597)* (Rome: Edizioni di storia e letteratura, 1959–67), 2:238n62.

88. Alfonso Cattanio to Ulisse Aldrovandi, 1 April 1559, BUB, Fondo Aldrovandi, Ms. 38² v. 2, f. 18r. The original letter, written from Ferrara, reads, "Ho sentito con mio gran dolore l'abbrugiamento delli lib[ri], s'io non conoscessi V[ostra] S[ignoria] nelle cose di questo mondo risorgere come fa la palma quanto più viene opressa gli porgerei qualche conforto ch'ha sentito per la perdita di tanti sui fatiche studii, e vigilie, et per q[uesto] il danno di tante, e sì rare opere di quali tanto vi sete affaticato di fornare il vostro studio."

89. There is also speculation that Aldrovandi may have found himself in trouble with the Roman Inquisition again in July 1558; however, documents from the Holy Office records conflict with other archival evidence about Aldrovandi from the period. *CCMS*, vol. 1, t. 2, p. 733.

90. Conrad Gessner, *De raris et admirandis herbis* (Tiguri: Andreas Gessner, 1555). Copy in BUB, A.4.F.7.52/2.

91. Baccio Puccini to M. Ioseff, 6 March 1559, BUB, Fondo Aldrovandi, 38.2 v. 4, f. 239r.

92. Baccio Puccini to Ulisse Aldrovandi, 25 April 1559, BUB, Fondo Aldrovandi, 38.2 v. 4, f. 240r. Also published in Alessandro Tosi, ed., *Ulisse Aldrovandi e la Toscana: Carteggio e testimonianze documentarie* (Florence: Olschki, 1989), 195: "Parte perché manchavono certi fogli et per ancora non si erano possuti havere per la indisposizone del tempo. Et di poi essere accaduto che è stato di bisognio per conto delle stampe mandarli in mano dello inquisitore insieme con assai libri di V[ostra] E[ccelenza]. Quanto a quella parte che mancha prego V[ost]ra E[ccelenza] che adesso che el tempo e commodo vegga con ogni opportuno modo quanto più presto si puo sia mandato da Venetia mi rendo certo che quella non sia per manchare accio e libro sia integro che altrimente non varrebe niente. Quanto a quelli che sono in mano dello inquisitor, V[ostra] E[ccelenza] intenda che ancora che in elli editti si contenesse tutte le cose che fossero tradocte o stampate in quelli luoghi et da quelli homini, li litterati sono stati per quanto si intende dal duca et così non si sono dati con questo però che se ne scripto a Roma."

93. For Guidoli, also known as Giovammaria Guidoni, as Aldrovandi's student, see Giovanni Fideli to Ulisse Aldrovandi, 23 September 1556, in Tosi, *Aldrovandi e la Toscana*, 201.

94. Gio. Mario Guidoli to Ulisse Aldrovandi, 15 July 1560, BUB, Fondo Aldrovandi, Ms. 38 v. I, f. 270 r; letter written from Masera.

95. On Donzellini, see Palmer, "Physicians and the Inquisition," 118–33; and Celati, "Heresy, Medicine and Paracelsianism," 5–37.

96. On Pietro Perna, see Leandro Perini, *La vita e i tempi di Pietro Perna* (Rome: Edizioni di storia e letteratura, 2002). Donzellini's letters were used against him in the first of his trials before the Venetian Inquisition. The first document in his folder is a

letter from Pietro Perna sent from Basel, 13 November 1550. ASV, Savi all'Eresia, busta 39, f. 1r.

97. *CCMS*, vol. 1, t. 1, pp. 683–704.

98. ASV, Savi all'Eresia, busta 39, f. 48r: "Et siccome in filosofia et medicina ho letto Averrois et Avicenna, li quali contengono molti errori, et ciò tirato dal comun corso dello schole, le quali leggono tai autori; così dall'impeto di questa nostra corrotta età son io stato tirato al legger si fatti libri. Et come che ho letto Averrois colla norma di san Thomaso, et Avicenna colla norma di Galeno, così li detti libri ho letto riguardando sempre alla dottrina sana et ortodossa della santa chiesa catolica."

99. ASV, Savi all'Eresia, busta 39, f. 173r: "Tre volumi di Conrado Gesneri de Animalium historia col nome dell'authore coperto di carta."

100. Chapter 6 includes a more thorough treatment of this phenomenon.

101. Schwarze, "Cornerus, Christoph," in *Allgemeine Deutsche Biographie* 4 (1876): 499.

102. ASV, Savi all'Eresia, busta 39, f. 173r: "ch'ella ci aspettava a fare questa visita."

103. ASV, Savi all'Eresia, busta 39, f. 106v: "qu[est]a licentia ho havuto in scriptas et e attacata ai libri propri."

104. On Donzellini's continued importation of prohibited books, see *CCMS*, vol. 1, t. 1, pp. 683–704; and John Jeffries Martin, *Venice's Hidden Enemies: Italian Heretics in a Renaissance City* (Berkeley: University of California Press, 1993), 81.

105. P. Grendler, *Roman Inquisition and Venetian Press*, 186–87.

106. P. Grendler, *Roman Inquisition and Venetian Press*, 188–89.

107. Gian Vincenzo Pinelli to Ulisse Aldrovandi, 11 April 1572, BUB, Fondo Aldrovandi, Ms. 38 v. I, f. 76.

108. Alfonso Cattanio to Ulisse Aldrovandi, 22 June 1567, BUB, Fondo Aldrovandi, Ms. 38 v. 2, f. 56r. "Io so' quasi di fantasia in questo anno leggere Theoph[rastus] *De causis plantarum* havendovi[havendomi?] visto sopra il Scaliger, desidero saper s'altro vi ha fatto che lui."

109. *CCMS*, vol. 1, t. 3, p. 2304.

110. Alfonso Cattanio to Ulisse Aldrovandi, 10 July 1567, BUB, Fondo Aldrovandi, Ms. 38 v. 2, f. 57r. "Desiderarei anco intendere se si potessero haver le cose del Fuchsio cio'è tutto quello che lui ha scritto del suo o commentato gal[en]o et quanto costarebbe per che ne so' pregato da un scolare mio amico, et darmene risposta."

111. BUB, Fondo Aldrovandi, Ms. 98 v. 2, 152v.

112. For Camerarius, see, e.g., BUB, Fondo Aldrovandi, Ms. 136 Tomo XXVI f. 44r. For Zwinger, see, e.g., BUB, Fondo Aldrovandi, 136 Tomo XVI, f. 149v, 151r. On Aldrovandi's German medical correspondents, see Mario Maragi, "Corrispondenze mediche di Ulisse Aldrovandi coi paesi germanici," *Pagine della storia della medicina* 13 (July–August 1969): 468–75.

113. Candice Delisle, "Letter," 173.

114. Girolamo Cardano, *Book of My Life*, 216–22.

115. Cardano, *Book of My Life*, 221.

116. Cardano, *Book of My Life*, 89.

117. On Grataroli, see Alessandro Pastore, "Guglielmo Grataroli," in *DBI*, 58:731–35; and *CCMS*, vol. 1, t. 2, pp. 1885–1904.

118. For an introduction to the literature on Cardano, see especially Siraisi, *Clock and Mirror*; and Grafton, *Cardano's Cosmos*. There is also an extensive literature on the censorship of Cardano: for example, Michaela Valente, *"Correzioni d'autore" e censure dell'opera di Cardano*, in *Cardano e la tradizione dei saperi*, ed. M. Baldi and G. Canziani (Milan: FrancoAngeli, 2003), 437–56.

119. *CCMS*, vol. 1, t. 3, pp. 2596–97.

120. For the documentation on Cardano in the ACDF, see *CCMS*, vol. 1, t. 2, pp. 1033–1472.

121. See Nancy G. Siraisi, "Mercuriale's Letters to Zwinger and Humanist Medicine," in Arcangeli and Nutton, *Girolamo Mercuriale*, 77–95; and Antonio Rotondò, *Studi e ricerche di storia ereticale italiana del Cinquecento* (Turin: Giappichelli, 1974) 287, 399–407, 546–48.

122. *CCMS*, vol. 1, t. 2, p. 1565. On Erastus, see Charles D. Gunnoe, *Thomas Erastus*.

123. *CCMS*, vol. 1, t. 2, p. 1565. We might wonder if Mercuriale's decision to eschew controversy in other public works was part of a deliberate strategy and model for how to maintain relationships in a fragmented republic of letters. See Durling, "Girolamo Mercuriale's *De modo studendi*," 182.

124. For a few examples of recent work in English on Paracelsus, see Charles Webster, *Paracelsus: Medicine, Magic and Mission at the End of Time* (New Haven, CT: Yale University Press, 2008); Gerhild Scholz Williams and Charles D. Gunnoe Jr., eds., *Paracelsian Moments: Science, Medicine, and Astrology in Early Modern Europe* (Kirksville, MO : Truman State University Press, 2002); and Philip Ball, *The Devil's Doctor: Paracelsus and the World of Renaissance Magic and Science* (London: William Heinemann, 2006).

125. On antiparacelsianism, see Tilman Walter, "New Light on Antiparacelsianism (c. 1570–1610): The Medical Republic of Letters and the Idea of Progress in Science," *Sixteenth Century Journal* 43, no. 3 (Fall 2012): 701–25; and Siraisi, *Communities of Learned Experience*, 34–36.

126. *ILI*, 9:720–21.

127. *CCMS*, vol. 1, t. 3, pp. 2166–96.

128. BUB, Fondo Aldrovandi, Ms. 38. v.I, f. 20v: "Io so che voi siete huomo di tanta authorità, che a dispetto suo haveste possuto far prohibire che tal poltroneria, come libello da infamia, et contra ogni legge, honestà, et politia, non solamente non si fosse publicata, ma che la fosse stata tolta dalle librarie, et abbrusciata quando pur non haveste possuto avviarte che la fusse uscita in luce, come cosa inicua et piena d'ogni tristizia."

129. *ILI*, 9:81, 143, 157, 599.

130. *ILI*, 9:632–33.

131. "Girolamo Caratto, Inquisitor of Asti, to Scipione Rebiba, Card. of Pisa, in Rome (Asti, 20 September 1574)," in *CCMS*, vol. 1, t. 3, p. 2616. Baldini and Spruit suggest three options for what this first text might be: *Compendiaria in artem medendi introductio* (1535), *De medendi methodo* (1539), and *De curandi ratione* (1542). The second text Caratto cites is *Paradoxa medicinae*, "derived from the 1530 edition of *Errata recentiorum medicorum*" (1535 and later editions). *CCMS*, vol. 1, t. 3, p. 2616n182.

132. *CCMS*, vol. 1, t. 3, p. 2616: "gli medici me protestano che loro sono confusi nel medicare senza il fucsio de medendis morbis, et paradoxa medicinalia. Se parra espediente all'Ill.ᵐᵃ S. S. che gli vega, corregga, et concedi lo faro. quando anco no, habbino paciencia."

CHAPTER 2

1. "Girolamo di Capugnano, Inquisitor of Vicenza, to Agostino Valier, Card. of Verona, in Rome." *CCMS*, vol. 1, t. 1, p. 620: "Con li stampatori di questa città procurano q[uest]i SS.ʳⁱ Medici, che vogliano stampare certi libri dello lor professione medicinale, come sarebbe tra questi Lionardo il Fucssio, et a me sopra di ciò hanno parlato ancora."

2. Prodi, *Il cardinale Gabriele Paleotti*, 2:234–243.

3. Prodi, *Il cardinale Gabriele Paleotti*, 2:228. See also *ILI*, vol. 6.

4. *CCMS*, vol. 1, t. 1, p. 113.

5. For a brief narrative of this complicated period, see *CCMS*, vol. 1, t. 1, pp. 111–23.

6. ASV, Savi all'Eresia (Sant'Ufficio) 156, f. 16r: "Essendo cosi, che io sto sempre in moto, et vado a diverse fiere, ne mi fermo in bottega non sapeva che vi fussero detti libri, anzi dico seben sono in Venetia non vado quasi mai in bottega, et quando furono espurgate le botteghe et magazeni de librari del S[an]to Officio. Io commisi tutto il carico ad un nostro giovene, che non sapria dir il nome, perche ne havemo mutati molti, ne so se al tempo dell'espurgation mi trovasse in Ven[eti]a ci credo più presto de no, che altramente."

7. Vittorio Frajese, *Nascita dell'Indice: La censura ecclesiastica dal Rinascimento alla Controriforma* (Brescia, It.: Morcelliana, 2006), 295.

8. *CCMS*, vol. 1, t. 3, p. 2616.

9. The content of the expurgations will be discussed in chapter 4.

10. This conclusion runs counter to a widely held view that the Counter-Reformation fundamentally clericalized Italian culture. See, for example, R. Po-Chia Hsia, *The World of Catholic Renewal, 1540–1770* (Cambridge: Cambridge University Press, 1998); Paolo Prodi, *The Papal Prince: One Body and Two Souls* (Cambridge: Cambridge University Press, 1987); and Paolo Prodi and Wolfgang Reinhard, eds., *Il concilio di Trento e il moderno* (Bologna: Il Mulino, 1996).

11. On the Counter-Reformation as negotiation, see Simon Ditchfield, *Liturgy, Sanctity and History in Tridentine Italy: Pietro Maria Campi and the Preservation of the Particular* (Cambridge: Cambridge University Press, 1995); Marc R. Forster, *The Counter-Reformation in the Villages: Religion and Reform in the Bishopric of Speyer, 1560–1720* (Ithaca, NY: Cornell University Press, 1992); Marc R. Forster, *Catholic Revival In the Age of the Baroque: Religious Identity in Southwest Germany, 1550–1750* (Cambridge: Cambridge University Press, 2001); David Gentilcore, *From Bishop to Witch: The System of the Sacred in Early Modern Terra d'Otranto* (Manchester: Manchester University Press, 1992); and Carvarzere, *La prassi della censura*. For the negotiation of Calvinist censorship in Geneva, see Ingeborg Jostock, *La censure négociée: Le contrôle du livre a Genève, 1560–1625* (Geneva: Droz, 2007).

12. "*Auctores damnati quos naturalis scientiae et medicinae studiosi desiderarent*

(Rome, ante 1590)," *CCMS*, vol. 1. t. 1, p. 603. Sabina Brevaglieri has also noted that the medical profession in Rome participated in censorship at the beginning of the seventeenth century and that utility was one of the reasons most often used to argue in favor of accessing a text. Sabina Brevaglieri, "Science, Books and Censorship in the Academy of the Lincei: Johannes Faber as Cultural Mediator," in Donato and Kraye, *Conflicting Duties*, 119.

13. *CCMS*, vol. 1. t. 1, p. 603.

14. *ILI*, 8:80n146: "Othonis Brunfelsi Onomasticon seu vocabularium necessarium, cum nullam habeamus nisi ipsam."

15. *CCMS*, vol. 1, t. 1, p. 603.

16. *CCMS*, vol. 1, t. 1, p. 605: "Catalogus plantarum nomina praeponens quae diversis gentibus in usu sunt. Eius historia animalium auctoribus pollet à quibus contracta est, et picturae quae in ea represantant animalia legentes multum iuvant."

17. Anthony Grafton, "Philological and Artisanal Knowledge Making in Renaissance Natural History: A Study in Cultures of Knowledge," *History of Humanities* 3, no. 1 (2018): 39–55.

18. *CCMS*, vol. 1, t. 1, p. 605: "Eiusdem indices librorum, commodam sufficerent studiosis, auctorum copiam, si purgati essent, neque bonos auctores vetitis interpellarent."

19. ACDF, Index II (Protocolli), P, f. 101r. This may have been the copy from which Antonio Posio, Secretary of the Congregation from 1571 to 1580, composed the set of expurgations held in the Vatican Library. *CCMS*, vol. 1, t. 2., pp. 1675–76n10.

20. ACDF, Index I, Diarii, v. 1, f. 145v, 7 April 1601.

21. *CCMS*, vol. 1, t. 1, p. 57; Mayer, *Roman Inquisition: Papal Bureaucracy*, 17–19.

22. Vicenzo Bonardi, "*Discorso intorno all'Indice da farsi de libri proibiti* (Rome, post February 1587, ante 1589)," *CCMS*, vol. 1, t. 1, pp. 249–52: "Ma specialmente si desiderano l'espurgationi del . . . Fuchsio, Cardano, Gesnero la Biblioteca, de animalibus, de Planctis, de fossilibus."

23. Agostino Valier, cardinal of Verona to vicar general of Naples, 3 December 1599, *CCMS*, vol. 1, t. 1, pp. 661–62: "Aspettano questi miei Ill.^{mi} SS.^{ri} della Cong[regatio]ne dell'Indice la censura de libri di Medicina et Philosofia cominciando da quelli, che sono espressi nel Indice donec expurgatur, seguitando nel resto de libri della medesima professione, che sono desiderati da molti et giudicati utili et hanno bisogno di censura."

24. "Solution of some Doubts (Rome, [post 1596])," *CCMS*, vol. 1, t. 1, p. 201: "Phisionomia, per quanto serve a giudicar le complessioni flematica, o colerica o sanguinea, si concede: per quanto si suole [vuole] abusar, ad indovinar, e, nel modo che si fà della chiromantia, è prohibita; nel resto si [servi] il tenor del indice." On physiognomy, see Jole Agrimi, *Ingeniosa scientia nature: Studi sulla fisiognomica medievale* (Florence: Sismel, 2002); and Nancy G. Siraisi, "The Fielding H. Garrison Lecture: Medicine and the Renaissance World of Learning," *Bulletin of the History of Medicine* 78, no. 1 (Spring 2004): 1–36.

25. Baldini, "Roman Inquisition's Condemnation of Astrology," 91. See also Germana Ernst and Guido Giglioni, introduction to *Il linguaggio dei cieli: Astri e simboli nel Rinascimento* (Rome: Carocci, 2012), 11–20.

26. Gaspare Mosca, canon of Salerno Cathedral, to cardinals of the Congregation of the Index, circa end 1596–beginning 1597, *CCMS*, vol. 1, t. 1, p. 528: "Domanda s'alcuni libri che trattano di detta Iudiciaria, se posseno permettere sotto pretesto che vogliono alcuni servirsene per l'agricoltura, navigatione, et medicina conforme à detta bolla, et quali si possono per detta causa permettere."

27. Giuseppe Rossi, bishop of L'Aquila, to Simone Tagliavia d'Aragona, cardinal of Terranova, 3 November 1599, *CCMS*, vol. 1, t. 1, p. 559: "Non sò se trà li libri di Astrologia può alcuni servirsi de quelli, che trattano della giuditiaria in materia di medicina, agricoltura, et navigatione, per esser connessi, et non potersi dividere questi trattati dall'altri, che sono nell'istessi libri intorno all'avvenimenti humani."

28. Agostino Valier, cardinal of Verona, to inquisitor of Cremona, 3 December 1599, *CCMS*, vol.1, t. 1, pp. 559–60: "procurando di trovar persone idonee, che seguitino la censura de libri d'Astrologia cavandone quel che ci è di buono in servitio della Nautica Agricultura e medicina." Baldini and Spruit note that no other documents relating to this situation exist in the ACDF. *CCMS*, vol. 1, t. 1, p. 560n370.

29. *CCMS*, vol. 1, t. 1, p. 98.

30. Gigliola Fragnito, "The Central and Peripheral Organisation of Censorship," in Fragnito, *Church, Censorship, and Culture in Early Modern Italy*, 13–49.

31. On Jewish students, see David B. Ruderman, *Jewish Thought and Scientific Discovery in Early Modern Europe* (New Haven, CT: Yale University Press, 1995).

32. P. Grendler, *Universities of the Italian Renaissance*, 37. See also R. Kagan, "Universities in Italy, 1500–1700," in *Les Universités européennes du XVIe au XVIIIe siècle: Histoire sociale des populations étudiantes*, ed. Dominique Julia, Jacques Revel, and Roger Chartier (Paris: Editions de l'Ecole des hautes études en sciences sociales, 1986).

33. Jonathan Woolfson, *Padua and the Tudors: English Students in Italy, 1485–1603* (Toronto: University of Toronto Press, 1998); Michael Tworek, "Learning Ennobles: Study Abroad, Renaissance Humanism, and the Transformation of the Polish Nation in the Republic of Letters, 1517–1605" (PhD diss., Harvard University, 2014); A. Francis Steuart, "The Scottish 'Nation' at the University of Padua," *Scottish Historical Review* 3, no. 9 (October 1905): 53–62.

34. Thomas Coryate and George Coryate, *Coryate's Crudities: Hastily Gobled up in Five Months . . .* (Glasgow: J. MacLehose, 1905), 297–98.

35. On Fabrici d'Aquapendente's theater, see Cynthia Klestinec, *Theaters of Anatomy: Students, Teachers, and Traditions of Dissection in Renaissance Venice* (Baltimore: Johns Hopkins University Press, 2011), 55–89.

36. On Pinelli, see especially Marcella Grendler, "Book-Collecting in Counter-Reformation Italy: The Library of Gian Vincenzo Pinelli, 1535–1601," *Journal of Library History* 16, no. 1 (1981): 143–51; and Angela Nuovo, "The Creation and Dispersal of the Library of Gian Vincenzo Pinelli," in *Books on the Move: Tracking Copies through Collections and the Book Trade*, ed. Robin Myers, Michael Harris, and Giles Mandelbrote (New Castle, DE: Oak Knoll Press; London: British Library, 2007), 39–68.

37. Marco Cornaro was bishop of Padua from 1594 until his death in 1625.

38. Giovanni Angeli and Antonino Poppi, eds., *Lettere del Sant'Ufficio di Roma all'Inquisizione di Padova, 1567–1660: Con nuovi documenti sulla carcerazione pa-*

dovana di Tommaso Campanella in appendice (1594) (Padua: Centro studi antoniani, 2013). On Barozzi, see Federico Barbierato, *Nella stanza dei circoli: Clavicula Salomonis e libri di magia a Venezia nei secoli XVII e XVIII* (Milan: Edizioni Sylvestre Bonnard, 2002), 113–16; Paul Rose, "A Venetian Patron and Mathematician of the Sixteenth Century: Francesco Barozzi (1537–1604), *Studi Veneziani* 1 (1977): 119–80; and *CCMS*, vol. 1, t. 4, p. 2800.

39. On the Jesuit College in Padua, see P. Grendler, *Universities of the Italian Renaissance*, 479–83; Christopher Carlsmith, "Struggling toward Success: Jesuit Education in Italy, 1540–1600," *History of Education Quarterly* 42, no. 2 (Summer 2002): 229–31; and Edward Muir, *Culture Wars of the Late Renaissance: Skeptics, Libertines and Opera* (Cambridge, MA: Harvard University Press, 2007), chap. 1.

40. On Cremonini's speech, see Nick Wilding, *Galileo's Idol: Gianfrancesco Sagredo and the Politics of Knowledge* (Chicago: University of Chicago Press, 2014), 21–25.

41. "Decree of the Congregation of the Index," *CCMS*, vol. 1, t. 1, p. 621: "Patavino Ep[iscop]o scriptum ut expurg[atio]nem librorum Philosophiae et Medicinae conficiat ex celebri illa Universitate adhibitis Consultoribus et Vicentini Inq[uisito]ris utatur opera." An outline of the following episode about the expurgation of books in Padua can be found in Ricci, *Inquisitori, censori, filosofi*, 363–76.

42. Agostino Valier, cardinal of Verona, to Marco Cornaro, bishop of Padua, 8 March 1597, *CCMS*, vol. 1, t. 1, p. 622: "Aspetto aviso che habbi dato // principio a si honorata impresa . . ."

43. Fragnito, "Central and Peripheral Organization of Censorship," 40.

44. As cited in Rodolfo Savelli, "The Censoring of Law Books," in Fragnito, *Church, Censorship, and Culture in Early Modern Italy*, 249.

45. *CCMS*, vol. 1, t. 1, p. 623: "aspetta conforme al valor' suo un'Indice Espurg[ato]rio de libri gravi e utili, essendosi in ciò molti anni affatigata V.R. et havendo copia costì et in Padua et in Venetia d'huomini dotti che l'aiuterano, faccia un'espurg[atio]ne di libri medicinali che costì si vogliano stampare."

46. Agostino Valier, cardinal of Verona, to Girolamo Giovannini, inquisitor of Vicenza, 8 March 1597, *CCMS*, vol. 1, t. 1, p. 623: "Non doverà però essere universalmente abbracciata, se p[rim]a non sarà approvata l'espurg[azio]ne da questa S. Congregatione et in ciò potrà comunicare con Mons.ʳ Vesc[ov]o di Padua, dandosi il carico a quella università di fare un'Indice Espurg[ato]rio di libri di Filosofia et Medicina, ma né l'espurg[atio]ni vostre, né della Università di Padua saranno universalmente abbracciate, se p[rim]a non si mandano a Roma e dalla nostra Congregazione sieno approvate etc."

47. Gigliola Fragnito observes that part of what made the development of an organized system for expurgatory censorship possible was that "all of the corrections would have to be subordinated to their [the Congregation of the Index's] approval." Although this centralized control may have ultimately been one of the reasons that expurgatory censorship failed, it was this centralized authority that made an expurgatory apparatus possible in the first place. Gigliola Fragnito, "Aspetti e problemi della censura espurgatoria," in *L'inquisizione e gli storici: Un cantiere aperto* (Rome: Accademia nazionale dei lincei, 2000), 168–69.

48. *CCMS*, vol. 1, t. 4, pp. 2833–34.

49. Aldo Stella, "L'età postridentina," in *Diocesi di Padova*, ed. Pierantonio Gios (Padua: Gregoriana libreria editrice, 1996), 231–37.

50. Marco Cornaro, bishop of Padua, to Agostino Valier, cardinal of Verona, 28 March 1597, *CCMS*, vol. 1, t. 1, pp. 623–24: "poiché ci sono attraversati alcuni impedimenti. Il Padre Inquisitore dice d'haver fatto non so che, et di darne conto a V.S. Ill.ma."

51. Cardinal Marcantonio Colonna to Marco Cornaro, bishop of Padua, 10 April 1597, *CCMS*, vol. 1, t. 1, pp. 624–25: "con tutto che si assicuriamo che V.S. con la prudenza e authorità sua sia per superare ogni difficoltà."

52. Marco Cornaro, bishop of Padua, to Cardinal Marcantonio Colonna, 25 April 1597, *CCMS*, vol. 1, t. 1, pp. 626: "sia certa V.S. Ill.ma ch'io non mancarò d'impiegarmi in tutto quello che saprò, e potrò, acciò un negotio cosi utile, et importante alla Christianità sortisca quel buon fine, che con tanta pieta desiderano."

53. On Pranzini, see Schwedt, "Gli inquisitori generali di Siena," XLVI–XLVII.

54. On April 4, 1597, Pranzini wrote to Valier that he had "burned 29 sacks of damned books." ACDF, Index IV, f. 125r.

55. Congregation of the Index to Felice Pranzini, inquisitor of Padua, 26 April 1597, *CCMS*, vol. 1, t. 1, pp. 627–28.

56. Congregation of the Index to Pranzini, *CCMS*, vol. 1, t. 1, pp. 627–28: "e per essere il negotio del'Indice non di minore importanza di quello del S. Offitio, dal quel deriva l'Indice, conviene anco che nell'essecutione di quanto dalla nostra Congregatione vien' ordinato, V.R. convenga unitamente per servitio di Dio con Mons. Vescovo in tutte le Congregationi che sopra cio si faranno alla presenza di Mons. e con quello medesimo zelo e solicitudine che si trattano li negotij del Santo Offitio, s'attenderà al'Indice espurgatorio de libri di Medicina, e Filosofia."

57. Schwedt suggests that Pranzini taught philosophy to his order at the convent of San Miniato in Tuscany, but this is never mentioned in letters with the Congregation of the Index. By contrast, the *uomini dotti* (learned men) of Padua are mentioned regularly. Schwedt, "Gli inquisitori generali di Siena."

58. Valier to Cornaro, 8 March 1597, p. 622: "cotesta tanto célèbre et ill[ust]re Università, la quale sempre è stata ripiena d'huomini Dotti"; Valier to Giovannini, 8 March 1597, p. 623. "essendoci in cio molti anni affaticata V.R. et havendo copia costi et in Padua et in Venetia d'huomini dotti che l'aiuteranno . . ."

59. *CCMS*, vol. 1, t. 4, p. 2932.

60. Ercole Sassonia to cardinals of the Congregation of the Index, 11 July 1597, *CCMS*, vol. 1, t. 1, pp. 630–31: "Non havessimo tardato tanto tempo à dar risposta alle lettere delle SS. VV. Ill.me e R.me . . . habbiamo nel nostro Collegio eletto dodeci Dottori alla correttione de i libri, cioè sei in Medicina . . . et sei in filosofia . . . i quali tutti hanno con molta prontezza abbracciato questo carico."

61. The biographical information provided in this paragraph can be found in the "Biographical Vademecum" in *CCMS*, vol. 1, t. 4, pp. 2783–2965.

62. The document actually lists Alessandro Piccolomini, but Baldini and Spruit suggest that this is a mistake and should be Francesco Piccolomini since Alessandro had been dead for nearly twenty years at this point. *CCMS*, vol. 1, t. 1, p. 630n119. On Piccolomini, see David A. Lines, "Latin and Vernacular in Francesco Piccolomini's

Moral Philosophy," in *"Aristotele fatto volgare": Tradizione aristotelica e cultura volgare nel Rinascimento*, ed. David A. Lines and Eugenio Refini (Pisa: Edizioni ETS, 2014), 169–99.

63. For a list of these philosophers, see *CCMS*, vol.1, t. 1, pp. 630–31.

64. P. Grendler, *Universities of the Italian Renaissance*, 40. Not only was Padua one of the most illustrious places to study medicine in sixteenth-century Italy, medical learning in Italy had a unique and widely respected position in Europe because of its professional, social, and institutional context and the debates among Italian scholars (Siraisi, *Medicine and the Italian Universities*, 143).

65. P. Grendler, *Universities of the Italian Renaissance*, 4–5.

66. Although Brevaglieri refers here specifically to Rome, this is also an important description of how Rome related to other cities and institutions on the Italian peninsula as exemplified by the outsourcing of censorship. Brevaglieri, "Science, Books and Censorship," 110.

67. This list is reproduced in *CCMS*, vol. 1, t. 1, pp. 638–49.

68. *CCMS*, vol. 1, t. 2, pp. 1473–80.

69. Simone Tagliavia d'Aragona, cardinal of Terranova, to Felice Pranzini, inquisitor of Padua, 16 July 1599, *CCMS*, vol. 1, t. 1, pp. 653–54: "Desiderano questi miei Ill.mi SS.ri che nella censura de libri di filosofia, et Medicina usi ogni diligenza con Monsignor Vescovo sollicitando il Collegio de Dottori con li Theologi assignati per Classi, con avertir, che il carico pricipale è del Vescovo, et del Inquisitore et non altrimenti del Collegio."

70. Fragnito, "Central and Peripheral Organization of Censorship," 22–23.

71. Fragnito, "Central and Peripheral Organization of Censorship," 23.

72. Felice Pranzini, inquisitor of Padua, to Agostino Valier, cardinal of Verona, 14 March 1598, *CCMS*, vol. 1, t. 1, p. 650: "E perche due giorni della settimana il lunedi et mercordi sono deputati à convenire insieme per attendere alle cause del Inquisitione, in tali giorni non si può attendere compitam[en]te ad altro."

73. On the economic functioning of the Inquisition, see Germano Maifreda, *I denari dell'inquisitore: Affari e giustizia di fede nell'Italia moderna* (Turin: Giulio Einaudi Editore, 2014).

74. Marco Cornaro, bishop of Padua, to Simone Tagliavia d'Aragona, cardinal of Terranova, 30 July 1599, *CCMS*, vol. 1, t. 1, p. 655: "Ma quando si è voluto metter il neg[oti]o in pratica, i Teologhi dissero di non haver i libri, ne i Medici parimente dicono di haverli; poiché sono libri di qualche importanza, et che converrà forsi à mandarli prender fuori d'Italia con qualche spesa."

75. Felice Pranzini, inquisitor of Padua, to Simone Tagliavia d'Aragona, cardinal of Terranova, 13 August 1599, *CCMS*, vol. 1, t. 1, p. 656: "non si manca di diligenza, ma da alcuni pochi authori in poi, fa bisogno di fare venire il resto de' libri da censurare, che importarà buona soma di denari; Io sono povero frate, et non posso per la mia povertà comp[er]are simili libri, ne so' dove voltarmi à chi sia per fare tale spesa etc."

76. It seems likely that Tagliavia is referring here to expurgations composed by Girolamo Rossi and Giovan Battista Codronchi, which will be discussed further in chapters 3 and 4. Simone Tagliavia d'Aragona, cardinal of Terranova, to Marco Cornaro,

bishop of Padua, 16 August 1599, *CCMS*, vol. 1, t. 1, pp. 658–59: "Son presentate alla nostra Congregatione diverse Censure de libri di Medicina e filosofia mandate da luoghi dove non è molta copia d'huomini dotti e vi è anco penuria de libri, onde par difficile, che in Padua son tanti huomini insigni, e tante librarie famose sia necessario far tanta spesa, e mandar in paesi lontani per libri e se dara una revista alle licenze date da V.S. e dal Inquisitore considerando la qualita de libri permessi troverà che vi saranno tanti libri di Medicina e filosofia da censurare che saranno sufficienti per occupar molti Consultori, e dar principio à cosi necessaria impresa altrimenti, non seguendo il frutto, che si sperava si revocaranno tutte le licenze date e la facoltà a Vescovi et Inquisitori di concederle il che potra V.S. farlo intimare à tutti prefigendo quel termine, che giudicherà oportuno."

77. Tagliavia to Cornaro, *CCMS*, vol.1, t. 1, p. 659.

78. Claudio Donati has suggested that some scholars may have agreed to expurgate texts as an act of subterfuge with the goal of influencing the process more easily from within it. Claudio Donati, "A Project of 'Expurgation' by the Congregation of the Index: Treatises on Duelling," in Fragnito, *Church, Censorship and Culture in Early Modern Italy*, 148.

79. Godman, *Saint as Censor*, 51.

80. Prospero Podiano, "Nota delli libri che si comprano per ordine della Sacra Congregazione dell'Indice," Rome, 7 August 1594, *CCMS*, vol. 1, t. 1, pp. 264–67.

81. "Decree of the Congregation of the Index," Rome, 14 August 1599, *CCMS*, vol. 1, t. 1, p. 657: "Comensi Episcopo comissa Censura operum Iovij, et Inq[uisito]ri Pisano tradita expurgatio librorum Medicinae et Philosophiae cum Vic[ari]o Archiepiscopali."

82. Simone Tagliavia d'Aragona, cardinal of Terranova, to vicar of Pisa, 16 August 1599, *CCMS*, vol. 1, t. 1, p. 658: "instituir una Cong[regatio]ne de varii Consultori a questo effetto, et andar censurando, e correggendo libri di Medicina, e filosofia con l'aiuto dell'Inquisitore e di tanti Ecc[ellen]ti huomini, che sono in cosi honorato studio."

83. On the priority of the Pisan botanical garden, see Findlen, "Death of a Naturalist," 156n3.

84. On ecclesiastical censorship in Pisa, see Adriano Prosperi, "Anime in trappola: Confessione e censura ecclesiastica all'università di Pisa fra '500 e '600," *Belfagor* 54, no. 321 (1999): 257–87.

85. Simone Tagliavia d'Aragona, cardinal of Terranova, to vicar of Naples, 22 September 1599, *CCMS*, vol. 1, t. 1, p. 660: "Aspettano questi miei Ill.ᵐⁱ Sig.ʳⁱ le Censure, e confidano nel molto valor vostro che usarete ogni diligenza in sollicitar che s'attenda a censurar varii libri di Medicina, e filosofia."

86. Agostino Valier, cardinal of Verona, to vicar of Pisa, 3 December 1599, *CCMS*, vol. 1, t. 1, pp. 660-61, and Agostino Valier, cardinal of Verona, to vicar general of Naples, 3 December 1599, *CCMS*, vol. 1, t. 1, pp. 661.

87. Greater detail on this collection of expurgations will be provided in chapter 4.

88. Cesare Speciano, bishop of Cremona, to the Congregation of the Index, 13 October 1603, ACDF, Index III, v. V, f. 326: "Alcuni de quali si sono dolusi gli sia stata assegnata la correttione dei medesimi libri di medicina che si sono corretti anche altrove nominando Padova et Milano."

89. In addition, there was the constant risk that the force of the Index and Inquisi-

tion would turn against the censors if their edits were seen as heterodox. Godman, *Saint as Censor*, 101.

90. Fragnito, "Central and Peripheral Organization of Censorship," 42. For another example of expurgation projects that were attempted several times but never successfully concluded, see Peter Godman's discussion of the censorship of Machiavelli's works in *From Poliziano to Machiavelli: Florentine Humanism in the High Renaissance* (Princeton, NJ: Princeton University Press, 1998), 303–33.

91. Felice Pranzini, inquisitor of Padua, and Camillo Peltrari to Agostino Valier, cardinal of Verona, 1 December 1600, ACDF, Index III, v. VII, f. 384. "Et se si havesse copia de' libri da espurgarsi di medicina, et philosophia, si farebbe molto più, ma noi per la nostra povertà non possiamo farne venire."

92. ASV, Savi all'Eresia (Sant'Ufficio) 27, 14r., 2 July 1568: "A Padoa al Bo. L'ho trovato nel loco dove andai a urinar."

93. As quoted in Charles Webster, *Health, Medicine, and Mortality in the Sixteenth Century* (Cambridge: Cambridge University Press, 1979), 351.

94. For the library as a whole, see Lucia Rosetti, "Le biblioteche delle 'nationes' nello studio di Padova," *Quaderni per la storia dell'Università di Padova* 2 (1969): 53–67; and the seventeenth-century list compiled by librarians of the time, *Bibliotheca medico-philosophico-philologica inclytae nationis germanae artistarum . . .* (Padua: Frambotti, 1685). See also the published acts of the German Nation in Antonio Favaro, *Atti della Nazione Germanica di Padova* (Venice: A spese della Società, 1911–12); and Lucia Rosetti, ed. *Acta Nationis Germanicae Artistarum (1616–1636)* (Padua: Antenore, 1967).

95. Fra Domenico Istriani da Pesaro, inquisitor of Mantua, to Congregation of the Index, 23 May 1596, ACDF, Index III, v. I, f. 287: "Il farsi questa un tal libro espurgatorio lo reputo difficilissmo e tengo di certo che qui nessuno vorrà pigliare questo carico."

96. Stefano da Cento, inquisitor of Bologna, to Congregation of the Index, 28 March 1597, ACDF, Index III. v. III, f. 49r: "La maggior parte disse di non potervi attender per il tempo . . . essendo impediti dalle predicationi confessioni et lettioni. . . . Con tutto niuno sia finito l'opera sua, et pochi incominciati."

97. Giovan Battista Lanci, inquisitor of Genoa, to Congregation of the Index, 12 April 1597, ACDF, Index III, v. III, f. 155r: "attendendo li dottori a fatti suoi, s'ha in cio pochissimo aiuto."

98. Vicenzo Castrucci, the inquisitor of Perugia, to Congregation of the Index, 7 February 1597, ACDF, Index III, v. III, f. 242r: "Hora perche questi Signori legisti diranno senza dubbio esser molto occupati in molte loro letioni, avocationi, et altri proprii negotii familiari, Monsignor Reverendissimo, che in questa opera farà particolare fatica, et io giudichiamo essere necessario che con una buona lettera sopra di cio al Collegio di questi signori scritta, et inviata a Monsignore Reverendissimo sieno eccitati et con emulatione di Bologna."

99. On the publication history of Zabarella, especially in Northern Europe, see Ian Maclean, "Mediations of Zabarella in Northern Germany, 1586–1623," in *Learning and the Market Place: Essays in the History of the Book* (Leiden, Neth.: Brill, 2009), 39–55. On the Paduan censor Francesco Piccolomini's disputes with Zabarella, see Nicholas Jardine, "Keeping Order in the School of Padua; Jacopo Zabarella and Francesco Piccolo-

mini on the Offices of Philosophy," in *Method and Order in Renaissance Philosophy of Nature: The Aristotelian Commentary Tradition* (Aldershot, UK: Ashgate, 1997).

100. Alfonso Soto to Congregation of the Index, 25 August 1601, ACDF, Index III, v. VII, f. 386r/v: "Questa è fatica d'importanza . . . ricerca una persona libera, e non una obligata come sono io. . . . bisognera mettere tutta l'opera sottosopra, con dar diverse interpretationi, e far diversa concatenatione, e porre discorsi fundamente; Ma se il negotio consistesse in notare solamente i luoghi repugnanti alla verita, quest'e cosa che si potrebbe fare da molti."

101. The ACDF does not hold any expurgations signed by that original committee of elected scholars. *CCMS*, vol. 1, t. 1, p. 602.

102. *CCMS*, vol. 1, t. 2, p. 1664.

103. For the denunciation, see Antonino Poppi, *Cremonini, Galilei e gli inquisitori del Santo a Padova* (Padua: Centro Studi Antoniani, 1993), 9–11. There is vast literature on Cremonini. See in particular Thomas F. Mayer, *The Roman Inquisition on the Stage of Italy* (Philadelphia: University of Pennsylvania Press, 2014), 124–34; Ezio Riondato and Antonino Poppi, *Cesare Cremonini: Aspetti del pensiero e scritti* (Padua: Accademia galileiana di scienze, lettere ed arti in Padova, 2000); Antonino Poppi, *Cremonini e Galilei inquisiti a Padova nel 1604: Nuovi documenti d'archivio* (Padua: Antenore, 1992); Leen Spruit, "Cremonini nelle carte del Sant'Uffizio Romano," in Riondato and Poppi, *Cesare Cremonini*, 193–205; and Muir, *Culture Wars*.

104. Muir, *Culture Wars*, 37.

105. See also John Jones, "The Censor Censored: The Case of Benito Arias Montano," *Romance Studies* 13 (1995): 19–29.

106. More literally, "Inwardly as you please, outwardly according to custom."

107. Luigi Firpo, "The Flowering and Withering of Speculative Philosophy—Italian Philosophy and the Counter Reformation: The Condemnation of Francesco Patrizi," in *The Late Italian Renaissance 1525–1630*, ed. Eric Cochrane (London: Macmillan, 1970), 266–84.

108. On Pomponazzi, see Martin L. Pine, *Pietro Pomponazzi: Radical Philosopher of the Renaissance* (Padua: Editrice Antinore, 1986); and Marco Sgarbi, ed., *Pietro Pomponazzi: Tradizione e dissenso* (Florence: Leo S. Olschki Editore, 2010).

109. On his response to the admonition, see Poppi, *Cremonini e Galilei inquisiti a Padova nel 1604*, IX.

110. Mayer, *Roman Inquisition on the Stage of Italy*, 124–25.

111. ASV, Riformatori dello Studio di Padova b. 168 and b. 420, unnumbered folios.

112. Mayer, *Roman Inquisition on the Stage of Italy*, 125.

113. Mayer, *Roman Inquisition on the Stage of Italy*, 150.

114. Spruit, "Cremonini," 202. During its 1611 investigation of Cremonini, the Congregation of the Index also ordered a search in their archives to see what they had on Galileo. See Antonio Favaro, ed., *Edizione nazionale delle Opere di Galileo Galilei*, 20 vols. (Florence: Giunti Barbèra, 1890–1909), 19:275 (hereafter cited as *OG* with volume and page numbers); and Thomas F. Mayer, *The Roman Inquisition: Trying Galileo* (Philadelphia: University of Pennsylvania Press, 2015), 7.

115. Marco Cornaro, bishop of Padua, to Agostino Valier, cardinal of Verona, 15

February 1602, ACDF, Index III, v. VII, f. 387: "Si sono fatte qui in Padova molte Congregationi per la espurgatione de libri, et molte opere si sono messe in mano di diversi, i quali per l'occasioni di lettioni, anco altri sono esercitii non hanno fatto altro che dare principio et le difficulta del progresso furono avisate da me alla sudetta Congregatione senza haverne havuto risposta."

116. Felice Pranzini, inquisitor of Siena, to Congregation of the Index, 26 October 1602, ACDF, Index III, v. VI, f. 117.

117. Charles B. Schmitt, "Philosophy and Science in Sixteenth-Century Universities: Some Preliminary Comments," in *Studies in Renaissance Philosophy and Science* (London: Variorum Reprints, 1981), 504.

118. For the Venetian Inquisition's use of physicians as expert witnesses in Inquisition trials, see Jonathan Seitz, *Witchcraft and Inquisition in Early Modern Venice* (Cambridge: Cambridge University Press, 2011). On the essential role of physicians in providing expertise in canonization cases, see Bouley, *Pious Postmortems*, esp. 30–37, 96–101.

119. On professional expertise in the late Renaissance, see George W. McClure, *The Culture of Profession in Late Renaissance Italy* (Toronto: University of Toronto Press, 2004).

CHAPTER 3

1. For a few examples, see *CCMS*; Paul Grendler, "The 'Tre Savii sopra Eresia' 1547–1605: A Prosopographical Study," *Studi veneziani* 3 (1979): 283–340; Nesvig, *Ideology and Inquisition*; Lynn, *Between Court and Confessional*; Mayer, *Roman Inquisition: Papal Bureaucracy*; and Herman H. Schwedt et al., *Prosopographie von Römischer Inquisition und Indexkongregation*.

2. *CCMS*, vol. 1, t. 4, p. 2926. On the opening of the Archive of the Congregation for the Doctrine of the Faith, see Schutte, "Palazzo del Sant'Uffizio," 25–28; and *L'Apertura degli archivi del Sant'Uffizio romano*.

3. The episode in Padua is discussed in Ricci, *Inquisitori, censori*, 363–76.

4. For basic biographical information on Rossi, see Primo Uccellini, *Dizionario storico di Ravenna e di altri luoghi di Romagna* (Ravenna: Tipografia del Ven. Seminario Arcivescovile, 1855), 416–17; and Pietro Paolo Ginanni, *Memorie storico-critiche degli scrittori ravennati*, vol. 2 (Faenza: Gioseffantonio Archi, 1769), 313–26.

5. The Collegio Ancarano was among the best schooling opportunities available in Bologna, considered second only to the Collegio di Spagna. Christopher Carlsmith, "'Cacciò fuori un bastone bianco': Conflicts between the Ancarano College and the Episcopal Seminary in Bologna," in *The Culture of Violence in Renaissance Italy*, ed. Samuel Kline Cohn and Fabrizio Ricciardelli (Florence: Le lettere, 2012), 194–95.

6. Ginanni, *Memorie storico-critiche*, 314.

7. Eric Cochrane places Rossi's work in his category of the "definitive histories" of the 1580s and 1590s. See Eric W. Cochrane, *Historians and Historiography in the Italian Renaissance* (Chicago: University of Chicago Press, 1981), 284–91.

8. On Rossi's research and sources, see Mario Pierpaoli, "Girolamo Rossi medico

e storico ravennate," in *Storie ravennati*, by Girolamo Rossi, ed. and trans. Mario Pierpaoli (Ravenna: Longo Editore 1996), xiv–xv.

9. Girolamo Rossi to [?] Caraffa, 11 March 1595, Biblioteca Comunale Classense, Ravenna (hereafter cited as BCRa), Fondo Manoscritti, Manoscritti vari di Girolamo Rossi, Mob. 3. 1 B, n. 5, f. 431v.: "Sono circa 30 anni ch'io vidi l'opera di Ricobaldo la quale è in foglio alta quattro dita, scritta a mano latinamente in carta pecora; si trovava nella libraria Vaticana, al sesto banco se mal non mi ricordo."

10. It is likely that Rossi was referring to the manuscript currently classified as part of Biblioteca Apostolica Vaticana (hereafter cited as BAV), Ott.lat.2072.

11. On the relationship between medicine and history writing in this period, see Siraisi, *History, Medicine*; and Pomata and Siraisi, *Historia*.

12. G. Rossi, *Storie ravennati*, XIV.

13. Ginanni, *Memorie storico-critiche*, 317.

14. BCRa, Fondo Manoscritti, Manoscritti vari di Girolamo Rossi, Mob. 3. 1 B, n. 1, see ff. 131r, 152r–154v, 156r.

15. On Rossi serving Cardinal Salviati and his family, see his correspondence detailing the management of Salviati properties and the sales of their crops. On the patronage relationship, see especially Rossi's letter to the cardinal's brother following the death of Cardinal Salviati in April 1602. BCRa, Fondo Manoscritti, Manoscritti vari di Girolamo Rossi, Mob. 3. 1 B, n. 5, f. 485v.

16. BUB, Fondo Aldrovandi. Ms. 136 vol. 32, f. 142r–143v. The treatise is dated November 18, 1602.

17. On the connection between the sixteenth-century study of history and the study of nature, see Pomata and Siraisi, *Historia*; and Siraisi, *History, Medicine*. The latter mentions Girolamo Rossi briefly on page 197.

18. There is surprisingly little research on Rossi's *On distillation*. See François Secret, "De Mésué à Hieronymus Rubeus en passant par Giovanni Mainardi et Jacques Dubois," *Chrysopoeia* 5 (1996): 453–66.

19. BCRa, Fondo Manoscritti, Manoscritti vari di Girolamo Rossi, Mob. 3. 1 B, n. 5, f. 436r, 443r.

20. Girolamo Rossi to Fabio Paolini, 18 May 1596, BCRa, Fondo Manoscritti, Manoscritti vari di Girolamo Rossi, Mob. 3. 1 B, n. 5, f. 443r: "Nondimeno mando un mio libro de Destillatione gia stampato in questa citta in quarto et ristampato in Basilea in ottavo: et perche quella ristampata fu senza mia saputta, non vi potevo aggiungere alcuni additioni che hora mando, desiderando di non spender più di quello che sia il mio credito, et pero o in ottavo o in 16mo o in qual altra forma . . . desiderarei bene che in ogni evento, forse ben corretto; et che da VS fosse giudicato al tutto non indegno d'esser ristampato."

21. Rossi to Paolini, BCRa, Fondo Manoscritti, Manoscritti vari di Girolamo Rossi, Mob. 3. 1 B, n. 5, f. 458r, 458v: "acio non sia troppo grande, ne troppo piccolo ma conviendo al foglio nel qual si stampe."

22. Massimo Firpo, "Giovanni Battista Ciotti," in *DBI*, 25:692–96. Note also that Ciotti was apprehended by the Venetian Inquisition in 1599 for importing prohibited books (P. Grendler, *Roman Inquisition and Venetian Press*, 280).

23. In the 1604 edition, "On the antiquity of distillation" appears as chapter 2

rather than chapter 4, chapter 9 is retitled, and page 37 is new content. In general, Rossi tended to add material to the ends of chapters (e.g., pp. 45, 48, 170–71), but he also added material within the body of the text (e.g., p. 126). Girolamo Rossi, *De destillatione liber* (Venice: apud Ioannem Baptistam Ciottum Senemsem, 1604).

24. In addition to his history of Ravenna and books on distillation, Rossi published widely on a number of other topics. Among his medical works, see his short pamphlet (really a collection of medical letters) about melons and asthma, *Diputatio de meloni-bus . . .* (Venice: Ioannem Baptistam Bertonum, 1607), and his commentary on Celsus dedicated to Scipione Borghese and published posthumously by his son Francesco, *Annotationes in libros octo Cornelii Celsi . . .* (Venice, 1616). Rossi exchanged treatises on the plague with Girolamo Mercuriale in the 1570s, which though not printed were certainly published in manuscript. Copies exist among Rossi's own papers in BCRa, Fondo Manoscritti, Manoscritti vari di Girolamo Rossi, Mob. 3. 1 B, n. 1, f. 162r–169v; and in the Biblioteca Ambrosiana, Milan, (hereafter cited as BA), Ms. Q. 117 sup. "Miscellaneo", f. 168r–171v. There is also an unpublished and likely incomplete work mostly in Rossi's hand preserved in the Vatican Library called *Rerum naturalium*, BAV, Vat.lat.5361.

25. Gian Antonio Grassi to Congregation of the Index, 21 December 1596, ACDF, Index III, v. I, f. 381r.

26. Fra Alberto Chelli was the inquisitor of Romagna from 1592 to 1603 and the inquisitor of Cremona from 1600 to 1603. *CCMS*, vol. I, t. 4, p. 2827. For an account of Chelli's arrest and accused conspiracy against Venice, see P. Grendler, *Roman Inquisition and Venetian Press*, 216–18.

27. Alberto Chelli to Congregation of the Index, ACDF, Index III, v. III, f. 108v. Also in *CCMS*, vol. 1, t. 1, p. 619.

28. These expurgations can be found in ACDF, Index, Protocolli, O (II.a.13) fols. 258r–264v; also published in *CCMS*, vol. 1, t. 1, pp. 607–18.

29. Carlo Colombero, "Giovan Battista Codronchi," in *DBI*, 26:604–5; and Luigi Angeli, *Sulla vita e su gli scritti di alcuni medici imolesi: Memorie storiche* (Imola, It., 1808), 153–69.

30. Giovan Battista Codronchi, *De christiana ac tuta medendi ratione libri duo varia doctrina referti* (Ferrara, 1591).

31. *CCMS*, vol. 1, t. 1, 607–18.

32. *CCMS*, vol. 1, t. 1, p. 618.

33. ACDF, Index V, f. 46r.

34. ACDF, Index III, v. III, f. 110r.

35. Lusitanus's *Centuriae* was published in seven installments between 1549 and 1570 and republished regularly into the seventeenth century. On Cardano's horoscope of Jesus, see Grafton, *Cardano's Cosmos*, 152, 154–55.

36. Girolamo Rossi to Alberto Chelli, inquisitor of Romagna, 15 February 1597, BCRa, Fondo Manoscritti, Manoscritti vari di Girolamo Rossi, Mob. 3. 1 B, n. 5, f. 448r.

37. Girolamo Rossi to Congregation of the Index, 27 March 1598, BCRa, Fondo Manoscritti, Manoscritti vari di Girolamo Rossi, Mob. 3. 1 B, n. 5, f. 455v.

38. ACDF, Index III, v. IV, f. 82r–83v, f. 191r.

39. CCMS, vol. I, t. I, pp. 559–60: "In questo mense potrà censurare qualche libro di Medicina, ò Philosofia, [. . .] ò pur da libri di Astrologia far scelta di quello che può servire alla Nautica, agricoltura, et Medicina, accio si possa resecar tutto il superfluo, è pernitioso, et metter insieme quanto ci è di buono nel uno, et nel altro."

40. ACDF, Index III, v. V, f. 450r.

41. ACDF, Index V, ff. 111v, 161v.

42. Girolamo Rossi to Congregation of the Index, 31 July 1602, ACDF, Index III, v. VI, f. 6r/v. Rossi was responding to a letter from the Congregation of the Index (in the person of Cardinal Terranova), written on 20 July 1602. ACDF, Index V, f. 164r: "Le espurgationi che io ho fatto in diversi tempi di alcuni libri di varie professioni et mass[im]e di medicina, io raccogliere in un corpo, [. . .] et quanto prima sarà arrivato il nuovo Vicario di Monsignor Arcivescovo Nostro, che s'aspetta in breve, ha detto di fare, insieme con esso lui, la deputatione de li tre Theologi che Vostra Signoria Illustrissima commanda i quali riveggano et sottoscrivano dette spurgationi."

43. BCRa, Fondo Manoscritti, Manoscritti vari di Girolamo Rossi, Mob. 3. I B, n. 3, "Hier. Rubei Correctiones 1) in Cardani opuscula 2) in Centurias Amati Lusitani 3) in librum de conservanda bona valetudine Arnoldi Novicomentis."

44. Fabio Tempestivo to Congregation of the Index, 2 February 1602, ACDF, Index III, v. VI, f. 7r: "Desidero, che resti servita d'avvisarmene, perché quando non fusse capitata se ne mandarà una copia, si come con questa mia mando la correttione, fatta dall'istesso Signore Gerollamo Rossi sopra la schola Salernitana, che aspettaro d'intender l'habbia recenta."

45. ACDF, Index V, f. 164r.

46. Stefano de Vicariis to Congregation of the Index, 24 October 1602, ACDF, Index III, v. V, f. 453r/v: "Questi dottori dicono che queste censure non si possono fare nelle città piccole ma nelle grandi e di studii generali."

47. Girolamo Rossi to Congregation of the Index, 7 August 1603, ACDF, Index III, v. VI, f. 11r/v: "Se a me solo stesse il totale compimento di mandare le censure fatte, già molto tempo, conforme al debito mio, sarebbono costì a servire Vostra Signoria Illustrissima ma poiché il più depende da altri, io non so che poter fare, salvo che da la parte mia compire perfettamente il detto servitio, copiando le censure, et in quello che a gli altri tocca[?] sollecitare si come ho già anco' hora fatto."

48. ACDF, Index VI, f. 11v.

49. ACDF, Index VI, f. 7r.

50. Girolamo Rossi to Congregation of the Index, 7 August 1603, ACDF, Index III, v. VI, f. 11r/v: "Nel resto che V.S. Ill[ustrissi]ma mi comanda, di notare in qualche Historia, massime di Oltamontani, che a le mani mi capitasse, quello che mi paresse degno di censura, obedire volontieri, ma in questo angolo dove mi trovo, rarissme volte arrivano libri nuovi."

51. ACDF, Index VI, f. 20r.

52. Pope Clement VIII had been periodically quite ill for several years when Rossi entered the scene, and the pope died of a stroke on March 5, 1605. See Ludwig von Pastor, *The History of the Popes, from the Close of the Middle Ages*, vol. 24 (London: K. Paul, Trench, Trübner, 1906), 430–34. On the ambiguity of the appointment of papal physicians, see Richard Palmer, "Medicine at the Papal Court in the Sixteenth

Century," in *Medicine at the Courts of Europe*, ed. Vivian Nutton (London: Routledge, 1990), 49–78.

53. Maria Pia Donato, "Scienze della natura," in *Dizionario storico dell'Inquisizione*, ed. Adriano Prosperi (Pisa: Scuola Normale Superiore, 2010), 1394–98.

54. These types of expurgations will be explored further in chapter 4.

55. *CCMS*, vol. 1, t. 2, pp. 1887, 1889–93.

56. Fabio Tempestivo to Congregation of the Index, 29 February 1601, ACDF, Index III, v. VI, f. 5r: "Signor Gerollamo Rossi huomo non meno ricco di zelo et pietà Christiana che di scienze."

57. The most recent study of Giovan Battista Rossi is the volume of conference proceedings *Giovanni Battista Rossi: Carmelitano ravennate* (Ravenna: Centro Studio "G. Donati," 1980).

58. Giovan Battista Rossi's visit to Avila and his meetings with Teresa are the subject of chapter 2 of the *Book of the Foundations of Saint Teresa*.

59. The others were Giovanni Morone, Marcantonio Arnulio, and Giuglielmo Sirleto. They are mentioned in, for example, Francesco Saverio de Feller, *Dizionario storico ossia storia compendiata* (Venice, 1834), 9:298.

60. The prefatory letter to which this story refers explained Erasmus's project of drawing on Christian antiquity to purify the Church. Stefano Passanzini, "Giovanni Battista Rossi Carmelitano: La famiglia, la patria, il personaggio," in *Giovanni Battista Rossi*, 36. Another story about Giovan Battista relates that he never opened a papal document without first putting the document on top of his own head as a symbol of his obedience and reverence for the office (Passanzini, "Giovanni Battista Rossi," 30; for this example, Passanzini cites Nicola Rouhier, BAV, Barb. Ist. Ms 2267, f. 2v). On Erasmus in Italy, see Seidel Menchi, *Erasmo in Italia*; Marcella T. Grendler and Paul F. Grendler, "The Survival of Erasmus in Italy," *Erasmus in English* 8 (1976): 2–22.

61. Giovanni Battista Rossi makes several appearances in Girolamo Rossi's *History of Ravenna*. See G. Rossi, *Storie Ravennati*, 8, 754–55, 758.

62. For a list of his children, see BCRa, Fondo Manoscritti, Manoscritti vari di Girolamo Rossi, Mob. 3. 1 B, n. 5, f. 507r.

63. Ginanni, *Memorie storico critiche*, 315: "magno in rei domesticae administratione ingenio, sed maiori in Deum pietate."

64. "De earum q[uae] elementis tribuuntur qualitatum, quantitate, Hieronymi Rubei Ravennatis Disputatio," BCRa, Fondo Manoscritti, Manoscritti vari di Girolamo Rossi, Mob. 3. 1 B, n. 4, f. 405r–414v.

65. BCRa, Fondo Manoscritti, Manoscritti vari di Girolamo Rossi, Mob. 3. 1 B, n. 4, f. 405r.

66. BCRa, Fondo Manoscritti, Manoscritti vari di Girolamo Rossi, Mob. 3. 1 B, n. 4, ff. 406r, 408v.

67. Paul Grendler, *Critics of the Italian World, 1530–1560: Anton Francesco Doni, Nicolò Franco and Ortensio Lando* (Madison: University of Wisconsin Press, 1969), 22. On Lando generally, see Simonetta Adorni Braccesi and Simone Ragagli, "Lando, Ortensio," in *DBI*, 63:451–59.

68. BCRa, Fondo Manoscritti, Manoscritti vari di Girolamo Rossi, Mob. 3. 1 B, n. 4, f. 410r. The original, showing the struck-through position, reads, "Hortensius Landus,

vi dicendi egregia, summa festivitate, et venustate coniuncta, suis paradoxa." There has been a renewed interest in Lando's work. For his *Paradossi,* see Ortensio Lando, *Paradossi: Cioè sentenze fuori del comun parere,* ed. Antonio Corsaro (Rome: Edizione di Storia e Letteratura, 2000); and Patrizia Grimaldi Pizzorno, *The Ways of Paradox from Lando to Donne* (Florence: Olschki, 2007).

69. BCRa, Fondo Manoscritti, Manoscritti vari di Girolamo Rossi, Mob. 3. 1 B, n. 4, f. 412r.

70. Maria Muccillo, "Da Monte, Giovanni Battista, dettò Montano," in *DBI,* 32:365–67. This entry is quite short; for more on Giovan Battista da Monte, see Daniela Mugnai Carrara, "Le epistole prefatorie sull'ordine dei libri di Galeno di Giovan Battista da Monte: Esigenze di metodo e dilemmi editoriali," in *Vetustatis indagatur, Scritti offerti a Filippo di Benedetto,* ed. Vicenzo Fera and Agostino Guida (Messina: Università degli Studi, 1999), 207–34; Silvia Ferretto, *Maestri per il metodo di trattar le cose: Bassiano Lando, Giovan Battista da Monte e la scienza della medicina nel XVI secolo* (Padua: Belzoni, 2012); and Nancy G. Siraisi, *Avicenna in Renaissance Italy: The "Canon" and Medical Teaching in Italian Universities after 1500* (Princeton, NJ: Princeton University Press, 1987), 194–202.

71. On the hermeneutics of reading early modern works in light of self-censorship, see Patterson, *Censorship and Interpretation.*

72. Michaela Valente, *Bodin in Italia: La "Démonomanie des sorciers" e le vincende della sua traduzione* (Florence: Centro Editoriale Toscano, 1999), 75–146. For other studies dealing with literary self-censorship, see Ugo Rozzo, "Gli 'Hecatommithi' all'Indice," *La Bibliofilia* 93 (1991): 21–51; and Luigi Firpo, "Correzioni d'autore coatte," in *Studi e problemi di critica testuale* (Bologna: Commissione per i Testi di Lingua, 1961), 143–57.

73. Girolamo Rossi, *De destillatione* (Ravenna,1582) ††v, ††2r–v, [††3]r. See the following examples and pages, though of course many of these authors appear repeatedly throughout the text: Thomas Erastus, 125; Arnald of Villanova, 124; Girolamo Cardano, 127.

74. Ugo Rozzo, *Biblioteche italiane del Cinquecento tra Riforma e Controriforma* (Udine, It.: Arti Grafiche Friulane, 1994), 28.

75. Ugo Rozzo has posited that beginning in the 1560s, the expurgation of literary texts was "defined and implemented with the uncoerced complicity of the many intellectuals who turned themselves into expurgators." While I agree that Rossi's participation was "uncoerced," it took place in a context of official expurgatory assignments around Italy. See Ugo Rozzo, "Italian Literature on the Index," in Fragnito, *Church, Censorship, and Culture in Early Modern Italy,* 222.

76. Darnton, *Censors at Work,* 234.

77. Christopher Black, *The Italian Inquisition* (New Haven, CT: Yale University Press, 2009), 207.

78. For a similar point about the role of censors, see Virgilio Pinto, "Censorship: A System of Control and an Instrument of Action," in *The Spanish Inquisition and the Inquisitorial Mind,* ed. Angel Alcala (Boulder, CO: Social Science Monographs, 1987), 307–11.

79. As another example, consider the University of Bologna's attempt to hire the Flemish humanist and political theorist Justus Lipsius in 1595. The Bolognese ambassador to Rome, Camillo Gozzadini, assured the Bolognese Senate that "it was true that the lords of the Holy Office ascribe a certain interpretation to some of his political works . . . but Lipsius corrected himself, when he judged that it was necessary, and was therefore not a man of scandal." ASB, Governo Misto, Assunteria di Studio 75, "Letter dell'Amb.re agli Assonti di Studio" 1571–1694. Unnumbered folios. This letter is dated 8 April 1595.

80. See Anthony Grafton, *The Culture of Correction in Renaissance Europe* (London: British Library, 2011).

81. On humanist reading with pen in hand, see Lisa Jardine and Anthony Grafton, "'Studied for Action': How Gabriel Harvey Read His Livy," *Past & Present* 129 (November 1990): 30–78.

82. For an introduction to the vast literature on early modern commonplacing, see Ann Moss, *Printed Commonplace-Books and the Structuring of Renaissance Thought* (Oxford: Clarendon Press, 1996); Blair, *Too Much to Know*, 62–116; Ann Blair, "Humanist Methods in Natural Philosophy: The Commonplace Book," *Journal of the History of Ideas* 53 (1992): 541–51.

83. I am grateful to Simon Reader for this turn of phrase.

84. *Indicis librorum expurgandorum in studiosorum gratiam confecti tomus primus* (Rome: Ex typographia R. Cam. Apost., 1607), 12–13 (hereafter cited as *Rome 1607*). The missing words removed from between these two indicators are "quia solicitus admodum erat circa divinas preces, quas Hebraei illis diebus noctu effundere ad Deum solent."

85. On willful ignorance and the art of forgetting, see Robert Proctor and Londa Schiebinger, eds., *Agnotology: The Making and Unmaking of Ignorance* (Stanford, CA: Stanford University Press, 2008). On memory and the role of archives in early modern Europe, see Alexandra Walsham, "The Social History of the Archive: Record-Keeping in Early Modern Europe," *Past & Present* 230, suppl. 11 (November 2016): 9–48.

CHAPTER 4

1. *ILI*, 9:20, 742, 750–51.

2. Gigliola Fragnito, "Un archivio conteso: Le 'carte' dell'Indice tra Congregazione e Maestro del Sacro Palazzo," *Rivista storica italiana* 119 (2007): 1298n76.

3. *Rome 1607*.

4. Fragnito, "Un archivio conteso," 1286n41.

5. For a similar conclusion about the censorship of Jean Bodin, see Sara Miglietti, "The Censor as Reader: Censorial Responses to Bodin's Methodus in Counter-Reformation Italy (1587–1607)," *History of European Ideas* 42, no. 5 (2016): 707–21.

6. ACDF, Index, Diari, 1, f. 77v, also in *CCMS*, vol. 1, t. 1, p. 2919. On book censorship in Naples, see Milena Sabato, *Il sapere che brucia: Libri, censure e rapporti Stato-Chiesa nel regno di Napoli fra '500 e '600* (Galatina, It.: Congedo, 2009).

7. *CCMS*, vol. 1, t. 4, pp. 2809–10, 2964, 2835, 2959–60.

8. ACDF, Index XXIII.

9. ACDF, Index V, f. 87v—13 May 1598; f. 92v—12 June 1599; f. 96r—26 July 1599; f. 102r—16 August 1599. For the receipt of the expurgations, see *CCMS*, vol. 1, t. 2, pp. 1382–84n921.

10. Rodolfo Savelli describes the expurgation of legal texts by this congregation in some detail in Rodolfo Savelli, *Censori e giuristi: Storie di libri, di idee e di costumi (secoli XVI–XVII)* (Milan: A. Giuffrè, 2011), 93–147, esp. 136–46.

11. The later decision to ask Neapolitan censors for expurgations of medical texts almost certainly reflected the knowledge that the expurgations already existed and was not an endorsement of local expertise.

12. Savelli has written about both of these groups—physicians and especially lawyers—in Savelli, *Censori e giuristi*; Savelli, "La biblioteca disciplinata"; and Savelli, "Censoring of Law Books," esp. 224, 249–51.

13. Savelli, *Censori e giuristi*, 1–46.

14. Savelli addresses some of the structural similarities in the prohibition of legal and scientific texts in *Censori e giuristi*, xxii–xxxii.

15. ACDF, Index V, f. 120v, and *CCMS*, vol. 1, t. 2, pp. 1382–84n921: "e piu grato sara il vederlo in stampa a publico benefitio."

16. *CCMS*, vol. 1, t. 2, pp. 1382–84n921.

17. Donato Favale to the Congregation of the Index, 24 March 1600, ACDF, Index III, v. VII, f. 227–28: "L'indice de libri prohibiti . . . e stato necessario alla Cristianità: per la porta dell'inferno s'è allargata, atteso con la curiosità humana non si mira alla prohibitione, et molt' anime la giù traboccano. Si potria facilmente rimediare, con publicarse un indice expurgatorio."

18. Fragnito, "Un archivio conteso," 1292.

19. *CCMS*, vol. 1, t. 1, p. 111.

20. Reading licenses will be discussed further in chapter 5.

21. See, for example, the entries for Conrad Gessner. *Index expurgatorius librorum, qui hoc seculo prodierunt [. . .]* (Antwerp, 1571), 46 (hereafter cited as *Antwerp 1571*).

22. For the censorship of scientific books in Spain, see José Pardo Tomás, *Ciencia y censura: La Inquisición Española y los libros científicos en los siglos XVI y XVII* (Madrid: Consejo superior de investigaciones científicas, 1991).

23. For the confusion about Theophrastus, see *Rome 1607*, 723.

24. David C. Lindberg, *The Beginnings of Western Science: The European Scientific Tradition in Philosophical, Religious, and Institutional Context, Prehistory to A.D. 1450* (Chicago: University of Chicago Press, 2007), 214, 273–74.

25. For an overview of early modern Catholic approaches to astrology, see Baldini, "Roman Inquisition's Condemnation of Astrology," and *CCMS*, vol. 1, t. 1, pp. 440–69. On Protestant debates about astrology, see Paola Zambelli, *Astrologi Hallucinati: Stars and the End of the World in Luther's Time* (Berlin: De Gruyter, 1986); and more recently, Robin Bruce Barnes, *Astrology and Reformation* (New York: Oxford University Press, 2016).

26. Rule IX quoted from "Ten Rules Concerning Prohibited Books Drawn Up by the Fathers Chosen by the Council of Trent and Approved by Pope Pius," in *Canons and*

Decrees of the Council of Trent, by Council of Trent, trans. H. J. Schroeder (1941; repr., Rockford, IL: Tan Books, 1978), 276.

27. *CCMS,* vol. 1, t. 1, p. 455.

28. *CCMS,* vol. 1, t. 1, pp. 155–57.

29. *CCMS,* vol. 1, t. 1, pp. 500–521, 524–25.

30. *CCMS,* vol. 1, t. 1, pp. 521–39.

31. *CCMS,* vol. 1, t. 1, pp. 465–66.

32. *CCMS,* vol. 1, t. 1, pp. 539–54.

33. *CCMS,* vol. 1, t. 1, p. 769.

34. *CCMS,* vol. 1, t. 1, p. 246–47: "Libri, qui utiliores et magis necessarii videntur, quo notantur ordine, essent corrigendi."

35. *CCMS,* vol. 1, t. 1, pp. 235, 237–38: "M.F. Alfonsi Ciaconis Sententia de expurgandis denuo aliquod libris catholicorum, qui cum multa utilitate expurgati legi possent vel quorum auctores suspecti sunt. . . . Opusculum de conservanda bona valetudine, quod Schola Salernitana medicorum ad quondam Anglorum regem transmisit, carmine confectum, ut facilius memoria retineri posset, liberum est ab omni errore ego enim puer legi, et usque adhuc mente teneo, verum fuerunt qui in ipsum commentaria ediderunt, quae revidenda essent, si quid habeant, quod purgationem admittat, est enim liber in medicina utilis."

36. *CCMS,* vol. 1, t. 1, p. 774: "De conservanda bona valetudine opusculum Scholae Salernitanae, non deberet interdici; nihil enim mali habet, sed essent interdicenda commentaria in opusculum de bona valetudine conservanda Scholae Salernitanae."

37. If this expurgation exists in the Roman archives, it is included only anonymously. *CCMS,* vol. 1, t. 1, p. 772.

38. *CCMS,* vol. 1, t. 3, pp. 1983–2050.

39. *CCMS,* vol. 1, t. 1, p. 783: "nel trattato delle visioni quae fiunt in somnis dice mille impertinenze."

40. *CCMS,* vol. 1, t. 1, pp. 787–88: "At Medici considerant somnia solum ad agnoscendos humores peccantes, non autem ad futurorum eventus certò predicendos."

41. *CCMS,* vol. 1, t. 1, p. 789: "Non ergo est ad adiuvandam artem medicam, ut fronte pollicetur, sed ad divinationem."

42. *CCMS,* vol. 1, t. 1, p. 786.

43. *CCMS,* vol. 1, t. 1, p. 790: "Ergo malè inscribitur de phisicis ligaturis, cum incantationes, et admirationes, et superstitiosas contineat ligaturas."

44. *CCMS,* vol. 1, t. 1, p. 785.

45. *CCMS,* vol. 1, t. 1, p. 787: "Cum manifeste faveant auguria, auspicia, sortilegia, incantationes, adiurationes, et magicas ligaturas, aperiant viam superstitioni, passim redoleant fatalem necessitatem, et confinia iudiciariae."

46. *CCMS,* vol. 1, t. 1, p. 787: "Qui ferè omnes iudicio meo potius essent expungendi, quam expurgandi; nam artem medicam non iuvant (ut profitetur Arnaldus) sed commiculant, et inficiunt, ideo a peritis Medicis nostri temporis praesertime à catholicis flocipendenteur."

47. *CCMS,* vol. 1, t. 1, p. 790: "Ibidem sequitur tractatus de Iudicijs Astronomiae ad artem medicam prorsus impertinens."

48. *CCMS*, vol. 1, t. 1, p. 791.

49. *CCMS*, vol. 1, t. 1, p. 793.

50. *CCMS*, vol. 1, t. 1, p. 793: "Erasmi auctoris primae classis, cuius de rebus sacris ioci, viam Lutero ad suas dissemiandas haereses straverunt: Unde vulgo Germania dicebatur Erasmus innuit, Luterus irruit: Erasmus parit ova, Luterus excludit pullos."

51. BCRa, Fondo Manoscritti, Mob. 3. 1 B, n. 3, c 319v.

52. Arnould Visser, "Irreverent Reading: Martin Luther as Annotator of Erasmus," *Sixteenth Century Journal* 48, no. 1 (2017): 87. Candida Carella points to Possevino's use of this line in "Antonio Possevino e la biblioteca 'selecta' del principe Cristiano," in *Bibliothecae selectae: Da Cusano a Leopardi*, ed. Eugenio Canone (Florence: L. S. Olschki, 1993), 509n17.

53. *CCMS*, vol. 1, t. 1, pp. 795–96: "Delerem etiam nomen 'Villanovani' in principio epistolae nuncupatoriae positum; tum ut piorum conscientiam consulatur, qui nullam in libri titulo Villanovani mentionem factam legerunt: tum quia revera, licet complura in hoc commentario, ex Villanovano commentarijs in scholam hanc, sint excerpta, tamen // ita mutatus est stilus, tam multa sunt audita, et pleraque secus enarrata, ut alterius quam Villanovani dici possit. Quamobrem tollerem etiam vocem 'antiqui' in libri titulo, cum Arnaldi antiqui non sit, sed iunioris cuiusdam auctoris, et forte Curionis: cuius esse evidens indicium possunt, recentes auctores, quos adducit, cuiusmodi sunt Erasmus, Fuchsius, Manardus, et alij." On Curione's heresy, see Lucio Biasiori, *L'eresia di un umanista: Celio Secondo Curione nell'Europa del Cinquecento*, ed. Studi Storici Carocci (Rome: Carocci Editore, 2015), 250.

54. See Grafton, *Culture of Correction*.

55. On the medieval past as a site of political negotiation in this period, see Ditchfield, *Liturgy, Sanctity and History*; Katrina Olds, *Forging the Past: Invented Histories in Counter-Reformation Spain* (New Haven, CT: Yale University Press, 2015); and Katherine van Liere, Simon Ditchfield, and Howard Louthan, eds., *Sacred History: Uses of the Christian Past in the Renaissance World* (Oxford: Oxford University Press, 2012).

56. José Pardo Tomás's work on the censorship of Fuchs focuses on the seventeenth century, the period after the publication of Fuchs's collected works. See Pardo Tomás, *Ciencia y censura*.

57. *CCMS*, vol. 1, t. 2, p. 1482; and *ILI*, 8:80n146.

58. *Antwerp 1571*, 48: "In IANI Cornarij commentariis, *parum est, quod offendat*."

59. *Antwerp 1571*, 48. The passage is in Janus Cornarius, *Universae rei medicae epigrafe seu enumeratio* (Basel: Froben, 1534), 40.

60. *Antwerp 1571*, 48: "An verò aliquid de religione habuerint, non meminimus nam visitatae nondum sunt."

61. See for example, Janus Cornarius, *Claudii Galeni Pergameni De compositione pharmacorum localium libri decem* (n.p., 1549), 2–3, 732. For the marginal notation, see *Index librorum prohibitorum et expurgatorum* (Madrid: Alfonsum Gomezium Regium Typographum, 1584; hereafter cited as *Spain 1584*), f. 149r: "*Neque id Ianus Cornarius reijcit*, addatur in margine. *Hoc vanum, & superstitiosum est*."

62. ACDF, Index XXIII, f. 12v.

63. A number of the proposed expurgations of Gessner's *Historiae animalium* (*Histories of Animals*) include line-by-line references to the names of heretics through-

out the text that need to be expunged. This work was laborious and never formalized in the 1607 Index but clearly reflected a preoccupation of Italian censors. See for example, *CCMS*, vol. 1, t. 2, pp. 1695–1730.

64. *Rome 1607*, 599, "Prefatio in Platonis opera . . . post interpretis mortem nihil habet, quod ad Religionis nostrae negocium pertineat"; "Eglog Iani Cornarii in Platonis Dialogos, nihil ad Religionem faciens immixtum habent"; "Prefatio Iani Cornarii in libros Constantini Caesaris de Agricultura . . . nihil Religionis, aut offendiculi habet"; "Prefatio . . . ad Sinesii Syrenaei . . . nihil mali habet."

65. José María López Piñero, *La traducción por Juan de Jarava de Leonhart Fuchs y la terminología botánica castellana del siglo XVI* (Valencia: Instituto de Estudios Documentales e Históricos sobre la Ciencia, 1994), 35–43.

66. *CCMS*, vol. 1, t. 2, p. 1638: "Leonardo fussio proibito nella prima classe già anticam[en]te, con tutto ciò si è trovato quasi in ogni parte permesso, come molti ne sono stati apresentati con sottos[critio]ne de Inq[uisito]ri et pare vi sia una espurgat[io] ne quale io non ho."

67. *CCMS*, vol. 1, t. 2, p. 1642.

68. *Rome 1607*, 649.

69. For more substantial expurgations of the prefatory letter, see Anonymous in *CCMS*, vol. 1, t. 2, p. 1635; Felice Pranzini in *CCMS*, vol. 1, t. 2, 1640–41; and the Neapolitan censors in *CCMS*, vol. 1, t. 2, p. 1653.

70. Anonymous expurgations from between 1574 and 1590 with specific and repeated references and descriptions of insults to Catholics, in *CCMS*, vol. 1, t. 2, pp. 1631–33.

71. *CCMS*, vol. 1, t. 2, p. 1661: "Censuit nihilominus congregatio ipsa mature tota praedicta dedicatoria considerate cum praecipue nil ad medicam facultatem pertineat, nihilque afferat utilitatis in totum esse [. . .] atque tollenda."

72. *CCMS*, vol. 1, t. 2, p. 1662: "eum debere toli, eo max[im]e quod hibil continet ad Artem medicine spectans."

73. *Antwerp 1571*, 50: "nam quod Deo Medicinae & stirpium inventionem assignandam dicat, non existimatur reprehensioni expositum esse."*Rome 1607*, 651: "ad religionem pertinens, nihil habent."

74. Anthony Grafton, *The Footnote: A Curious History*, rev. ed. (Cambridge, MA: Harvard University Press, 1997); and Blair, *Too Much To Know*.

75. The literature on the social construction of medical and natural historical knowledge is vast. For a start, see Findlen, *Possessing Nature*; Ogilvie, *Science of Describing*; Siraisi, *Communities of Learned Experience*; and Grafton, "Philological and Artisanal Knowledge Making."

76. *CCMS*, vol. 1, t. 2, p. 1731: "Est enim auctor diligentissimus et multa scripsit utilia quae facillime expurgari possunt. Nam libri de animalibus nihil continent offensivum praeter citationes auctorum haereticorum."

77. *CCMS*, vol. 1, t. 2, p. 1695: "De Gesnero jo ho visto fino al foglio 217 exclusive." Gessner's work is *Historiae animalium Liber III. Qui est de Avium natura* (Zurich: apud Christoph. Froschoverum, 1555).

78. *CCMS*, vol. 1, t. 3, p. 2168.

79. On Paracelsus's name, see Charles Webster, *Paracelsus: Medicine, Magic and Mission at the End of Time* (New Haven, CT: Yale University Press, 2008), 39–43.

80. *Indicis librorum expurgandorum*, 723–74; *Spain 1584*, 192r–v. The anonymous expurgation reproduced in Baldini and Spruit is a copy from the Spanish Index and was likely compiled in the preparation of the 1607 Index. *CCMS*, vol. 1, t. 3, p. 2195.

81. Reading licenses will be discussed further in chapter 5.

82. Girolamo Rossi, *De destillatione liber* (Basel, 1585), 171–72.

83. *CCMS*, vol. 1, t. 2, p. 1574. Rossi's expurgation reads, "171.7. 'soni articulati' addendum est 'naturaliter loquendo' propter verba sacramentorum, accedit enim verbum ad elementum, et fit sacramentum."

84. Lyke de Vries and Leen Spruit, "Paracelsus and Roman Censorship: Johannes Faber's 1616 Report in Context," *Intellectual History Review* 28, no. 2 (2018): 225–54, https://doi.org/10.1080/17496977.2017.1361060.

85. Vries and Spruit, "Paracelsus and Roman Censorship," 26. This is their translation.

86. On Grataroli, see Pastore, "Guglielmo Grataroli"; Ian Maclean, "Heterodoxy in Natural Philosophy and Medicine: Pietro Pomponazzi, Guglielmo Gratarolo, Girolamo Cardano," in Brooke and Maclean, *Heterodoxy in Early Modern Science and Religion*, 1–30; and Alessandra Celati, "Heretical Physicians in Sixteenth-Century Italy: The Fortunes of Girolamo Massari, Guglielmo Grataroli, and Teofilo Panarelli," *Societate Si Politica* 12, no. 1 (2018): 11–31.

87. Cardano, *Book of My Life*, 89.

88. *ILI*, 9:818.

89. *CCMS*, vol. 1, t. 2, p. 1892.

90. BCRa, Int. V, f. 507r. Rossi also composed a poem in 1571 about the Christian victory against the Turks. Girolamo Rossi, *Canzone sopra la felicissima vittoria de' christiani contra turchi* (Venice, 1571).

91. *CCMS*, vol. 1, t. 2, p. 1890: "Ianuenses ut plurimum talis sunt figurae et naturae, et plurimi cucullati, quos religiosos vocant, praesertim Bigotti et hypocritae."

92. *CCMS*, vol. 1, t. 2, p. 1890: "Quae cum famam laedant status religiosi, et proximorum, delenda sunt."

93. *CCMS*, vol. 1, t. 2, p. 1891: "Non enim auspicial a christiano homine exercenda sunt."

94. *CCMS*, vol. 1, t. 2, p. 1891: "187. in fine, cum ait 'daemonicos morbos' tollatur, quia Daemoniaci, ut communiter accipiuntur, non fiunt ab aeris qualitate, ut dixi in expurgatione Cardani."

95. The work Rossi expurgated was Girolamo Cardano, *Libelli duo. Unus, de supplemento almanach. Alter, de restitutione temporum et motuum coelestium. Item geeniturae LXVII* [. . .] (Nuremberg, 1543).

96. *CCMS*, vol. 1, t. 2, p. 1358: "quia Daemones cum sint incorporei naturaliter non subijciuntur astris, sed quia humoribus abutuntur in corpore existentibus, id saepe melancholia faciunt. Hinc videmus exorcistas vacuare tales et vomitu, et secessu."

97. *CCMS*, vol. 1, t. 2, p. 1358: "Verùm Daemonem dicimus facile arripere melancholicos, et melancholia abuti, quod humor ille max[im]e omnium aptus est in insaniam et desperationem adducere homines quia max[im]e mala, mali Daemones optant: imo ipsi, qui corporum norunt et humorum dispositionem, invadunt paratos ad huiusmodi morbos, et Dei permissu agitant, et ad [actum] deducunt."

98. *CCMS*, vol. 1, t. 2, p. 1904: "Pag. 70. ʃʃ *Caput pineatum* dele, *Ianuenses usque, et hypocritae.*"

99. *Indicis librorum expurgandorum*, 559: "Pag. 244. lin. 1. dele, *ac cognitione sincera invito Sathana, & membris eius.*"

100. *Indicis librorum expurgandorum*, 559: "Christi Servatoris nostri, dicatur Christi Salvatoris nostri, ut vitentur prophaneae vocum novitates."

101. Guglielmo Grataroli, *Opuscula* [. . .] (Lyon: apud G. Coterium, 1558), 243–44: "ex quibus quos memoria teneo brevissime recensebo, ut iuniores in familia nostra ad imitationem accendam, non quod inde Gloria affectem, cum in sola Christi servatoris nostri cruce ac cognitione syncera (invito sathana & membris eius) glorier."

102. *CCMS*, vol. 1, t. 2, pp. 745–46. Licentiousness surrounding medical images is a separate and related issue. Whether a work was anatomical or titillating lay in part in the eye of the beholder or in the reader's interpretation of censorship laws. Images of sexual organs in anatomical texts had a medical purpose that was not sexual. Anatomical images of genitalia were not automatically licentious, though some readers of Vesalius, especially of copies held in monastic libraries, have obliterated or removed images of genitalia. Dániel Margócsy, Mark Somos, and Stephen N. Joffe, *The Fabrica of Andreas Vesalius: A Worldwide Descriptive Census, Ownership, and Annotations of the 1543 and 1555 Editions* (Leiden, Neth.: Brill, 2018), 121–30.

103. Dov Front has touched on some of this material already: see Dov Front, "The Expurgation of Medical Books in Sixteenth-Century Spain," *Bulletin of the History of Medicine* 75, no. 2 (Summer 2001): 290–96; and Dov Front, "The Expurgation of the Books of Amatus Lusitanus: Censorship and the Bibliography of the Individual Book," *Book Collector* 47, no. 4 (1998): 520–36.

104. *CCMS*, vol. 1, t. 1, p. 785.

105. See Pomata and Siraisi, *Historia*; and Siraisi, *History, Medicine*.

106. *CCMS*, vol. 1, t. 1, p. 786.

107. For a sample of the literature, see Jonathan Regier, "Reading Cardano with the Roman Inquisition: Astrology, Celestial Physics and the Force of Heresy," *Isis* 10, no. 4 (December 2019): 661–79; Valente, *"Correzioni d'autore"*; Ugo Baldini, "Cardano negli archivi dell'inquisizione e dell'indice: Note su una ricerca," *Rivista di storia della filosofia*, no. 4 (1998): 761–66; and Ugo Baldini and Leen Spruit, "Cardano e Aldrovandi nelle lettere del Sant'Uffizio romano all'Inquisitore di Bologna (1571–73), *Bruniana & Campanelliana* 6 (2000): 145–63; and *CCMS*, vol. 1, t. 2, pp. 1033–1472.

108. *CCMS*, vol. 1, t. 2, pp. 1037–39.

109. *CCMS*, vol. 1, t. 2, p. 1171: "Hoc solum advertens libros hos hieronymi Cardani si purgentur, utilissimos futuros cunctis philosophis, medicis, mathematicis, astronomis, architectis, agricolis, nautis, curam rei famliaris gerentibus, et universis demum artificibus." On this expurgation see Valente, *"Correzioni d'autore,"* 447–49.

110. Rossi cites Cardano's *De subtilitate* in Rossi, *De destillatione* (1585), 85, 99, 203, 245.

111. *CCMS*, vol. 1, t. 2, pp. 1370–71: "Quae dixisse volui, ut ostenderem quantum pertinet ad hunc locum, quomodo de Demoniacis loqui Astrologi et Medici possint, sed multum sobriè, ne ad omnes Daemoniacos trahatur, et quod est a Daemone, aliquis putaret esse ab humore; ideo melius est talia expungere, praesertim quia, quod aliquis

a Daemone arripiatur, licet sit melancholicus, non est ab astris, sed a Deo, eiusque permissu."

112. *CCMS*, vol. 1, t. 2, pp. 1396–97.

113. This occurs in only one other place in the Index: the beginning of the entry for the expurgation of Robert Estienne's Bible. *Indicis librorum expurgandorum*, 46.

114. *Indicis librorum expurgandorum*, 559–60: "Lectori salutem. Ut tuae amice lector commoditati consuleremus, excogitavimus rationem qua facilius ea, quae in Cardani operibus expurganda sunt invenires; quandoquidem sua scripta in libros tantum sine ulla capitum partitione digessit. Singulis igitur locorum animadversionibus certas voces ex indice ipsius libri corrigendi petitas adiecimus, quae quidem voces loca, quae ex praescripto huius censurae emendanda sunt, quota pagina reperiantur, quasi digito commonstrabunt. Curavimus autem, ut omnibus impressionibus (quantum fieri potuit) responderent. Vale, & hanc nostram diligentiam boni consule."

115. Cardano himself was so deeply interested in the concept of utility that he wrote a book *De utilitate ex adversis capienda* (*On Gaining Advantage from Misfortunes*; [Rome?], 1561), which brought together a number of his reflections on happiness and misfortune in life.

116. Kathleen M. Crowther, "Sacred Philosophy, Secular Theology: The Mosaic Physics of Levinus Lemnius (1505–1568) and Francisco Vallés (1524–1592)," in *Nature and Scripture in the Abrahamic Religions: Up to 1700*, ed. Jitse M. van der Meer (Leiden, Neth.: Brill, 2008), 407–38.

117. On Vallés in particular, see Concetta Pennuto, "Francisco Vallés' *De Sacra Philosophia: A Medical Reading of the Bible*," in *Lay Readings of the Bible in Early Modern Europe*, ed. Erminia Ardissino and Élise Boillet (Leiden, Neth.: Brill, 2019), 235. More generally, see Ann Blair, "Mosaic Physics and the Search for a Pious Natural Philosophy in the Late Renaissance," *Isis* 91, no. 1 (2000): 50; and Tricia M. Ross, "Sacred Medicine and the Bible: Thomas Bartholin's *On Biblical Diseases* (1672)," *Early Science and Medicine* 24, no. 1 (May 2019): 101–3.

118. *CCMS*, vol. 1, t. 3, p. 2437.

119. *CCMS*, vol. 1, t. 3, p. 2440. "catholicè loquatur."

120. *CCMS*, vol. 1, t. 3, pp. 2439–44.

121. Compare *CCMS*, vol. 1, t. 3, pp. 2442–44; and *Index librorum expurgandorum*, 727–29.

122. *CCMS*, vol. 1, t. 3, p. 2444.

123. These works were *De astrologia* (1554), *De miraculis occultis naturae* (1559), *De habitu et constitutione corporis* (1561), and *Herbarum atque arborum quae in Bibliis passim obviae sunt* (1566). *CCMS*, vol. 1, t. 3, p. 1915. These works appeared in many editions.

124. *CCMS*, vol. 1, t. 1, p. 258: "est enim vir catholicus et pius . . . et postea inter catholicos eius naturae miracula expurganda proponuntur."

125. *CCMS*, vol. 1, t. 3, pp. 1951–52, 1965.

126. *CCMS*, vol. 1, t. 3, p. 1952.

127. *CCMS*, vol. 1, t. 3, p. 1952.

128. *CCMS*, vol. 1, t. 3, pp. 1962, 65.

129. For example, da Asola proposes changing a marginal note from "Deus col-

lapsae Naturae instaurator" to "Deus collapsae Naturae per sacramenta instaurator," and elsewhere reminds readers of the sacrament of the penance and reconciliation, or confession. *CCMS*, vol. 1, t. 3, pp. 1953, 1955, 1966–68.

130. There are many examples; see, for instance, *CCMS*, vol. 1, t. 3, p. 1953.

131. *Indicis librorum expurgandorum*, 652; *CCMS*, vol. 1, t. 3, p. 1918.

132. *CCMS*, vol. 1, t. 3, p. 1919.

133. Levinus Lemnius, *Occulta naturae miracula* (Antwerp, 1564), 373.

134. Grataroli, *Opuscula*, 241: "& receptissimum est carmen illud vulgatum, Optandum est ut sit mens sana in corpore sano. Neque enim divitiarum ulla gratia, aut liberorum aut regnorum, demumque virtutes nec utiles, nec conferentes esse absque sanitate possunt." Thank you to Thomas Wilson for pointing out that this is a slight misquotation of Juvenal 10.356.

135. *CCMS*, vol. 1, t. 2, p. 1892: "tollatur, quia temperentia, patientia, et aliae etiam in aegroto utiles sunt."

136. *CCMS*, vol. 1, t. 1, p. 795: "Post principium, versus hic qui ibi scribitur 'Regula praesbiteri iubet hoc pro lege teneri' ['quod bona sint ova candida, longa, nova.'] delendus est, quia aliquo modo deridere videtur, et ledere famam sacerdotum, qui cum eam rationem qua sunt sacerdotes, medicini non sint, ut de his possint sanitatis gratia consulere, videntur eo versu taxari, quasi palato studeant, et animum habeant in patinis. Reponi autem forsan ita posset. 'Regula sic medici, iubet hoc pro lege teneri.'"

137. Cardano, *De subtilitate*, 682: "Haec igitur summa, quam homini Deus tribuere aut potuit, aut voluit, felicitas est."

138. Cardano *De subtilitate*, 682–83: "Ad hanc habendam sapientiam delectu in studiis opus est. . . . Hinc ad quam voluerimus artem profiteri, ut medicam, aut iurisprudentiam, aut Theologiam transibimus. Atque hic ordo scientiarum."

139. *CCMS*, vol. 1, t. 2, p. 1395.

140. *CCMS*, vol. 1, t. 2, p. 1395: "sed tantum de speculativa, quae tamen non acquiritur per artem medicam, ut de hac exempli gratia loquar, de qua et ipse scribit Cardanus."

141. *CCMS*, vol. 1, t. 2, p. 1395: "Quae volui hic attexere, ut etiam id confirmarem, quod supra dixi, Veram philosophiam, non repugnare fidei."

142. *Expurgatio quorundam librorum medicorum prohibitorum, videlicet* (Turin: Io. Antonium Seghinum, 1610), identified in the list of Index editions as *Turin 1610*.

143. In Naples, the expurgation of law books was such a desirable outcome for booksellers that they offered to finance the project! See Savelli, "Censoring of Law Books," 251.

144. Thomas James, *Index generalis librorum prohibitorum a Pontificiis* (Oxford: William Turner, 1627).

CHAPTER 5

1. The literature that deals with this subject directly is limited. See Hannah Marcus, "Bibliography and Book Bureaucracy: Reading Licenses and the Circulation of Prohibited Books in Counter-Reformation Italy," *Papers of the Bibliographical Society of America* 110, no. 4 (December 2016): 433–57; *CCMS*, vol. 1, t. 3, pp. 2567–2779, and

t. 4, pp. 3084–3100; Ugo Baldini, "Il pubblico della scienza nei permissi di lettura di libri proibiti delle Congregazioni del Sant'Ufficio e Dell'Indice (secolo XVI): Verso una tipologia professionale e disciplinare," in *Censura ecclesiastica e cultura politica in Italia tra Cinquecento e Seicento*, ed. Cristina Stango (Florence: Olschki, 2001), 171–201; Vittorio Frajese, "Le licenze di lettura tra vescovi ed inquisitori: Aspetti della politica dell'indice dopo il 1596," *Società e storia* 86 (October 1999): 766–818; Vittorio Frajese, *Nascita dell'Indice*, 208–20. There is also discussion of reading licenses in Gigliola Fragnito, *La Bibbia al rogo*; and in the scholarship on monastic libraries including M. M. Lebreton and L. Fiorani, *Codices Vaticani Latini: Codices 11266–11326 Inventari di biblioteche religiose italiane alla fine del Cinquecento*, (Rome: Typis Polyglottis Vaticanis, 1985); Rosa Marisa Borraccini Verducci and Roberto Rusconi, *Libri, biblioteche e cultura degli ordini regolari nell'Italia moderna attraverso la documentazione della Congregazione dell'Indice* (Vatican City: Biblioteca Apostolica Vaticana, 2006); Francesco De Luca, *Biblioteche monastiche in Puglia nel Cinquecento* (Lecce, It.: Conte, 1996); and Flavia Bruni, *Erano di molti libri proibiti: Frate Lorenzo Lucchesi e la censura libraria a Lucca alla fine del Cinquecento* (Rome: Marianum, 2009).

2. On the bureaucracy of issuing licenses, see Marcus, "Bibliography and Book Bureaucracy," and on the organization of the Holy Office archive, see Francesco Beretta, "L'archivio della Congregazione del Sant'Ufficio: Bilancio provvisorio della storia e natura dei fondi d'antico regime," in *L'Inquisizione romana: Metodologia delle fonti e storia istituzionale*, ed. Andrea Del Col and Giovanna Paolin (Trieste: Edizioni Università di Trieste, 2000), 119–44.

3. Deborah Shuger's work on Tudor censorship has similarly tried to position the regulation of writing and language alongside regulations of "nonverbal behavior." See Deborah Shuger, *Censorship and Cultural Sensibility: The Regulation of Language in Tudor-Stuart England* (Philadelphia: University of Pennsylvania Press, 2006), 5.

4. The Index of Prohibited Books was not dissolved until Vatican II (1962–65), so the ACDF contains records of reading licenses through the twentieth century.

5. Frajese, "Le licenze di lettura," 770–71, 794–95. This issue arose less often with other kinds of licenses, though by contrast, on February 28, 1641, the Holy Office granted the bishop of Kefalonia and Zakinthos permission to grant his own licenses or even to deputize the process of allowing Jewish physicians to treat Christians in his diocese. ACDF, S.O. st. st. Q 1 h, f. 13v.

6. *OG*, 16:445: "atteso che il dar licenza di leggergli è ridotto a tale strettezza, che S. S.tà la riserba in sè solo."

7. Mayer, *Roman Inquisition*, 143.

8. University of Toronto, Thomas Fisher Rare Book Library, MSS gen 25, item 7.

9. For a few examples, see ACDF, Index V, f. 72v, 121v, 122v.

10. Cardinal Millini to the inquisitor of Florence, 11 October 1613, Archivio Arcivescovile di Firenze (hereafter cited as AAF), Florence, Fondo Sant'Ufficio 22, busta 32, f. 8.

11. Archivio di Stato di Faenza (hereafter cited as ASF), Archivi delle Congregazioni Religiose, Domenicani di San Andrea, X.84, unnumbered folios: "Die 21 Januarii 1632. Prefata licentie legendi il tesoro politico di Filippo Onorii et ut supra copia transcripta fuit ex suo originali exhibito mihi. ab adm R P Inq Generali Romagne Mag'o Fratre Toma de Tabia et verbaliter registrata in p'nti folio die et prout supra. In quor'. Ego

Fr' Hipp. M.a a Sacra Inq.nis Fav.a Not.s." On the archival sources for the inquisition remaining in Faenza, see Adriano Prosperi, "Faenza," in Prosperi, *Dizionario storico dell'Inquisizione*, 2:573–75.

12. Archivio di Stato di Udine (hereafter cited as ASU), Famiglia Caimo, b. 79, fasc. 7. Harvard Countway Library, 1.MX.123. no 4 (1555).

13. *CCMS*, vol. 1, t. 3, p. 2757.

14. Marcus, "Bibliography and Book Bureaucracy," 443–50.

15. On censorship in Naples, see Sabato, *Il sapere che brucia*; and Pasquale Lopez, *Inquisizione, stampa e censura nel Regno di Napoli tra '500 e '600* (Naples: Edizioni del delfino, 1974).

16. ACDF, Index IX, f. 184: "le assicura non intender di servirsene per studio di dottrina perniciosa, o vana, ma solo per quanto concerne la sua professione di Medicina."

17. Simonetta Adorni Braccesi, "Il dissenso religioso nel contesto urbano lucchese della Controriforma," in *Città italiane del '500 tra Riforma e Controriforma* (Lucca, It.: Maria Pacini Fazzi editore, 1988), 231.

18. For a case study of prohibited books in Lucca, see Bruni, *Erano di molti libri proibiti*.

19. Carlo M. Cipolla, *Public Health and the Medical Profession in the Renaissance* (Cambridge: Cambridge University Press, 1976), 118–24.

20. All licenses are in the ACDF. For Aggiunti's license, see Index VIII, f. 218. For Ottonaio's: S.O. st. st. Q 1 b, f. 211r; S.O. st. st. Q 1 c, f. 105v; and S.O. st. st. Q 1 d, f. 213v. S. For Punta's: S.O. st. st. Q 1 b, f. 65v. For Ronconi's: S.O. st. st. Q 1 g.39r.

21. The physician who held the license was Panfilo Topi.

22. Angela Nuovo, "A proposito del carteggio Pinelli-Dupuy," *Bibliotheca: Rivista di studi bibliografici* 2 (2002): 110n70.

23. Paolo Gualdo's *Vita Ioannis Vincentii Pinelli* (Augustae Vindelicorum: [Christophorus Mangus], 1607) is full of references to Pinelli's many supporters within the Catholic hierarchy.

24. Nuccio Ordine, "Caravage et Bruno: Les relations de Gian Vincenzo Pinelli avec Della Rovere, Paolo Gualdo et les frères Del Monte," in *Une Traversée des Savoirs: Mélanges Offerts à Jackie Pigeaud*, ed. Philippe Heuzé, Yves Hersant, and Eric Van Der Schueren (Québec: Presses de L'Université Laval, 2008), 618.

25. Ordine, "Caravage et Bruno," 618.

26. Chapter 7 will have more on the license and the fate of the Pinelli collection at the Biblioteca Ambrosiana.

27. There are many studies of prohibited books in Venice. See for example, P. Grendler, *Roman Inquisition and Venetian Press*; Barbierato, *La rovina di Venetia in materia de' libri prohibiti*; and Barbierato, *Inquisitor in the Hat Shop*.

28. S.O. st. st. Q 1 d, f. 65r. There is also another record of this license in S.O. st. st. Q 1 d, f. 27r.

29. S.O. st. st. Q 1 d, f. 65r: "e se non in tutto in parte che di tutto."

30. Pope Urban VIII in particular took an active role in granting reading licenses. In my database, of the fifty-three total licenses granted by popes, all but nine of them were granted by Urban VIII.

31. Maria Gemma Paviolo, *I testamenti dei cardinali: Carlo Gaudenzio Madruzzo*

(1562–1629) (self-pub., 2015), 49. See also ACDF, S.O. st. st. Q 1 d: fol. 109r, and S.O. st. st. Q 1 h: fol. 122r.

32. On Canevari, see Rodolfo Savelli, "La critica roditrice dei censori," in *Saperi e meraviglie: Tradizione e nuove scienze nella libraria del medico genovese Demetrio Canevari*, ed. L. Malfatto and E. Ferro. (Genoa: Sagep, 2004), 41–62; and George Sarton, "The Strange Fame of Demetrio Canevari, Philosopher and Physician, Genoese Patrician (1559–1625)," *Journal of the History of Medicine and Allied Sciences* 1, no. 3 (July 1946): 398–418.

33. ACDF, S.O. st. st. Q 1 c: fols, 152v–153r: "Nicolai Copernicum De revolutionibus, sublatis Capitibus, in quibus terram moveri secundum antiquos docet."

34. Baldini, "Il pubblico della scienza."

35. Sabina Brevaglieri, "Libri e circolazione della cultura medico-scientifica nella Roma del Seicento: La biblioteca di Johannes Faber," *Mélanges de L'Ecole française de Rome: Italie et Méditerranée* 120, no. 2 (2008): 427.

36. Brevaglieri, "Science, Books, and Censorship," 147–52. On the prohibited books in Faber's library, see G. Miggiano, "Libri prohibiti: Qualche appunto dalle carte di Johannes Faber Lynceus Bambergensis," in *L'organizzazione del sapere: Studi in onore di A. Serrai*, ed. M. T. Biagetti (Milan: S. Bonnard, 2004), 245–73.

37. Silvia De Renzi, "Medical Competence, Anatomy and the Polity in Seventeenth-Century Rome," *Renaissance Studies* 21, no. 4 (2007): 557.

38. BANL, Fondo Faber, tome 413, fol. 344r.

39. The Catholic Church also sought this kind of ecumenical expertise to head the Vatican Library, which was regularly staffed by converts and scholars from Northern Europe.

40. ACDF, S.O. st. st. Q 1 c, f. 170v.

41. Parma, Giovanni Comiti, ACDF, S.O. st. st. Q 1 e, f. 26r, S.O. st. st. Q 1 g, ff. 29r–30r, S.O. st. st. Q 1 g, f. 69v.

42. ACDF, S.O. st. st. Q 1 d, 20v–21r.

43. ACDF, S.O. st. st. Q 1 d, f. 200r.

44. Rossi is famous for, among other things, claiming that Galileo was born illegitimately. See Paolo Galluzzi, "The Sepulchers of Galileo: The 'Living' Remains of a Hero of Science," in *The Cambridge Companion to Galileo*, ed. Peter Machamer (Cambridge: Cambridge University Press, 1998), 427. For Rossi's time in Padua, see Francesca Zen Benetti, "La libreria di Girolamo Fabrici d'Acquapendente," *Quaderni per la storia dell'Università di Padova* 9–10 (1976–77): 165n11.

45. ACDF, S.O. st. st. Q 1 b, f. 56r, S.O. st. st. Q 1 b, ff. 173v–174v, S.O. st. st. Q 1 c, f. 121r, S.O. st. st. Q 1 c, f. 152v, S.O. st. st. Q 1 d, f. 117, Index IX, f. 579r.

46. Zen Benetti, "La libreria di Girolamo Fabrici d'Acquapendente," 165.

47. ACDF, S.O. st. st. Q 1 b: 173v–174v: "Cum conditione ut corrigat eos quorum extat correctione in Indice Expurgatione edito Romae anno 1607 et nomina hereticorum deleat, et in aliis quidquid interlegendum bonis moribus, et veritati Catholice contrarium repererit, deleat, et in scriptis notet, et Santo Officio quamprimi referat."

48. Marco Bizzarini, *Luca Marenzio: The Career of a Musician between the Renaissance and the Counter-Reformation*, trans. James Chater (Burlington, VT: Ashgate, 2003), 230. This is Chater and Bizzarini's translation from the Latin.

49. For Vincenzi: ACDF, Index IX. f. 62, 63. For Regolino: ACDF, S.O. st. st. Q 1 g, f. 36v.

50. ACDF, S.O. st. st. Q 1 c, 238r.

51. ACDF, Index IX, f. 517r. This request is undated, but in it Pisano states that he has held a license since September 2, 1626.

52. ACDF, S.O. st. st. Q 1 e, f. 32v.

53. Calefilippo's request is ACDF, S.O. st. st. Q 1 d, f. 192v. Squillace's can be found in Index IX, f. 554r.

54. Apothecaries in Venice were certainly reading prohibited books. See Sabina Minuzzi, *Sul filo dei segreti: Farmacopea, libri e pratiche terapeutiche a Venesia in età moderna* (Milan: Edizioni unicopli, 2016); Richard Palmer, "Pharmacy in the Republic of Venice in the Sixteenth Century," in *The Medical Renaissance in the Sixteenth Century,* ed. Andrew Wear, R. K. French, and Ian M. Lonie (Cambridge: Cambridge University Press, 1985), 100–117; Filippo de Vivo, "Pharmacies as Centres of Communication in Early Modern Venice," *Renaissance Studies* 21, no. 4 (September 2007): 505–21; Paula Findlen, "Aristotle in the Pharmacy: The Ambitions of Camilla Erculiani in Sixteenth-Century Padua," foreword to *Letters on Natural Philosophy: The Scientific Correspondence of a Sixteenth-Century Pharmacist, with Related Texts,* by Camilla Erculiani, ed. Eleonora Carinci, trans. Hannah Marcus (Toronto: Iter Press, 2020).

55. ACDF, S.O. st. st. Q 1 b, f. 194v. In 1621 he received another similar license, S.O. st. st. Q 1 c, f. 82v.

56. ACDF S.O. st. st. Q 1 c, ff. 217v–218r.

57. ACDF S.O. st. st. Q 1 g.22v, 76v. Ian Maclean points out that the Venetian printer Giovanni Battista Ciotti had this book printed in Frankfurt by Johann Wechel with a false title page in order to avoid prepublication censorship (Maclean, "Mediations of Zabarella," 42).

58. *CCMS,* vol. 1, t. 3, p. 2669.

59. Vincenzo Malacarne, *Delle opere de' medici, e de' cerusici che nacquero, o fiorirono prima del secolo XVI negli stati della real casa di Savoja monumenti* (n.p.: Stamperia Reale, 1786), 57. See also Andrea De Pasquale, "La biblioteca di Orlando Fresia di Moncalvo, medico del Duca di Savoia," *Platano indici,* 18 (2002): 23–40.

60. As quoted in De Pasquale, "La biblioteca," 24.

61. ACDF, S.O. st. st. Q 1 d, f. 206v; S.O. st. st. Q 1 e, ff. 24r–24v: "Giammaria Mazzuchelli, Gli scrittori d'Italia, cioè notizie storiche e critiche intorno alle vite e . . . ," v. 2, 2463.

62. Scotto requested at least three different licenses. The 1633 license is ACDF, S.O. st. st. Q 1 e, f. 31v. The other two can be found at Index IX, f. 103; and S.O. st. st. Q 1 g, f. 14v.

63. On physicians for the Holy Office and their role as experts, see Federico Barbierato, "Il medico e l'inquisitore: Note su medici e perizie mediche nel tribunale del Sant'Uffizio veneziano fra sei e Settecento," in *Paolo Zacchia: Alle origini della medicina legale, 158–1659,* ed. Alessandro Pastore and Giovanni Rossi (Milan: Franco Angeli, 2008), 266–85; and Seitz, *Witchcraft and Inquisition.*

64. On licenses granted to read Bruno, see Leen Spruit, "Roman Reading Permits for the Works of Bruno and Campanella," *Bruniana & Campanelliana* 18, no. 2 (2012): 571–78.

65. Niccolò Toppi, *Biblioteca Napoletana et apparato agli huomini illustri in lettere di Napoli* (Naples, 1678), 90.

66. ACDF, Index IX.164r–166v.

67. ACDF, Index IX.164r–166v. On Bellanti, see Cesare Vasoli, "Bellanti, Lucio," in *DBI*, 7: 597–99.

68. ACDF, S.O. st. st. Q 1 e, 144v, 159v.

69. ACDF, S.O. st. st. Q 1 g, 25r.

70. *Doppio ritratto con Stefano Coli*, by Pietro Paolini. On this portrait, see F. Baldassari, "Per Pietro Paolini: Un importante ritrovamento," *Paragone* 42, no. 26–27 (1991): 73–77; Eva Struhal, "Pittura e poesia a Lucca nel Seicento: Il caso di Pietro Paolini," in *Lucca città d'arte e i suoi archivi*, ed. Max Seidel and Romano Silva (Venice: Marsilio, 2001): 389–404.

71. I agree with Eva Struhal that Stefano Coli is the younger figure, though F. Baldassare has suggested that he is the older man, leaving the younger man unidentified. See Struhal, "Pittura e Poesia"; and F. Baldassare, "Per Pietro Paolini."

72. ACDF, S.O. st. st. Q 1 g, 75r/v.

73. ACDF, S.O. st. st. Q 1 d, 57r/v.

74. ACDF, Index IX, f. 344–45.

75. Coli's assumption that John Caius was prohibited is a classic example of an Italian scholar assuming that Northern European scholars were all Protestants.

76. My data set includes requests from four other physicians from Lucca in the 1620s and 1630s: Sebastiano Pardino, Ludovico Berlingzano, Francesco Maria Fiorentini, and Gherardo Riccarola. None of their licenses include the quantity or range of subjects that Girolamo Coli's requests do.

77. ACDF, S.O. st. st. Q 1 g, f. 75r/v.

78. Rozzo, "Italian Literature on the Index," 194–222. On the censorship of Ariosto, see Jennifer Helm, "Literary Censorship: The Case of the Orlando Furioso," in *Dimensioni e problemi della ricerca storica*, no. 1 (2012): 193–214.

79. Giovanni Sforza, *F. M. Fiorentini ed i suoi contemporanei lucchesi: Saggio di storia letteraria del secolo xvii*, Florence: F. Menozzi, 1879), 328.

80. ACDF, S.O. st. st. Q 1 g.75r/v.

81. Panfilo Topi in Montepulciano took the same steps by including his son Anselmo in his reading licenses. ACDF, S.O. st. st. Q 1 c, f. 174r.

82. With the same profession as his father, Stefano Coli would have been in a strong position to keep the books. The record shows that in May 1631, Sebastiano Cocioli da Canthiano, a philosopher and law professor, was granted permission to keep the works of nine authors that he found in the "library of his father who was a philosopher and physician." ACDF, Index IX, f. 548. Even when the expertise was not the same, a father's authority could still be transferrable.

83. ACDF, S.O. st. st. Q 1 g, f. 75r/v: "Dictionarium trium linguarum Latinae, Grecae, et Hebraicae cuius nomen est addeo deletum, ut nullibi inveniri possit."

84. ACDF, S.O. st. st. Q 1 e, f. 106r–107v: "cuius in' nomen fuit abrasum et praefatione lacerata, et multa loca hactenus correcta." On Costantini, see Vera Lettere, "Costantini, Toldo," in *DBI*, 30:304–6.

85. ACDF, S.O. st. st. Q 1 e, 158v: "Julii Pollucis Onomasticon, hoc est instruc-

tissima rerum et synonimorum Dictionarium, nunc latinitate donatum a . . . [qui e stato stato tagliato il nome, e messoci carta bianca] . . . Basilee apud Roberti Winter, 1541." The title page reads slightly differently than Bulgarini recorded it: "Julii Pollucis Onomasticon, hoc est instructissimum rerum et synonymorum Dictionarium, nunc primum latinitate donatum, Rodolpho Gualthero Tigurino interprete." Julius Pollox, *Onomasticon* (Basel, 1541).

86. BAV, Vat.lat.11286, f. 156r: "Nella prima carta di questo libbro ci e la licentia del P. inquisitore scritta di sua mano del 1587."

87. ACDF, Index IX, f. 579: "settimane a dietro sono stati revisti e corretti dall'inquisitore di Firenze."

88. ACDF, Index IX, f. 90: "Historia Germanica di diversi autori in sei tomi in foglio, corretta dall' Inquisitione di Spagna."

89. Andrea Ottone, "Pastoral Care and Cultural Accuracy: Book Collections of Secular Clergy in Three Southern Italian Dioceses," in *Documenting the Early Modern Book World: Inventories and Catalogues in Manuscript and Print*, ed. Malcolm Walsby and Natasha Constantinidou (Leiden, Neth.: Brill, 2013), 232.

90. Marcus, "Bibliography and Book Bureaucracy."

91. Eamon, *Science and the Secrets of Nature*, 273–78. More recently on secrets, see Elaine Leong and Alisha Michelle Rankin, eds., *Secrets and Knowledge in Medicine and Science, 1500–1800* (Burlington, VT: Ashgate, 2011).

92. For "slippery," see, Bruce T. Moran, *Distilling Knowledge: Alchemy, Chemistry, and the Scientific Revolution* (Cambridge, MA: Harvard University Press, 2005), 209. Tilmann Walter has argued that the same is true of "Antiparacelsianism" (Walter, "New Light on Antiparacelsianism," 701–25). On Paracelsianism in general, see Ole Peter Grell, ed., *Paracelsus: The Man and His Reputation, His Ideas and Their Transformation* (Leiden, Neth.: Brill, 1998); and Webster, *Paracelsus*, 34–36.

93. While these authors all dealt with chemistry and alchemy, they polemically debated all aspects of the subjects. See Owen Hannaway, *The Chemists and the Word: The Didactic Origins of Chemistry* (Baltimore: Johns Hopkins University Press); Bruce T. Moran, *Andreas Libavius and the Transformation of Alchemy: Separating Chemical Cultures with Polemical Fire* (Sagamore Beach, MA: Science History Publications, 2007).

94. On chemical medicine in Italy, see Antonio Clericuzio, "Chemical Medicines in Rome: Pietro Castelli and the Vitriol Debate (1616–1626)," in *Conflicting Duties: Science, Medicine and Religion in Rome 1550–1750*, ed. Maria Pia Donato and Jill Kraye (London: Warburg Institute, 2009), 281–302.

95. ACDF, S.O. st. st. Q 1 e, f. 142r.

96. ACDF, S.O. st. st. Q 1 b, f. 65v.

97. ACDF, S.O. st. st. Q 1 c, f. 171r.

98. ACDF, S.O. st. st. Q 1 b, f. 200r. On Gualterotti, see Paolo Galluzzi, "Motivi paracelsiani nella Toscana di Cosimo II e di Don Antonio dei Medici: Alchemia, medicina 'chimica' e riforma del sapere," in *Scienze, credenze occulte, livelli di cultura* (Florence: L. S. Olschki, 1982), 52–55. Eileen Reeves discusses Gualterotti in *Painting the Heavens: Art and Science in the Age of Galileo* (Princeton, NJ: Princeton University Press, 1997), 51–53, 78–82.

99. For Mino: ACDF, S.O. st. st. Q 1 b, f. 257v; S.O. st. st. Q 1 c, f. 234r; S.O. st. st. Q 1 e, f. 5v; S.O. st. st. Q 1 g, f. 27v. For Condero: S.O. st. st. Q 1 c, f. 227v.

100. ACDF, S.O. st. st. Q 1 f, f. 126r/v.

101. ACDF, S.O. st. st. Q 1 d, f. 113v.

102. For Trivellino: ACDF, S.O. st. st. Q 1 e, ff. 38v–39r. For Fererolo: ACDF, S.O. st. st. Q 1 e, 41r. For Soncino: ACDF, S.O. st. st. Q 1 e, 41v. For Schipano: S.O. st. st. Q 1 d, f. 38r. For Coccapani: S.O. st. st. Q 1 d, f. 115v. For Guarino: S.O. st. st. Q 1 d, f. 303v.

103. For Liceti: ACDF, S.O. st. st. Q 1 f, ff. 126r/v. For Rolando: ACDF, S.O. st. st. Q 1 c, ff. 217v–218r. For Caballo: ACDF, S.O. st. st. Q 1 d, f. 200r.

104. For secretary to the Duke of Savoy: ACDF, S.O. st. st. Q 1 e, f. 76r. For Arconati: ACDF, S.O. st. st. Q 1 g, f. 32r. For Balneo: ACDF, S.O. st. st. Q 1 g, f. 24v.

105. For Ferrari: ACDF, Index IX, f. 153r. For Vitelleschi: ACDF, S.O. st. st. Q 1 g, f. 12v.

106. ACDF, S.O. st. st. Q 1 e, f. 173v.

107. CCMS, vol. 1, t. 1, p. 603.

108. This finding suggests that enterprising Basel printers were well aware of which publications would sell well across professional divides.

109. ACDF, S.O. st. st. Q 1 d, f. 315r.

110. ACDF, S.O. st. st. Q 1 d, 149r.

111. ACDF, S.O. st. st. Q 1 d, 150r/v.

112. ACDF, Index IX, f. 301r.

113. ACDF, S.O. st. st. Q 1 d, f. 50r.

114. For the censorship of legal books, see Savelli, *Censori e giuristi.*

115. ACDF, S.O. st. st. Q 1 d, 247r.

116. CCMS, vol. 1, t. 3, pp. 2758–59.

117. CCMS, vol. 1, t. 4, p. 3086.

118. AAF, SU. 23. b 4.1: "con l'auttorità di communicarla ad un'altra persona a sua elettione."

119. ACDF, S.O. st. st. Q 1 c, f. 237r/v.

120. ACDF, Index IX, f. 339r.

121. Lorenzo Ghiradelli, *Il memorando contagio seguito in Bergamo l'anno 1630* (Bergamo, 1631), 15, 22.

122. Apothecaries also relied on many of these same books. See, for example, the appendix in Minuzzi, *Sul filo dei segreti,* 300–309.

123. ASV, Savi all'Eresia, 160, unnumbered folios. Letter dated 9 March 1559: "di nascosto li vanno prestando di mano in mano." See especially Ugo Rozzo, "Le 'biblioteche proibite' nel Friuli del Cinquecento," in Rozzo, *Biblioteche italiane del Cinquecento,* 1–58. See also Martin, *Venice's Hidden Enemies;* and Barbierato, *La rovina di Venetia in materia de' libri prohibiti.*

124. ACDF, Index IX, f. 152: "desiderando avanzarsi nella sua professione piu sia possibile et arrivassi a conoscere alcune cose tratte diffusamente dall'infrascritti authori." Undated, but from between 1628 and 1632.

125. ACDF, Index IX, f. 370r: "intendendo di servirsene a fine lecito o buono o per salute del parocchia."

126. ACDF, Index IX, f. 217: "per salute de gli infermi." Undated, but from between 1628 and 1632.

127. ACDF, Index IX, f. 61: "ad honor di Dio e benefitio publico." And ACDF, Index IX, f. 17: "per uso della sua professione et per servitio publico."

128. ACDF, Index IX, f. 5. Undated.

129. ACDF, Index IX, f. 271: "per suo gusto et per utile che ne cava in servitio publico per agiustamento di pace, come fa di continuo, non le desiderando ad altro effetto."

130. ACDF Index Diari 9, f. 6r/v: "le Università istesse fanno fede, che nessuno medicato dall'oratore è morto, percio si è risoluto con un Libro che ha fatto di scoprire al mondo la maniera che tiene per beneficio comune e perche per amplificarlo, li sarebbe necessario di leggere in Libri Astrologici che trattano di Medicina, percio supplica la Licenza." Many thanks to Andreea Badea for sharing this citation with me.

131. On Vincenzo Modena, see Maria Iolanda Palazzolo, "Modena, Vincenzo," in *DBI*, 75:200–203.

132. ACDF, S.O. Fac. L.P. 2. 1, Packet 5, "legere ac retinere, sub custodia tamen ne ad aliorum manus perveniant, libros prohibitos de re Medica, Anatomica, Chymica, et Chirurgica. Item grammaticos, rhetoricos, loaticos [sic? read as "logicos"], Philosophicos, Mathematicos, Astronomicos, et Historicos profanos; Exceptis Operibus Dupry per come p[rim]a."

CHAPTER 6

1. John Tedeschi, "The Dispersed Archives of the Roman Inquisition," in *The Prosecution of Heresy: Collected Studies on the Inquisition in Early Modern Italy* (Binghamton, NY: Medieval and Renaissance Texts and Studies, 1991), 23–45. On Italian archives, see also Filippo De Vivo, Andrea Guidi, Alessandro Silvestri, *Archivi e archivisti in Italia tra medioevo ed età moderna* (Rome: Viella, 2015).

2. *L'Apertura degli archivi del Sant'Uffizio romano.*

3. Hannah Marcus, "Expurgated Books as an Archive of Practice," *Archive Journal*, August 2017, http://www.archivejournal.net/essays/expurgated-books-as-an-archive-of-practice/. On the archive as the memory of the Inquisition, see Adriano Prosperi, "'Damnatio memoriae': Nomi e libri in una proposta della Controriforma," in *L'Inquisizione romana: Letture e richerche* (Rome: Edizioni di storia e letteratura, 2003), 385–411.

4. A recent article by Margherita Palumbo that examines the expurgation of Estienne's *Thesaurus Graecae Linguae* also concludes that expurgated copies are each unique—no two censors exacted the same punishment on the copies she consulted. Margherita Palumbo, "*Lexica malvagia et perniciosa*: The Case of Estienne's *Thesaurus graecae linguae*," *Lexicon Philosophicum: International Journal for the History of Texts and Ideas* 3 (2015): 1–22.

5. BAV, RG Medicina IV.3824 (int.1).

6. ACDF, S.O. st. st. Q 1 g, f. 55r–56r: "Di questi libri, e delli retroscritti, so che molti non sono prohibiti per altro che per essere stati composti da autori heretici, e condannati in prima classe, e che non trattando di fede, molti de essi col solo scancellar

il nome dell'autore, si potrebbere senza pecca[to], e pericolo tenere, e leggere; ma io non vorrei rovinar i libri e voglio piu presto non haverli, che haverli mal trattati."

7. See especially Blair, *Too Much to Know*.

8. I am very grateful to Fabiano Zambelli at the Biblioteca Comunale Manfrediana di Faenza (hereafter cited as BCMF) in Faenza for sharing this uncatalogued treasure with me.

9. For Erasmus, see in particular Silvana Seidel Menchi, *Erasmo in Italia*.

10. See the copy of Thomas Erastus, *Disputationum de noua Philippi Paracelsi medicina pars tertia* (Basileae: per Petram Pernam, 1572), held in Biblioteca Antica del Seminario Vescovile di Padova (hereafter cited as BSVP) in Padua, call number: 500. ROSSA.SUP.C.6.-43.3.

11. Janus Cornarius, *Hippocratis coi medicorum longe principis, opera quae ad nos extant omnia* (Basel: Froben, 1546), held in BCMF, call number: CINQ. 004 002 014.

12. For publications from exhibitions featuring expurgated books, see Roger E. Stoddard, *Marks in Books, Illustrated and Explained* (Cambridge, MA: Houghton Library, Harvard University, 1985); Eric Marshall White, *"Heresy and Error": The Ecclesiastical Censorship of Books 1400–1800* (Dallas: Bridwell Library, Perkins School of Theology, Southern Methodist University, 2010); Vittoria Bonani, ed., *Dal torchio alle fiamme: Inquisizione e censura* (Salerno: Biblioteca Provinciale, 2005); and Ada Palmer, *Censorship and Information Control* (Chicago: Swift Impressions, 2018). See also the online exhibit Biblioteca Panizzi, "Libri proibiti: Stampa e censura nel Cinquecento," accessed January 19, 2020, http://panizzi.comune.re.it/Sezione.jsp?idSezione=106. For expurgation in a Protestant context, see Martha W. Driver, *The Image in Print: Book Illustration in Late Medieval England and its Sources* (London: British Library, 2004), 185–214.

13. Savelli, "La biblioteca disciplinata," 856–944, quotation at 869. Savelli also published a shorter version of this article a few years earlier: see Savelli, "La critica roditrice dei censori," 41–76.

14. Silvana Seidel Menchi, "Sette modi di censurare Erasmo," in *La censura libraria nell'Europa del secolo 16.: Convegno internazionale di studi*, ed. Ugo Rozzo (Udine: Forum, 1997), 177–206.

15. Peter Stallybrass has worked to classify the material interventions of censorship in editions of Petrarch. See Peter Stallybrass, "Petrarch and Babylon: Censoring and Uncensoring the Rime, 1559–1651," in Blair and Goeing, *For the Sake of Learning: Essays in Honor of Anthony Grafton*, (Leiden, Neth.: Brill, 2015), 581–601. Francisco Malta Romeiras is in the process of systematically examining how scientific and medical books were expurgated in Portugal to better understand the readership of prohibited books in Portuguese monastic collections.

16. Fragnito, *La Bibbia al rogo*. On burning the Talmud, see Marvin J. Heller, *Printing the Talmud: A History of the Earliest Printed Editions of the Talmud* (Brooklyn: Im Hasefer, 1992), 201–40; Kenneth R. Stow, "The Burning of the Talmud in 1553, in the Light of Sixteenth Century Catholic Attitudes toward the Talmud," *Bibliothèque d'Humanisme et Renaissance* 34, no. 3 (1972): 435–59; and Amnon Raz-Krakotzkin, *The Censor, the Editor, and the Text: The Catholic Church and the Shaping of the Jewish Canon in the Sixteenth Century* (Philadelphia: University of Pennsylvania Press, 2007). I would also like to thank David Stern for generously sharing a draft of the chapter

"The Burning of the Talmud," from his forthcoming cultural biography of the Talmud as a book.

17. F. Vincenzo Castrucci inquisitor of Perugia to Congregation of the Index, 20 March 1597, ACDF, Index III, v. 3, f. 243r.

18. Dalma Frascarelli, "The Salone Sistino: The Iconographic Plan," in *The Vatican Library*, ed. Ambrogio M. Piazzoni et al. (Milan: Jaca Book, 2012), 202, 264.

19. On the layout of the frescoes, see Frascarelli, "Salone Sistino," 206–7. The caption under the fresco recounts, "EX DECRETO CONCILII CONSTANTINUS IMP. LIBROS ARIANORUM COMBURI IUBET" (By decree of the Council of Emperor Constantine the books of the Arians were ordered burned). See Frascarelli, "Salone Sistino," 232.

20. The caption reads, "EX DECRETO CONCILII BASILIUS IMP. CHIROGRAPHA PHOTII, EIUSQ. CONCILIAB. ACTA COMBURI IUSSET [*sic*]" (By decree of the Council of Emperor Basil the writings of Photius and the acts of his synods were ordered burned). See Frascarelli, "Salone Sistino," 236.

21. ACDF, Index III, v. 7, f. 190r: "Che se bene sono state levate alcune parole, tuttavia pero il senso de i Capitoli intieri resta obsceno, come facilmente si vede che gli legge: e percio a giudicio mio non si possino espurgar se non col fuoco."

22. On burning as ritual purification, see Richard Burt, "(Un)Censoring in Detail: The Fetish of Censorship in the Early Modern Past and the Postmodern Present," in *Censorship and Silencing: Practices of Cultural Regulation*, ed. Robert C. Post (Los Angeles: Getty Research Institute for the History of Art and the Humanities, 1998), 17, 36.

23. Bishop of Cagli to Congregation of the Index, 7 September 1597, ACDF, Index III, v. 3, f. 79r: "Mi fu commesso molti giorni sono da cotesta sacra Cong.ne ch'io insieme con la mia Cong.ne volessi rivedere et censurare la Christeide di Gio. Maria Velmatio, et questo per alcuni errori, ch'erano stati notati da noi nel vedere questo Poema: nel quale si sono trovati tanti, e tanti errori, et tante lascivie, che non so qual libro profano possi piu dissolutamente parlare; oltre molti errori che consistono ne la fede Catholica et siamo tutti concorsi in un parere che non solo si debba abbrusciare questo libro, ma che si dovesse anco abbrusciare l'auttore se fosse vivo."

24. ACDF, Index XVIII, f. 79r. By contrast, there were other books that ecclesiastical officials so badly wanted to burn that they used remarkable means to do so. In 1598 the inquisitor of Novarra was so surprised that no one had turned over a copy of Bodin's *De Republica* to him or the bishop that he took three copies of it from the inquisitor in Milan so that he could burn it. ACDF, Index III, v. 2, f. 46r.

25. ACDF, Index III, v. 5, f. 17r: "et per dar conto in parte della diligenza fin hora usata nelli *Paradossi della Medicina* del Fuchsio, il [. . .] quel ch'asserisce impiamente contra i Cattolici nell'Epistola dedicatoria nel primo libro a cap. 29 v'e un luogo della scrittura nel terzo capo dell' Epistola ad Romanos esposto a simili, quasi per passaggio nel qual' applica il suo perverso sentimento luterano et heretico, che gia e in poter mio, quali luoghi se bene sono stati scancellati . . . ma crederei fussi bene abbrugiarlo." The passage in Romans to which De Rubeis refers is likely the section on righteousness through faith, Romans 3:21–26. However, this passage is not in book 1, chapter 29 of Fuchs's work.

26. ACDF, S.O. st. st. Q 1 d, f. 158v: "qui e stato stato tagliato il nome, e messoci carta bianca."

27. BCMF, Otto Brunfels, *Herbarum vivae eicones* (1532), call number: 004 003 013.

28. See, for example, the copy of Leonhart Fuchs, *Institutionum medicinae* [. . .] (Venice: Ex Officina Ersamiana, Vincentii Valgrisi, 1556) at the Library of the New York Academy of Medicine (hereafter cited as NYAM) in New York City.

29. BVSP, Erastus, *Disputationum de nova philippi paracelsi medicina pars tertia. In qua dilucida et solida verae medicinae assertio, & falsae, vel Paracelsica Confutatio continentur . . .* , 1572, call number: 500.ROSSA.SUP.C.6.-43.3.

30. Copy of *Historia plantarum* printed by Melchiore Sessa in Venice, 1541. Held at BCMF, CINQ. 002 002 018, f. A4r.

31. For an example of scraping off from the binding, see Biblioteca Universitaria di Padova (hereafter cited as BUP), Fuchs, 89.a.184, and for cutting the name out of the binding, see BUP, Fuchs, 96.b.46.

32. Stallybrass, "Petrarch and Babylon." For an intriguing parallel to the professionalism of censors, see Laurie Nussdorfer, *Brokers of Public Trust: Notaries in Early Modern Rome* (Baltimore: Johns Hopkins University Press, 2009).

33. The copies are held in NYAM, Leonhart Fuchs, *De humani corporis fabrica* (1551); and BANL, Leonhart Fuchs, *De historia stirpium* (1542), call number: 139=I.8=.

34. For an example in a medical text see BSVP, Leonhart Fuchs, *Paradoxorum medicinae* (Venice: apud haeredes Petri Ravani et socios, 1547), call number: 500. ROSSA.SUP.APP.-5.2.—8.a/b.

35. A complete list of libraries and call numbers follows. BUP: Leonhart Fuchs, *De historia stirpium* (1542), call number 99.a.7; Leonhart Fuchs, *De humani corporis fabrica* (1551), call number 93.a.129; Leonhart Fuchs, *Paradoxorum medicinae* (1555), call number 88.a.199; Leonhart Fuchs, *Plantarum et stirpium icones* (1595), call number 93.a.136. BAV: Leonhart Fuchs, *De curandi ratione libri octo* (1548), call number RG Medic. V.1406; Amatus Lusitanus, *Curationum medicinalium . . .* (1620), call number Stamp. Barberini M.VIII.72; Leonhart Fuchs, *De sanandis totius humani corporis . . .* (1543), call number RG Medic. V. 1984; Leonhart Fuchs, *De curandi ratione libri octo* (1548), call number Stamp. Barberini N.VI.133; Leonhart Fuchs, *Plantarum eefigies* [sic] (1551), call number Stamp. Chigi VI.1603. BNCR, Theophrastus Paracelsus, *Chirurgia Magna* (1573), call number 12.26.H.26. NYAM: Leonhart Fuchs, commentary on Galen, *De sanitate tuenda libri sex* (Tubingen: Apud Ulricum Morhardum, 1541). Department of Special Collections, Stanford University Libraries (hereafter cited as SUL): Conrad Gessner, *Historiae animalium* (1551), call number QL41 .G37 1551 F V.1.

36. Leonhart Fuchs, *De humani corporis fabrica* (Lugduni: apud Ioannem Frellonium, 1551). Copy held at BUP, call number: 93.a.129.

37. I am grateful to Pietro Gnan, who facilitated this process for me at the Biblioteca Universitaria di Padova.

38. Lodovico Ariosto, *Orlando Furioso*, ed. Lanfranco Caretti (Turin: Einaudi, 1992), 75.

39. Lodovico Ariosto, *Orlando Furioso, The Frenzy of Orlando: A Romantic Epic*, ed. and trans. Barbara Reynolds (New York: Penguin, 1975), 1:178.

40. Francesco Petrarca and Robert M. Durling, *Petrarch's Lyric Poems: The Rime Sparse and Other Lyrics* (Cambridge, MA: Harvard University Press, 1976), 372–73.

41. ASV, SU, b. 156, f. 38r: "Int: se cendono li Galeni con la epistola de Slapner

[Slasner?]? Res: Io l'ho cassata con la sbiaca per non guastar il volume, et se piace cosi a V. P. Io faro incolar sopra anche una carta sopra la Epistola. Altrimente non li vendere."

42. See Grafton, *Culture of Correction*, 107;and Sarah Werner, "Correcting with Cancel Slips," *The Collation: Research and Exploration at the Folger* (blog), Folger Shakespeare Library, April 14, 2015, https://collation.folger.edu/2015/04/correcting-with -cancel-slips/.

43. ASV, SU, b. 39, f. 173r.

44. As cited in Rozzo, "Italian Literature on the Index," 201.

45. BNCR, Conrad Gessner, *Historie animalium* (1551), β[1]r: "Henrichus Petri Candido Lectori."

46. Biblioteca Roberto Caracciolo (hereafter cited as BRC), Lecce, It., Desiderius Erasmus, *D. Hieronymi Operum* (Basileae: Froben, 1565), call number: G-4-IV-12. Tome 1, page 201 shown here.

47. Stallybrass, "Petrarch and Babylon."

48. See, for example, the NYAM copy of Fuchs's *De historia stirpium* (1542).

49. Hippocrates, *Hippocratis medicorum omnium principis . . .* (Basel: Isingrin, 1537), Biblioteca Nazionale Centrale di Roma (hereafter cited as BNCR), 55. 7.G.7.

50. On using manuscript marginalia to understand readers' engagement with texts, see William H. Sherman, *Used Books: Marking Readers in Renaissance England* (Philadelphia: University of Pennsylvania Press, 2008).

51. Exodus 32:32–33. Douay-Rheims translation: "strike me out of the book thou hast written."

52. On violent persecution as a form of worship, see Christine Caldwell Ames, *Righteous Persecution* (Philadelphia: University of Pennsylvania Press, 2009), and also her article, "Does Inquisition Belong to Religious History?" *American Historical Review* 110, no. 1 (2005): 11–37.

53. For recent work on Roman practices of *damnatio memoriae*, see especially Eric R. Varner, *Mutilation and Transformation: Damnatio Memoriae and Roman Imperial Portraiture* (Leiden, Neth.: Brill, 2004); and Harriet I. Flower, *The Art of Forgetting: Disgrace and Oblivion in Roman Political Culture* (Chapel Hill: University of North Carolina Press, 2006). Flower is careful to remind us that Roman memory sanctions have an important precedent in Greek culture.

54. Most scholarship cites Friedrich Vittinghoff, *Der Staatsfeind in der römischen Kaiserzeit: Untersuchungen zur "damnatio memoriae"* (Berlin: Juncker und Dünnhaupt, 1936), 64–74. P. Stewart dates the expression to a 1689 publication by Christoph Schreiter and J. H. Gerlach, *Dissertationem juridicam de damnatione memoriae* (Leipzig: Christophori Fleischeri, 1689), 184n.3.

55. Fernando Martins de Mascarenhas, *Index auctorum damnatae memoriae* (Lisbon, 1624), identified in the list of Index editions as *Lisbon 1624*.

56. *Nouus index librorum prohibitorum et expurgatorum* (Seville: ex typographaeo Francisci de Lyra, 1632; hereafter cited as *Spain 1632*.)

57. *Spain 1632*, 19.

58. Ettore Capecelatro, *Selectiorum consultationum iuris in variis, ac frequentioribus facti contingentiis* (Naples: typis Iacobi Gaffari, 1643), e.g., 232.

59. There have also been several recent studies of practices of *damnatio memoriae*

in the late antique, medieval, and early modern period. See in particular Charles W. Hedrick Jr., *History and Silence: Purge and Rehabilitation of Memory in Late Antiquity* (Austin: University of Texas Press, 2000); Isa Lori Sanfilippo and Antonio Rigoni. *Condannare all'oblio: Pratiche della damnatio memoriae nel medioevo* (Ascoli Piceno, It.: Istituto superiore di studi medievali Cecco d'Ascoli, 2010); Tracy E. Robey, "Damnatio memoriae: The Rebirth of Condemnation of Memory in Renaissance Florence," *Renaissance and Reformation* 36, no. 3 (2013): 5–32.

60. Deut. 25:19, Douay-Rheims translation: "Remember what Amalec did to thee in the way when thou camest out of Egypt: How he met thee: and slew the hindmost of the army, who sat down, being weary, when thou wast spent with hunger and labour, and he feared not God. Therefore when the Lord thy God shall give thee rest, and shall have subdued all the nations round about in the land which he hath promised thee: thou shalt blot out his name from under heaven. See thou forget it not." I would like to thank Tali Winkler and Yair Mintzker for bringing my attention to this passage.

61. In the copy of Leonhart Fuchs, *Institutionum medicinae libri quinque* (1605), Biblioteca Civica Romolo Spezioli Fermo (hereafter cited as BCF), call number: 1P2/7647. This original license is dated July 24, 1610. I do not have a reference to this license from the ACDF.

62. See Michel Foucault, "What Is an Author," in *The Foucault Reader*, ed. Paul Rabinow (New York: Pantheon Books, 1984); and Roger Chartier's response, "Foucault's Chiasmus: Authorship between Science and Literature in the Seventeenth and Eighteenth Centuries," in *Scientific Authorship: Credit and Intellectual Property in Science*, ed. Mario Biagioli and Peter Galison (New York: Routledge, 2003), 13–31.

63. BAV, Leonhart Fuchs, *Plantarum eefigies* [sic] (1551), call number: Stamp.Chigi. VI.1603.

64. Romedio Schmitz-Esser, "The Cursed and the Holy Body: Burning Corpses in the Middle Ages," *Journal of Medieval and Early Modern Studies* 45, no. 1 (January 2015): 131–57. See also Dyan Elliott, "Violence against the Dead: The Negative Translation and *damnatio memoriae* in the Middle Ages," *Speculum* 92, no. 4 (October 2017): 1020–1055.

65. ACDF, Index XVIII, f. 209r. Charles Hedrick makes a similar argument in *History and Silence*, 93–94.

66. Seidel Menchi, "Sette modi di censurare Erasmo," 190.

67. Roger Chartier has described how agencies of control have sought to "reduce this diversity by postulating the existence of a work that remains the same regardless of its form." Roger Chartier, *Inscription and Erasure: Literature and Written Culture from the Eleventh to the Eighteenth Century*, trans. Arthur Goldhammer (Philadelphia: University of Pennsylvania Press, 2007), ix–x.

CHAPTER 7

1. John Churchill and Awnsham Churchill, eds., *A Collection of voyages and travels* [. . .] (London, 1732), 6:575.

2. For an exploration of similar issues in the Jesuit libraries of Perugia, see Natale Vacalebre, *Come le armadure e l'armi: Per una storia delle antiche biblioteche della*

Compagnia di Gesù; Con il caso di Perugia (Florence: Leo S. Olschki editore, 2016); and Christopher Black, "Perugia and Post-Tridentine Church Reform," *Journal of Ecclesiastical History* 35 (1984): 429–451. As an introduction to the extensive literature on scientific collecting in this period, see Krzysztof Pomian, *Collectors and Curiosities: Paris and Venice, 1500–1800*, trans. Elizabeth Wiles Portier (London: Polity, 1990); Giuseppe Olmi, *L'Inventario del mondo: Catalogazione della natura e luoghi del sapere nella prima età moderna* (Bologna: Il Mulino, 1992); Paula Findlen, "Anatomy Theaters, Botanical Gardens, and Natural History Collections," in Park and Daston, *Early Modern Science*, 283–89; and Findlen, *Possessing Nature*.

3. Adriano Prosperi, "Presentazione," in *La Biblioteca antica dell'Osservatorio Ximeniano, Catalogo*, ed. Mauro Guerrini (Florence: Regione Toscana, 1994). With thanks to Adriano Prosperi for finding me an image of this incredible stove in the monthly magazine of the Salzburg Museum; see Salzburg Museum, "Der Spottofen im SMCA," *Kunstwerk des*, February 1989. On anxieties about creating a universal library, see also Roger Chartier, *The Order of Books*, translated by Lydia Cochrane (Cambridge: Polity Press, 1994), 61–88.

4. Luigi Balsamo, *Bibliography: History of a Tradition*, trans. William A. Pettas (Berkeley, CA: Bernard M. Rosenthal, 1990), 5–6, 36. On the *bibliotheca* in ancient Rome as a place, a piece of furniture, and also a book collection, see Stephanie Frampton, "What to Do with Books in the *De finibus*," *Transactions of the American Philological Society* 146, no 1 (Spring 2016): 117–47.

5. Rozzo, *Biblioteche italiane del Cinquecento*, esp. chapter 1, "Le 'Biblioteche proibite" nel Friuli del Cinquecento."

6. On the problems with drawing a strict dichotomy between *Bibliotheca universalis* and *Bibliotheca selecta*, see Helmut Zedelmaier, *Bibliotheca universalis und bibliotheca selecta: Das Problem der Ordnung des gelehrten Wissens in der frühen Neuzeit* (Cologne: Böhlau, 1992); and Ugo Rozzo, "Biblioteche e censura: Da Conrad Gesner a Gabriel Naudé" *Bibliotheca: Rivista di studi bibliografici* 2 (2003): 33–72. On Kircher navigating the difficult terrain of universal and prohibited knowledge, see Daniel Stolzenberg, "Utility, Edification, and Superstition," 336–54.

7. This field is developing rapidly thanks to the Andrew W. Mellon Fellowship in Critical Bibliography at Rare Book School. For an example, see Michael J. Suarez, *The Reach of Bibliography*, Lyell Lectures in Bibliography at Oxford University (Oxford: Oxford University Press, 2017).

8. Pamela M. Jones, *Federico Borromeo and the Ambrosiana: Art, Patronage, and Reform in Seventeenth-Century Milan* (Cambridge: Cambridge University Press, 1993), 44.

9. As quoted in Pamela Jones, *Federico Borromeo and the Ambrosiana*, 44.

10. BA, G. 310. inf. 21. References to books "tolto dall'inquisitore" can be found on folios 1v, 2v, and 9r/v.

11. BA, G. 310. inf. 21.

12. On the acquisition of Pinelli's collection by the Ambrosiana, see Anthony Hobson, "A Sale by Candle in 1608," *Library* 26, no. 3 (1971): 215–33; Roberta Ferro, "Per la storia del Fondo Pinelli all'Ambrosiana: Notizie dalle lettere di Paolo Gualdo," in *Tra i fondi dell'Ambrosiana: Manoscritti italiani antichi e moderni*, ed. Marco Ballani et

al. (Milan: Cisalpino, 2008); Angela Nuovo, "Dispersione di una biblioteca privata: La biblioteca di Gian Vincenzo Pinelli dall'agosto 1601 all'ottobre 1604," in *Biblioteche private in età moderna e contemporanea*, ed. Angela Nuovo (Milan: Ed. Bonnard, 2005), 43–54; and Massimo Rodella, "Fortuna e sfortuna della biblioteca di Gian Vincenzo Pinelli: La vendita a Federico Borromeo," *Biblioteca. Revista di studi bibliografici* 2 (2003): 87–125.

13. For the contents of Pinelli's collection, see M. Grendler, "Book Collecting in Counter-Reformation Italy"; and Nuovo, "Creation and Dispersal of the Library of Gian Vincenzo Pinelli."

14. The letter draft is undated, but the other items in the volume in which it appears are all dated 1605. BA, G. 194, f. 119r. Several of the following letters are also transcribed in Cesare Pasini, "Il progetto biblioteconomico di Federico," in *La Biblioteca Ambrosiana: Tra Roma, Milano e l'Europa*, ed. Franco Buzzi and Roberta Ferro (Rome: Bulzoni, 2005), 260–70.

15. BA, G. 194, f. 119r/v: "Io ne vorrei un'breve, overo due brevi, per quiete della mia conscienza, la prima gratia, e che nella libraria, che si va facendo, si possino comprare, et tenere insieme con li altri libri, et non in loco distinto dagl'altri, ne rinchiuso, i libri, che sono prohibiti, over che si prohibitaranno nell'indice romano, overo altri indici, con questa nota, et conditione, donec expurgent[ur], et il medesimo sia de i sospesi, et non affatto prohibiti."

16. On debates about the censorship of Lull, recall the discussion in chapter 4 and see also *CCMS*, vol. 1, t. 3, pp. 1983–2050.

17. BA, G. 194, f. 119r/v and 119ar. Antonio Seneca wrote to Federico Borromeo from Rome December 10, 1605, to say that it would not be difficult to have Olgiati's first requests granted but that the works of Lull might still be a problem. BA, G., f. 194 bis infer., f. 196r/v.

18. The decree registers of the Holy Office in Rome record regular updates about the status of this arrest and trial, though none that I have found indicate the initial cause of his arrest. I will cite them individually as necessary, but they can be found in the following volumes: ACDF, S.O. Decreta 1619–20, f. 92r, 202v, 341r, 359v–360r, 356r, 359r, 388r, 411v, 418r, 441v, 443v; ACDF, S.O. Decreta 1620, f. 225; ACDF, S.O. Decreta 1621–22, 1v, 18r, 24v, 28r, 39r, 49r, 71r, 78r, 187v, 246r, 250v, 262v.

19. ACDF, S.O. Decreta 1619–20, f. 441: "Caroli Josephi Origoni de Mediolano carceri in hoc Santo Officio ob retensionem scriptorum hereticorum et p'nsam prolationem heretum proposita causa & Illustrissimi Dr' auditis notis' decreverunt, ut torqueatur pro ulteriori veritae; super usie, complicibus, et Intentione et si nihil supervenias abiuret de vehementi, damnetur ad carcerem per tempus arbitrio Sac. Congregationis, deinde relegetur in Urbe, ubi observetur, et se p'ntet in Santo Officio, multetur in scutis 500 applicari extinctioni debitorum Santi Officii Mediolani, cuius Inquisitione certioret, an muleta excedat vires eius facultatum."

20. ACDF, S.O. Decreta 1619–20, f. 441: "super usie, complicibus, et Inten[tion]e et si nihil supervenieas abiuret de vehementi."

21. On decree registers as a source, see Mayer, *Roman Inquisition: Papal Bureaucracy*, 26–37.

22. BA, G. 254 inf., f. 238r: "In questo mio ritorno da Roma ho commeso all'Olgiato

Bibliotecario della libreria Ambrosiana di giustificarsi di certa imputatione, che si e in-
tesa da piu bande, gli vien data da quel Carlo Gioseffo Origoni, d'haver lasciato ricavar
alcune cose, che si stimano prohibite, da libro di questa Biblioteca."

23. BA, G. 254 inf., f. 239r: "Quei manuscritti gli havevo fatti io di mia mano, parte
composti da me, et parte cavati da altri . . . Quell'oratione del Agrippa de Incertitudini,
io la cavai da un libro dell'Agrippa, che era nella libraria del Sig.r Cardinale ch'e corretto, quale e un tomo separato, che non tratta di Maggia . . . Quel trattato che comincia
chiromantia satir . . . credo di haverlo cavato da un libro di Girolamo Cardano, e non mi
raccordo se tal libro fosse nella libraria del Signor Cardinale o in altro luogho."

24. BA, G. 254 inf., f. 240r: "Percio io come Prefetto di essa, ho giudicato necessario
di mostrare, che della parte della libraria et de suoi Ministri, non e stato commesso
alcun fallo colpevole."

25. James Hankins, *Repertorium Brunianum: A Critical Guide to the Writings of
Leonardo Bruni* (Rome: Istituto storico italiano per il medioevo, 1997–), vol. 1, index.

26. The text the oration was with is the *Scriptores historiae augustae*, describing
the lives of six emperors, which we now know to have been an ancient forgery. Thank
you to Anthony Grafton for help identifying this text.

27. BA, G. 254 inf., f. 240r: "Et io confesso, che la detta oratione si ritrova congionta
alle vite de gl'imperatori scritte da Lampridio, et da certi altri con le note di Battista
Egnatio stampate in Venetia appresso Aldo Manutio. Ma aggiongo ancora, non ritrovarsi
libraria o publica o privata, che non habbia, over che non possa tenere lecitamente
questo libro."

28. BA, G. 254 inf., f. 240v: "Poiche diligentemente, et rigorosamente e stato corretto nella congregatione dell'indice, con l'intervento di molti eccellenti Theologi, et
dell'istesso P[ad]re Inquisitore."

29. BA, G. 254 inf., f. 241r.

30. I was unable to confirm that these books were expurgated because the titles
are among the forty thousand volumes from the Ambrosiana that burned in the World
War II bombings of August 15–16, 1943. See Olgiati's to-do list in BA, ms. Z.142 bis sup.,
f. 14 r/v, "Libri corretti dalla congregatione." Cesare Pasini dates this as 1604–5, but it is
in several distinct inks, so Olgiati clearly added to it over a period of time.

31. BA, G. 254 inf., f. 240v: "Le quali regole sunt in viridi observantia, perche Molineo, Fuchsio, Mustero, Zwingero, Roberto Stefano, et altri sono posti all'indice nella
p[rim]a classe, e nondimeno sono statti emendati, et le loro emendationi sono registrate
nell'Espurgatorio Romano stampato l'anno 1608." Olgiati here mentions the Expurgatory Index printed in Rome in 1608. This Index was printed in 1607, and we should
assume that Olgiati made a mistake.

32. See, for example, Roland Barthes, "From Word to Text," in *Textual Strategies:
Perspectives in Poststructural Criticism*, ed. J. V. Harari (Ithaca, NY: Cornell University
Press, 1979); Stanley Fish, *Is There a Text in This Class? The Authority of Interpre-
tive Communities* (Cambridge, MA: Harvard University Press, 1980); Roger Chartier,
"Texts, Printings, Readings," in *The New Cultural History*, ed. Lynn Hunt (Berkeley:
University of California Press, 1989), 154–75. In the context of the inquisition, see espe-
cially Ginzburg, *Cheese and Worms*.

33. BA, G. 254 inf., f. 246r: "Questi miei Illustrissimi Signori Cardinali Colleghi

sono restati a pieno sodisfatti di detta Libraria, e della diligenza de' suoi Ministri, e non credono che da essa possino seguire simili inconvenienti, stante in particolare il singolar' zelo, e prudenza di Vostra Signoria Illustrissima."

34. BA, G. 254 inf., f. 247r: "essendo restati sodisfatti della relatione data dal sudetto Bibliotecario."

35. BA, G. 254 inf., f. 248r, f. 274–75, f. 320r: "Non e stato possibile spuntare il Molineo, per essere apunto quell'opera che principalmente accese Clemente a farne quella severa prohibitione, che ne anco Cardinali lo potessero tenere." Cardinals were usually permitted to read prohibited books without seeking permission. On the censorship of Dumoulin, see Savelli, *Censori e giuristi.*

36. See Luigi Balsamo, "How to Doctor a Bibliography: Antonio Possevino's Practice," in Fragnito, *Church, Censorship, and Culture in Early Modern Italy,* 50–78.

37. Antonio Manfredi, ed., *Le origini della Biblioteca Vaticana tra umanesimo e Rinascimento, 1447–1534,* vol. 1 of *Storia della Biblioteca Apostolica Vaticana* (Vatican City: Biblioteca Apostolica Vaticana, 2010).

38. See Romeo De Maio, "La Biblioteca Vaticana nell'età della Controriforma," in *Riforme e miti nella Chiesa del Cinquecento* (Naples: Guida, 1973), 313–63; Massimo Ceresa, ed., *La Biblioteca Vaticana tra riforma cattolica, crescita delle collezioni e nuovo edificio (1535–1590),* vol. 2 of *Storia della Biblioteca Apostolica Vaticana* (Vatican City: Biblioteca Apostolica Vaticana, 2012); Claudia Montuschi, ed., *La Vaticana nel Seicento (1590–1700),* vol. 3 of *Storia della Biblioteca Apostolica Vaticana* (Vatican City: Biblioteca Apostolica Vaticana, 2014).

39. Anthony Grafton, *Rome Reborn: The Vatican Library and Renaissance Culture* (Washington, DC: Library of Congress, 1993), 45.

40. The Vatican Library was not only the repository of all Catholic knowledge; it was also a reference tool for inquisitors and censors who needed to make rulings on the acceptability of many texts that were prohibited elsewhere. ACDF, Index I (Diarii), v. 1, f. 173.

41. Jill Bepler, "Vicissitudo Temporum: Some Sidelights on Book Collecting in the Thirty Years' War," *Sixteenth Century Journal* 32, no. 4 (Winter 2001): 955–57. See also Fabien Montcher, "Early Modern Bibliopolitics," *Pacific Coast Philology* 52, no. 2 (2017): 206–18; and Prosperi, "'Damnatio memoriae,'" 397.

42. Leone Allacci, *Relazione sul trasporto della Bibliteca Palatina da Heidelberg a Roma* (Florence: Tipografia dei Fratelli Bencini, 1882), 18–19: "Nel cammino per inanzi era pericoloso, questo era l'istesso pericolo, poichè non eravamo sicuri, nè fuori di campagna, nè dentro l'habitato; nella campagna dai ladri e soldati, che tutta la scorrevano, nell'habitato per la gente del paese, che quando vedevano forastieri, se conoscevano la loro *(sic)* non facendo distintione di persone, tutti gli ammazzavano."

43. Allaci, *Relazione sul trasporto della Biblioteca Palatina,* 21.

44. Allaci, *Relazione sul trasporto della Biblioteca Palatina,* 21. Elmar Mittler's catalog to the Palatine exhibition of 1986 suggests that the following could be references to some of this material that Allacci's *Relazione* mentions in passing: Cod. Pal. Lat 1824–27, 1959 and in the university library, for example, Cod. Pal. Germ. 40: A sermon of Luther's on bringing children to school. Elmar Mittler, *Bibliotheca Palatina:*

Catalogue to the Exhibition from July 8th to November 2nd 1986 in the Heiliggeist Church, Heidelberg (Heidelberg: Edition Braus, 1986), 49.

45. Rozzo, "Biblioteche e censura," 66.

46. Mittler, *Bibliotheca Palatina*, 48–49.

47. Leonhart Fuchs, *Institutionum medicinae* [. . .] *libri quinque* (Basel: per Paulum Quercum, 1566), 23: "Medicina vitiatam corporis humani constitutionem emendat, corrigit, instaurat, atque adeo sanitatem facit, eamque iam praesentem conservat." BAV, Stamp. Pal. 912. V. 996.

48. The library of the Holy Office was transferred to the Vatican Library. Some of the books are still part of the Sant'Uffizio collection, but most of the books related to medicine or science were integrated into other collections in the larger Raccolta Generale series. The library of the Congregation of the Index had been held in the offices of the Secretary of the Congregation of the Index. Since the secretary was always a Dominican, the collection was therefore held at Santa Maria Sopra Minerva. A decree dated August 21, 1775, dictated that the library would be moved to the Biblioteca Casanatense.

49. Many thanks to Daniel Stolzenberg for pointing me to this list. BAV, Barb. Lat. 3131. The latest book publication date on the list is 1632, so the list was compiled sometime thereafter and makes it possible that the list was compiled by either Francesco or Antonio Barberini. It seems unlikely that it was compiled by Lucas Holstenius as suggested by Peter Rietbergen, *Power and Religion in Baroque Rome: Barberini Cultural Policies* (Leiden, Neth.: Brill, 2006), 205–31.

50. BAV, Barb. Lat. 3131, unpaginated flyleaf: "+ Libri qui nihil habent haereticum, et hoc tempore per omnes officinas librarias Romae impune venduntur."

51. BAV, Barb. Lat. 3131, unpaginated flyleaf: "++ SS. Patrum scripta ab haereticis edita, quae praeter interpretis aut editoris nomen nihil mali continent."

52. BAV, Barb. Lat. 3131, unpaginated flyleaf: "// Auctores antiqui profani, vel scripta recentium auctorum de rebus et materiis nihil ad fidem pertinentibus."

53. BAV, Barb. Lat. 3131, unpaginated flyleaf: "/ Libri haeretici, vel suspensi, ab historiam, vel eruditionem vel aliam noticiam tamen utiles."

54. BAV, Barb. Lat. 3131, f. 26r/v.

55. BAV, Barb. Lat. 3131, f. 36r.

56. BAV, Barb. Lat. 3131, f. 53v–54v.

57. On the Biblioteca Marciana, see Marino Zorzi, *La libreria di San Marco: Libri, lettori, società nella Venezia dei Dogi* (Milan: A. Mondadori, 1987).

58. Petrarch's donation took place in 1362. On Aldus Manutius's contribution to print and humanism, see Martin Lowry, *The World of Aldus Manutius: Business and Scholarship in Renaissance Venice* (Oxford: Blackwell, 1979).

59. Lotte Labowsky, *Bessarion's Library and the Biblioteca Marciana: Six Early Inventories* (Rome: Edizione di storia e letterature, 1979).

60. Caius published this volume with Froben in Basel in 1544. On Caius, see Vivian Nutton, *John Caius and the Manuscripts of Galen* (Cambridge: Cambridge Philological Society, 1987); and Grafton, "Philological and Artisanal Knowledge Making," 39–55.

61. Giacopo Morelli's history of the library is *Della pubblica libreria di San Marco*

in Venezia: Dissertazione storica di D. Jacopo Morelli sacerdote veneziano (Venice: presso Ant. Zatta, 1774), 64–65. Morelli's manuscript additions to the work can be found in the copy at the Biblioteca Nazionale Marciana (hereafter cited as BNM), Riservati 71, *Dissertazione della Pubblica Libraria di San Marco*, Jacopo Morelli, 1774.

62. Morelli, *Della pubblica libreria di San Marco*, 64–65; Sabba, *La Bibliotheca universalis*, 83.

63. ASV, Riformatori dello Studio di Padova, 420, Scritture di Rectori e pubblici Rappresentanti ed altre persone private dirette ai Riformatori e Senato dal 16-9-1622, unnumbered folios: "Attesto colla presente d'haver veduto tutto poco regolato, et molti libri dell'Illustrissimo Signor Cardinale Bessarion asportati, et levati dalli ladri, dove solevano stare colla soleta catenuccia di ferro per sicurezza ligati, molti de quelli del Secretario Vianello bagnati et guasti, et altri del corrotti et dalle pioggie malissimo trattati." Despite the date on the volume, many of the documents date from before 1622.

64. A copy of this edict can be found in ASV, Procuratori di San Marco, Procuratori "de supra," Chiesa, 68: "Siano etiandio obligati tutti quelli che stamperanno alcun libro, cosi in questa Città come fuori nello Stato Nostro consignar il primo di cadauna sorte de Libri, che stamparanno legato in Bergamina alla Libraria Nostra di San Marco, nè possano principiar a vender quel tal libro se non haveranno una Fede del Bibliotecario di detta Libraria di haverlo consignato. Et la essecutione della presente parte sia specialmente commessa alli Reformatori del Studio sopradetti per l'intiera, et inviolabile sua essecutione." On this decree, known as the *Parte* of 1603, see Angela Nuovo, *The Book Trade in the Italian Renaissance* (Leiden, Neth.: Brill, 2013), 218–19; Marino Zorzi, "La produzione e la circolazione del libro," in *Storia di Venezia dalle origini alla caduta della Serenissima*, vol. 7, ed. Gino Benzoni and Gaetano Cozzi (Rome: Istituto della Enciclopedia Italiana, 1998), 928; and Mario Infelise, "Deposito legale e censura a Venezia (1569–1650)," *La Bibliofilía* 109 (2007): 71–77.

65. For the list of expenses, see ASV, Procuratori di San Marco, Procuratori "de supra," Chiesa, 68, f. 13r. The inventory of the books can be found in the same volume, f. 15r–33r. There is another copy of the inventory of Wieland's books in ASV, Miscellanea di carte non appatenenti ad alcun archivio, 15, fasc. 1. For more on Wieland's books, see G. E. Ferrari, "Le opere a stampa del Guilandino: Per un paragrafo dell'editoria scientifica padovana del pieno Cinquecento," in *Libri e stampatori in Padova* (Padua: Tip. del Seminario, 1959), 377–463; and the forthcoming work to be published from Silvia Pugliese's doctoral thesis "Melchiorre Guilandino, 'Bazzaro Venetoteutonico' alla guida dell'Orto botanico di Padova: Studi su una biblioteca scientifica del Cinquecento" (PhD diss., Università degli Studi di Udine, 2014). There was another major medical donation to the Biblioteca Marciana in 1622, when Girolamo Fabrizio d'Acquapendente donated books and colored images to the library. The donation is found in d'Acquapendente's will dated 1615, but due to a legal battle, it was not carried out until 1622. On the *Lascito Acquapendente*, see Maurizio Rippa Bonati and José Pardo-Tomás, eds., *Il teatro dei corpi: Le pitture colorate d'anatomia di Girolamo Fabrici d'Acquapendente* (Milan: Mediamed, 2004); and Benetti, "La libreria di Girolamo Fabrici d'Acquapendente."

66. ASV, Procuratori di San Marco, Procuratori "de supra," Chiesa, 68, f. 21v–25r.

67. A list of mathematics books can be found in ASV, Procuratori di San Marco, Procuratori "de supra," Chiesa, 68, f. 30v–31v.

68. These books are all in chest 17, labeled *"facolta diverse,"* found on the inventory in ASV, Procuratori di San Marco, Procuratori "de supra," Chiesa, 68, f. 26v–27v.

69. Laurent Pinon notes that the prohibited works that Aldrovandi listed in the 1558 bibliography were not present in the 1583 list. Pinon suggests that the missing titles in the 1583 catalog indicate that Aldrovandi did not have official permission for the works or that he was not to publish that he possessed them; I find the latter argument far more convincing. Pinon, "Clématite bleue," 485n24.

70. For a series of the importation documents that Sozomeno undersigned and released from Venetian customs to circulate in the city, see ASV, Riformatori dello Studio di Padova, 285, Revisione libri che arrivano alla Dogana, 1609–1622.

71. Wipertus Hugues Rudt De Collenberg, "Les "custodi" de la Marciana Giovanni Sozomenos et Giovanni Matteo Bustron," *Miscellanea Marciana* 5 (1990): 9–76.

72. This volume was printed as an octavo in Venice by Roberto Meietti in 1617. On the elegy, see BNM, Riservati 71, *Dissertazione della Pubblica Libraria di San Marco,* Jacopo Morelli, 1774, 93.

73. BNM, Consultazione Catal. Mss. Marc. 1 E, Catalogus Librorum Manuscriptorum ex Legato Revendissimi Cardinalis Bessarionis.

74. On Donzellini, see Palmer, "Physicians and the Inquisition," 118–33; and Celati, "Heresy, Medicine and Paracelsianism," 5–37.

75. BNM, Consultazione Catal Mss. Marc. 1 E, 195. This section of the catalog continues through page 205.

76. BNM, Consultazione Catal Mss. Marc. 1 E, 195–96. Various editions of Gessner's *Biblioteca* were also listed as humanities books.

77. Blair, *Too Much To Know,* 142.

78. These titles are all scattered across the pages under the heading "Diversi in foglio, quarto, octavo, 12 & 16," BNM, Consultazione Catal Mss. Marc. 1 E, 195–204.

79. They also modeled themselves on institutions from antiquity. See for example, Justus Lipsius's *De bibliothecis* of 1602, recently edited and translated by Thomas Hendrickson, *Ancient Libraries and Renaissance Humanism: The "De bibliothecis" of Justus Lipsius* (Leiden, Neth.: Brill, 2017).

80. Marco Navoni, "Gli uomini di Federico Borromeo," in Buzzi and Ferro, *La Biblioteca Ambrosiana,* 296n39.

81. For the original request see ASV, Riformatori dello Studio di Padova, 64, Lettere de gli ecc'mi Riformatori dello studio scritte a diversi Ill'mi Rettori, 1601–1622, unnumbered folios: "Desideriamo havere informationi diligenti del modo et ordini con che sono tenuti i libri nelle librarie del Gran Duca . . . et quanto piu copiosa, distinta, e particolare ci veniva maggiormenti ci riuscira caro."

82. ASV, Riformatori dello Studio di Padova, 65 ([1611–22], Lettere de gli ecc'mi Riformatori dello studio scritte a diversi Illmi Rettori), unnumbered folios: "Ho veduto anche l'indice, che è un libro con nota particolare de tutti i libri, che vi sono, fatto colla debita distintione delle materie, et delle lingue."

83. ASV, Riformatori dello Studio di Padova, 65, unnumbered folios.

84. ASV, Riformatori dello Studio di Padova, 420 (Scritture di Rectori e pubblici Rappresentanti ed altre persone private dirette ai Rifomatori e Senato dal 16-9-1622), unnumbered folios: "Tutta l'administratione della Biblioteca Vaticana e porta in mano di un Cardinale di Santa Chiesa, che per tale Offitio e detto Bibliotecario, dal quale dipendono li Ministri et Offitiali."

85. Virgilio Pinto has suggested that this proliferation of prohibited books in Spanish libraries (which he dates to the eighteenth century) was due to "inquisitorial prodigality in conceding licenses for reading prohibited books" (Pinto, "Censorship," 311).

86. On library architecture and infrastructure, see John Willis Clark, *The Care of Books: An Essay on the Development of Libraries and their Fittings, from the Earliest Times to the End of the Eighteenth Century* (Cambridge: Cambridge University Press, 1902).

87. On Spezioli, see Fabiola Zurlini, *Romolo Spezioli (Fermo, 1642—Roma, 1723): Un medico fermano nel XVII secolo a Roma* (Rome: Vecchiarelli Editore, 2000); Fabiola Zurlini, *Cultura scientifica, formazione e professione medica tra la Marca e Roma nel Seicento: Il caso di Romolo Spezioli* (Fermo: Litografica Com, 2009); Annarita Franza, "Romolo Spezioli, Andrea Vesalio ed il *manuum munus*, 'Il dono delle mani' nella practica medica moderna," in *La formazione del medico in età moderna (secc. XVI–XVIII)*, ed. Roberto Sani and Fabiolo Zurlini (Macerata: Eum, 2012), 139–48; Francesca Coltrini, "Romolo Spezioli (1642–1723) medico, collezionista e committente d'arte fra Roma e Fermo," in Sani and Zurlini, *La formazione del medico in età moderna*, 183–228.

88. Filippo Raffaelli, *La Biblioteca Comunale di Fermo* (Recanati, It.: R. Simboli, 1890), 6–7.

89. On the volumes marked as prohibited in Spezioli's collection, see Zurlini, *Romolo Spezioli*, 147–91.

90. My source here is Romolo Spezioli's catalog of his library. There are several manuscript copies. See Zurlini, *Romolo Spezioli*, 90–102.

91. BCF, ms. Spezioli 1.0, 4 N 3/1. Catalogo primo.

92. See BCF, ms. Spezioli 2.0 4 N 3/2. Catalogo secondo.

93. I would like to thank Paula Findlen for noticing this architectural detail while consulting a microfilm reader at the Corsiniana.

94. Aby Warburg also kept "prohibited" books in his library on a shelf that he called the "poison cabinet." He did not want young people reading these books, but he was also unwilling to ban them outright, stating that people had to fight the devil with his own weapons. Ron Chernow, *The Warburgs: The Twentieth-Century Odyssey of a Remarkable Jewish Family* (New York: Random House, 1993), 124. I am grateful to Alexander Bevilacqua for this citation.

95. Gabriel Naudé, *Advice on Establishing a Library* (Berkeley: University of California Press, 1950). On Naudé as physician, see Siraisi, *History, Medicine*, 127–33.

96. "I think it neither an absurdity nor a danger to have in a library (under the restrictions, nevertheless, of license and permission obtained from the responsible authority) all the works of the most learned and famous heretics, such as Luther, Melancthon, Pomeranus, Bucer, Clavin, Beza, Daneau, Gaulther, Hospinian, Pare, Bullinger, . . ." Naudé, *Advice*, 28.

97. Naudé, *Advice*, 21–22.

98. BUB, Fondo Aldrovandi, Ms. 38 v. I, f. 53v.

99. BNM, Consultazione Catal. Mss. Marc. 1 E, Catalogus Librorum Manuscriptorum ex Legato Reverendissimi Cardinalis Bessarionis, 138.

EPILOGUE

1. See Lindberg, *Beginnings of Western Science*, 149–50; David C. Lindberg, "The
Medieval Church Encounters the Classical Tradition: Saint Augustine, Roger Bacon,
and the Handmaiden Metaphor," in *When Science and Christianity Meet*, ed. David C.
Lindberg and Ronald L. Numbers (Chicago: University of Chicago Press, 2008), 7–32.

2. As quoted in Lindberg, "Medieval Church Encounters Classical Tradition,"
14–15.

3. Nicolaus Copernicus, *De revolutionibus orbium coelestium libri VI* (Nuremberg:
Apud Ioh. Petreium, 1543).

4. Andreas Vesalius, *De humani corporis fabrica* (Basel: Ex Officina Ioannis
Oporini, 1543, mense Iunio).

5. On the early reception of Copernicanism, see Robert S. Westman, "The Melanchthon Circle, Rheticus, and the Wittenberg Interpretation of the Copernican Theory,"
Isis 66, no. 2 (1975): 165–93; and, more recently, Westman, *Copernican Question*.

6. Edward Rosen, "Was Copernicus' *Revolutions* Approved by the Pope?" *Journal of
the History of Ideas* 36, no. 3 (July–September 1975): 531–42.

7. Copernicus himself declared this in his preface to Pope Paul III. See Nicolaus
Copernicus, *On the Revolutions*, trans. Edward Rosen (Baltimore: Johns Hopkins
University Press, 1978), 5–6. On the uses of *De revolutionibus*, see Edoardo Proverbio,
"Francesco Giuntini e l'utilizzo delle tavole copernicane in Italia nel XVI secolo," in
La diffusione del Copernicanesimo in Italia, 1543–1610, ed. Massimo Bucciantini and
Maurizio Torrini (Florence: Olschki, 1997), 37–56.

8. Owen Gingerich, *The Book Nobody Read: Chasing the Revolutions of Nicolaus
Copernicus* (New York: Penguin Books, 2005). Gingerich borrows the phrase from Arthur Koestler, *The Sleepwalkers* (New York: Macmillan, 1968), 191–95.

9. For a detailed account of Galileo's telescope and his early telescopic discoveries,
see Massimo Bucciantini, Michele Camerota, and Franco Giudice, *Galileo's Telescope:
A European Story* (Cambridge, MA: Harvard University Press, 2015).

10. For an introduction to the vast literature on Galileo, see Stillman Drake, *Galileo
at Work* (Chicago: University of Chicago Press, 1978); Mario Biagioli, *Galileo Courtier:
The Practice of Science in an Age of Absolutism* (Chicago: University of Chicago Press,
1993); William Shea and Mariano Artigas, *Galileo in Rome* (Oxford: Oxford University
Press, 2003); John Heilbron, *Galileo* (Oxford: Oxford University Press, 2010); David
Wootton, *Galileo: Watcher of the Skies* (New Haven, CT: Yale University Press, 2010);
Giorgio De Santillana, *The Crime of Galileo* (Chicago: University of Chicago Press,
1955); Michele Camerota, *Galileo Galilei e la cultura scientifica nell'età della Controriforma* (Rome: Salerno, 2004); and Antonio Beltrán Marí, *Talento y poder: Historia de las
relaciones entre Galileo y la Iglesia católica*, 2nd ed. (Pamplona: Laetoli, 2007).

11. See Albert Van Helden, introduction to *Sidereus Nuncius, or the Starry Messenger*, by Galileo Galilei, 2nd. ed. (Chicago: University of Chicago Press, 2016), 1–26.

12. On manuscript publication, see Nick Wilding, "Manuscripts in Motion: The Diffusion of Galilean Copernicanism," *Italian Studies* 66, no. 2 (July 2011): 221–33. On the recent discovery of a copy of Galileo's letter to Castelli, see Michele Camerota, Franco Giudice, and Salvatore Ricciardo, "The Reappearance of Galileo's Original Letter to Benedetto Castelli," *Notes and Records: The Royal Society Journal of the History of Science* 73, no. 1 (2019): 11–28.

13. Maurice Finocchiaro offers a succinct overview in "Science, Religion, and the Historiography of the Galileo Affair: On the Undesirability of Oversimplification," *Osiris* 16 (2001): 120–22. For a more detailed account, see Alfredo Damanti, *Libertas philosophandi: Teologia e filosofia nella lettera alla granduchessa Cristina di Lorena di Galileo Galilei* (Rome: Edizioni di storia e letteratura, 2010).

14. Stefania Tutino, *Empire of Souls: Robert Bellarmine and the Christian Commonwealth* (Oxford: Oxford University Press, 2010), 274–80.

15. On truth, interpretation, and the encounter between Bellarmine, Galileo, and Foscarini, see Richard J. Blackwell, *Galileo, Bellarmine, and the Bible* (Notre Dame, IN: University of Notre Dame Press, 1991); Ugo Baldini, "Bellarmino tra vecchia e nuova scienza," in *Legem impone subactis: Studi su filosofia e scienza dei Gesuiti in Italia, 1540–1632* (Rome: Bulzoni, 1992), 305–31; Ugo Baldini, "L'astronomia del cardinale," in *Legem impone subactis*, 285–303.

16. ACDF, Index I (Diarii), v. 2, f. 89v/90r.

17. Josef Metzler, "Francesco Ingoli, der erste Sekretär der Kongregation (1578–1649)," in *Sacrae Congregationis de Propaganda Fide Memoria Rerum* (Rome: Herder, 1971): 197–243; and G. Pizzorusso, "Ingoli, Francesco," in *DBI*, 62:388–91.

18. Massimo Bucciantini, *Contro Galileo: Alle origini dell'affaire* (Florence: Olschki, 1995), 86n26. For new examples of Ingoli's involvement in the natural philosophical community in Rome, see Sabina Brevaglieri, *Natural desiderio di sapere: Roma barocca fra vecchi e nuovi mondi* (Rome: Viella, 2019), 224–25, 348, 378.

19. This manuscript is titled *De situ et quiete terrae contra copernici systema disputatio* and is published in *OG*, 5:403–12. See also Maurice A. Finocchiaro, *Defending Copernicus and Galileo* (Dordrecht: Springer Netherlands, 2009), 72–76.

20. Camerota, *Galileo Galilei e la cultura scientifica*, 335.

21. ACDF, Index, Diarii v. II, f. 128r: "qualiter liber Copernici iam prohibitus (eo quod falso asserat terram mobilem et coelum immobilem) valde utilis et necessarius ad Astronomiam erat, atque ob id ab omnibus desideratus." On Ingoli's role in the affair and for edited versions of some key documents, see Bucciantini, *Contro Galileo*.

22. On the sections expurgated from Copernicus's *De revolutionibus*, see Maurice Finocchiaro, ed., *The Galileo Affair: A Documentary History* (Berkeley: University of California Press, 1989), 200–202 and 350–51; and Owen Gingerich, "The Censorship of Copernicus's *De Revolutionibus*," *Annali dell'Istituto e museo di storia della scienza di Firenze* 7 (1981): 52–56.

23. Finocchiaro, *Galileo Affair*, 350.

24. ACDF, Index, Diarii v. II, f. 128r.

25. This is Richard Blackwell's translation, published in Blackwell, *Galileo, Bellarmine, and the Bible*, 204, 206. The letter was written on February 16, 1615.

26. On the conflict between Dominican and Jesuit authority and how it played

out in light of the Galileo affair, see Rivka Feldhay, *Galileo and the Church: Political Inquisition or Critical Dialogue?* (Cambridge: Cambridge University Press, 1995).

27. On the status of Grienberger as an intermediary in this dispute, see Mordechai Feingold, "The Grounds for Conflict: Grienberger, Grassi, Galileo, and Posterity," in *The New Science and Jesuit Science: Seventeenth Century Perspectives*, ed. Mordechai Feingold (Dordrecht, Neth.: Kluwer Academic Publishers, 2003), 121–57.

28. Here I am using Stefania Tutino's translation in Tutino, *Empire of Souls*, 277: "Nihilominus, quia in iis multa sunt reipublicae utilissima, unanimi consensu in eam iverunt sententiam: ut Copernici opera . . . permittenda essent . . . iis tamen correctis iuxta subiectam emendationem, locis in quibus non ex hypothesi, sed asserendo de situ, & motu terrae disputat."

29. For the license, see ACDF, S.O. st. st. Q 1 d: 1v–2v. For Balbi's correspondence with Galileo, see *OG*, 13:218–19.

30. ACDF, Index III, v. III, f. 223r.

31. Quotation from Galileo's letter to Dini as reproduced in Blackwell, *Galileo, Bellarmine, and the Bible*, 209.

32. For the letter, see *OG*, 6:509–61; and Finocchiaro's translation in Maurice Finocchiaro, ed., *The Galileo Affair: A Documentary History* (Berkeley: University of California Press, 1989) 154–97. The most extensive treatment of Ingoli and Galileo can be found in Bucciantini, *Contro Galileo*.

33. Finocchiaro, *Galileo Affair*, 156.

34. As quoted in Finocchiaro, *Galileo Affair*, 196.

35. Here I am using Stillman Drake's translation, in Galileo Galilei, *Dialogue Concerning the Two Chief World Systems*, ed. Stephen Jay Gould, trans. Stillman Drake, paperback ed. (New York: Modern Library, 2001), 5. The Italian reads, "Non mancò chi temerariamente asseri quel decreto essere stato parto, non di giudizioso esame, ma di passione troppo poco informata, e si udirono querele, che Consultori totalmente inesperti delle osservazioni Astronomiche no[n] dovevano con proibizione repentina tarpar l'ale a gl'intelletti speculativi."

36. On Galileo's engagement with literary tradition, see Crystal Hall, *Galileo's Reading* (Cambridge: Cambridge University Press, 2013); and Eileen Adair Reeves, *Galileo's Glassworks: The Telescope and the Mirror* (Cambridge, MA: Harvard University Press, 2008).

37. On strategic and encrypted communication in Galileo's writing, see Hannah Marcus and Paula Findlen, "Deciphering Galileo: Communication and Secrecy before and after the Trial," *Renaissance Quarterly* 72 (2019): 953–95.

38. Galileo, *Dialogo*, f. [π4]v. On the role of consultors, see Mayer, *Roman Inquisition*, 17–19.

39. Again, I am following the Stillman Drake translation; see Galileo, *Dialogue*, 4. The Italian reads, "Accettila dunque l'A.V. con la sua solita benignità; e se ci troverrà cosa alcuna, onde gli amatori del vero possan trar frutto di maggior cognizione, e di giovamento; riconoscala, come propria di Se medesima, avvezza tanto a giovare, che però nel suo felice Dominio non ha niuno."

40. On the publication and prohibition of the *Dialogue*, see the published and translated documents in Finocchiaro, *Galileo Affair*. For an introduction into the exten-

sive secondary literature, see Richard S. Westfall, "Patronage and the Publication of the *Dialogue,*" in *Essays on the Trial of Galileo* (Notre Dame, IN: University of Notre Dame Press and Vatican Observatory Publication, 1989), 58–83; and Paula Findlen and Tara Nummedal, "Words of Nature Scientific Books in the 17th Century," in *Thornton and Tully's Scientific Books, Libraries and Collectors,* ed. Andrew Hunter (Brookfield, VT: Ashgate, 2000), 216–57.

41. *OG,* 16:445: "atteso che il dar licenza di leggergli è ridotto a tale strettezza, che S. S.tà la riserba in sè solo."

42. ACDF, S.O. st. st. Q 1 f, f. 126r/v.

43. ACDF, S.O. st. st. Q 1, f. 70r/v: "omnes et quoscumque damnatorum Authorum aut alias quomod.le prohibitos libros etiam in Indice Romano librorum prohibitorum contentos, humaniorum litterarum ac Philosophie, exceptis tamen libris ex professo de Religione et Astrologia Iudiciaria, ac male de Iurisdictione Pontificia quovis modo tractantibus, nec non operibus damnate mem[oriae] Caroli Molinei, Macchiavelli, ac libro Galilei de Galileis de motu terrae, ac stabilitate Coeli."

44. Antonio Favaro, *Documenti inediti per la storia dei manoscritti Galileiani nella Biblioteca Nazionale di Firenze* (Rome, 1886) 115, 117.

45. Galileo Galilei and Giuseppe Toaldo, *Opere* (Padua, 1744).

46. Renée Jennifer Raphael, *Reading Galileo: Scribal Technologies and the Two New Sciences* (Baltimore: Johns Hopkins University Press, 2017).

BIBLIOGRAPHY

LIBRARY AND ARCHIVE ABBREVIATIONS

AAF—Archivio Arcivescovile di Firenze, Florence
ACDF—Archive of the Congregation for the Doctrine of the Faith, Vatican City
ASB—Archivio di Stato di Bologna, Bologna
ASF—Archivio di Stato di Faenza, Faenza
ASU—Archivio di Stato di Udine, Udine
ASV—Archivio di Stato di Venezia, Venice
BAMi—Biblioteca Ambrosiana, Milan
BANL—Biblioteca dell'Accademia Nazionale dei Lincei e Corsiniana, Rome
BAV—Biblioteca Apostolica Vaticana, Vatican City
BCRa—Biblioteca Comunale Classense, Ravenna
BCF—Biblioteca Civica Romolo Spezioli, Fermo, Italy
BCMF—Biblioteca Comunale Manfrediana di Faenza, Faenza, Italy
BMF—Biblioteca Marucelliana, Florence
BNCR—Biblioteca Nazionale Centrale di Roma, Rome
BNM—Biblioteca Nazionale Marciana, Venice
BRC—Biblioteca Roberto Caracciolo, Lecce, Italy
BSVP—Biblioteca Antica del Seminario Vescovile di Padova, Padua
BUB—Biblioteca Universitaria di Bologna, Bologna
BUP—Biblioteca Universitaria di Padova, Padua
NYAM—Library of the New York Academy of Medicine
SUL—Department of Special Collections, Stanford University Libraries

FREQUENTLY CITED REFERENCE WORKS

CCMS—Baldini, Ugo, and Leen Spruit, eds. *Catholic Church and Modern Science: Documents from the Archives of the Roman Congregations of the Holy Office and the Index.* Vol. 1. Rome: Libreria Editrice Vaticana, 2009.
DBI—*Dizionario Biografico degli Italiani.* 94 vols. Rome: Istituto della Enciclopedia Italiana, [1960–].

ILI—Bujanda, Jesús Martínez de. *Index des Livres Interdits.* 12 vols. Sherbrooke, QC: Centre d'études de la Renaissance, Editions de l'Université de Sherbrooke, 1984–96.

OG—Favaro, Antonio, ed. *Edizione nazionale delle opere di Galileo Galilei.* 20 vols. Florence: Giunti Barbèra, 1890–1909.

EDITIONS OF THE INDEX CITED

Antwerp 1571—*Index expurgatorius librorum, qui hoc seculo prodierunt [. . .].* Antwerp, 1571.

Spain 1584—*Index librorum prohibitorum et expurgatorum.* Madrid: Alfonsum Gomezium Regium Typographum, 1584.

Rome 1607—*Indicis librorum expurgandorum in studiosorum gratiam confecti tomus primus.* Rome: Ex typographia R. Cam. Apost., 1607.

Turin 1610—*Expurgatio quorundam librorum medicorum prohibitorum, videlicet.* Turin: Io. Antonium Seghinum, 1610.

Spain 1612—*Index librorum prohibitorum et expurgatorum.* Madrid: Ludovicum Sanchez Typographum Regium, 1612.

Lisbon 1624—*Index auctorum damnatae memoriae.* Lisbon, 1624.

Spain 1632—*Nouus index librorum prohibitorum et expurgatorum.* Seville: Ex typographaeo Francisci de Lyra, 1632.

EARLY MODERN EDITIONS

The following sources are listed alphabetically by author (or title if author unknown) and then chronologically for each author.

Adam, Melchior. *Vitae germanorum medicorum.* Frankfurt: heredum Jonae Rosae, 1620.

Bibliotheca medico-philosophico-philologica inclytae nationis germanae artistarum [. . .]. Padua: Frambotti, 1685.

Bock, Hieronymus. *De stirpium, maxime earum, quae in Germania nostra nascuntur.* Argentorati [Strasbourg]: excudebat Wendelinus Rihelius, 1552.

Brunfels, Otto. *Herbarum vivae eicones.* Argentorati [Strasbourg]: apud Ioannem Schottum, 1532.

Capecelatro, Ettore. *Selectiorum consultationum iuris in variis, ac frequentioribus facti contingentiis.* Naples: typis Iacobi Gaffari, 1643.

Cardano, Girolamo. *Libelli duo. Unus, de supplemento almanach. Alter, de restitutione temporum et motuum coelestium. Item geniturae LXVII* [. . .]. Nuremberg: apud Iohan. Petreium, 1543.

Cardano, Girolamo. *De subtilitate libri XXI.* Lyon: apud Philibertum Rolletium, 1554.

Cardano, Girolamo. *De utilitate ex adversis capienda libri IIII.* Basel: per Henricum Petri, 1561.

Churchill, John, and Awnsham Churchill. *A Collection of Voyages and Travels* [. . .]. 6 vols. London, 1732.

Codronchi, Giovan Battista. *De christiana ac tuta medendi ratione libri duo varia doctrina referti*. Ferrara: apud Benedictum Mammarellum, 1591.

Copernicus, Nicolaus. *De revolutionibus orbium coelestium libri VI*. Nuremberg: Apud Ioh. Petreium, 1543.

Copernicus, Nicolaus. *De revolutionibus orbium coelestium libri VI*. Basel: Ex Officina Henricpetrina, 1566.

Cornarius, Janus. *Universae rei medicae epigrafe seu enumeratio*. Basel: Froben, 1534.

Cornarius, Janus. *Hippocratis coi medicorum longe principis, opera quae ad nos extant omnia*. Basel: Froben, 1546.

Cornarius, Janus. *Claudii Galeni Pergameni de compositione pharmacorum localium libri decem*. Lyon: apud Gulielmum Rovillium,1549.

Epistolae medicinales diversorum authorum. Lyon: Apud haeredes Jacobi Juntae, 1557.

Erasmus, Desiderius. *D. Hieronymi Operum*. Basel: Froben, 1565.

Erastus, Thomas. *Disputationum de noua Philippi Paracelsi medicina pars tertia*. Basel: per Petram Pernam, 1572.

Fracastoro, Girolamo. *Syphilidis, sive Morbi Gallici*. Verona, 1530.

Fuchs, Leonhart. *Libri IIII, difficilium aliquot quaestionum, et hodie passim controversarum explicationes continents*. Basel: Robert Winter, 1540.

Fuchs, Leonhart. Commentary on Galen, *De sanitate tuenda libri sex*. Tubingen: Apud Ulricum Morhardum, 1541.

Fuchs, Leonhart. *De historia stirpium commentarii insignes*. Basel: in Officina Isingriniana, 1542.

Fuchs, Leonhart. *De sanandis totius humani corporis malis, libri V*. Venice: Apud Andream Arrivabenum, 1543.

Fuchs, Leonhart. *Paradoxorum medicinae*. Venice: apud haeredes Petri Ravani et socios, 1547.

Fuchs, Leonhart. *De curandi ratione libri octo*. Lyon: apud Guliel. Rovillium sub scuto Veneto, 1548.

Fuchs, Leonhart. *Methodus seu ratio compendiaria perveniendi ad culmen Medicinae, nunc denuo . . . recognita*. Lyon: Guliel. Rovillium, 1548.

Fuchs, Leonhart. *De humani corporis fabrica . . . .* Lyon: apud Ioannem Frellonium, 1551.

Fuchs, Leonhart. *Plantarum eefigies [sic] . . . quinque diversis linguis redditae*. Lyon: apud Balthazarem Arnoulletum, 1551.

Fuchs, Leonhart. *Institutionum medicinae ad Hippocratis, Galeni, aliorumque veterum scripta recte intelligenda mire utiles libri quinque*. Venice: Ex Officina Ersamiana, Vincentii Valgrisi, 1556. Fuchs, Leonhart. *Institutionum medicinae ad Hippocratis, Galeni, aliorumque veterum scripta recte intelligenda mire utiles libri quinque*. Basel: per Paulum Quercum, 1566.

Fuchs, Leonhart. *Institutionum medicinae libri quinque* [. . .]. Basel: ex officina Oporiniana, per Hieronymum Gemusaeum, 1594.

Fuchs, Leonhart. *Plantarum et stirpium icons*. Lyon, 1595.

Fuchs, Leonhart. *Institutionum medicinae libri quinque* [. . .]. Basel: typis Conradi Waldkirchii, 1605.

Galilei, Galileo. *Dialogo sopra i due massimi sistemi del mondo.* Florence: per Gio. Battista Landini, 1632.

Galilei, Galileo, and Giuseppe Toaldo. *Opere.* 4 vols. Padua, 1744.

Gessner, Conrad. *Conrad Gessner, Bibliotheca Uniuersalis, siue Catalogus omnium scriptorum locupletissimus.* [. . .]. Tiguri [Zurich]: Apud Christophorum Froschouerum, 1545. Zentralbibliothek Zürich, DrM 3, https://doi.org/10.3931/e -rara-16206.

Gessner, Conrad. *Pandectarum sive partitionum universalium . . . Libri XXI.* Zurich: excudebat Christophorus Froschoverus, 1548.

Gessner, Conrad. *Partitiones theologicae, Pandectarum universalium liber ultimus.* Zurich: Christophorus Froschoverus excudebat, 1549.

Gessner, Conrad. *Historiae animalium.* Zurich: apud Chist. Froschouerum, 1551. Copy consulted at Stanford University Library, call number QL41.G37 151 F copy 1.

Gessner, Conrad. *Thesaurus Euonymi Philiatri, De remediis secretis, liber physicus, medicus, et partim etiam chymicus* [. . .]. Zurich, 1552.

Gessner, Conrad. *De raris et admirandis herbis.* Zurich: Andreas Gesner, 1555.

Gessner, Conrad. *Historiae animalium Liber III. Qui est de Avium natura.* Zurich: apud Christoph. Froschoverum, 1555.

Ghiradelli, Lorenzo. *Il memorando contagio seguito in Bergamo l'anno 1630.* Bergamo, 1631.

Ginanni, Pietro Paolo. *Memorie Storico-critiche Degli Scrittori Ravennati.* Faenza: Gioseffantonio Archi, 1769.

Grataroli, Guglielmo. *Opuscula* [. . .] *ab ipso autore denuo correcta et aucta.* Lyon: apud G. Coterium, 1558.

Gualdo, Paolo. *Vita Ioannis Vincentii Pinelli.* Augsburg: Christoph Mangus, 1607.

Hippocrates. *Hippocratis medicorum omnium principis* [. . .]. Basel: Isingrin, 1537.

James, Thomas. *Index generalis librorum prohibitorum a Pontificiis.* Oxford: William Turner, 1627.

Le Coq (Gallus), Paschal. *Bibliotheca medica siue catalogus illorum, qui ex professo artem medicam in hunc usque annum scriptis illustrarunt.* Basel: per Conradum Waldkirch, 1590.

Lemnius, Levinus. *Libelli tres perlegantes ac festivi . . . De astrologia liber unus.* Antwerp: apud Martinum Nutium, 1554.

Lemnius, Levinus. *Occulta naturae miracula.* Antwerp: apud Gulielmum Simonem, 1559.

Lemnius, Levinus. *De habitu et constitutione corporis.* Antwerp: Guglielmum Simonem, 1561.

Lemnius, Levinus. *Occulta naturae miracula.* Antwerp: apud Guilielmum Simonem ad insigne Scuti Basiliensis, 1564.

Lemnius, Levinus. *Herbarum atque arborum quae in Bibliis passim obviae sunt.* Antwerp: Guglielmum Simonem, 1566.

Malacarne, Vincenzo. *Delle opere de' medici, e de' cerusicis che nacquero, o fiorirono prima del secolo XVI negli stati della real casa di Savoja monumenti.* [Turin]: Stamperia Reale, 1786.

Mattioli, Pietro Andrea. *Epistolarum medicinalium libri quinque.* Prague, 1561. Copy held at the Biblioteca Nazionale Marciana, Venice. Call number 18. D. 20.

Mazzuchelli, Giammaria. *Gli scrittori d'Italia cioé notizie storiche, e critiche intorno alle vite, e agli scritti dei litterati italiani.* Brescia, It.: Presso a G. Bossini, 1753.

Morelli, Giacopo. *Della pubblica libreria di San Marco in Venezia: Dissertazione storica di D. Jacopo Morelli sacerdote veneziano.* Venice: presso Ant. Zatta, 1774.

Paracelsus, Theophrastus. *Chirurgia Magna, in duos tomos digesta* [. . .]. 2 vols. Argentorati [= Basel: Petrus Perna], 1573.

Pollox, Julius. *Onomasticon.* Basel: Robertum Winter, 1541.

Rossi, Francesco. *Annotationes in libros octo Cornelii Celsi* [. . .]. Venice: Ioannem Guerilium, 1616.

Rossi, Girolamo. *Canzone sopra la felicissima Vittoria de' Christiani contra Turchi.* Venice, 1571.

Rossi, Girolamo. *De destillatione liber.* Ravenna: ex typographia Francisci Thebaldini, 1582.

Rossi, Girolamo. *De destillatione liber.* Basel: per Sebastianum Henricpetri, 1585.

Rossi, Girolamo. *De destillatione liber.* Venice: apud Ioannem Baptistam Ciottum Senemsem, 1604.

Rossi, Girolamo. *Diputatio de melonibus* [. . .]. Venice: Ioannem Baptistam Bertonum, 1607.

Schreiter, Christoph, and J. H. Gerlach. *Dissertationem juridicum de damnatione memoriae.* Leipzig: Christophori Fleischeri, 1689.

Spach, Israel. *Nomenclator scriptorum medicorum.* Frankfurt: ex typographica Martini Lechleri, impensis Nicolae Bassaei, 1591.

Toppi, Niccolò. *Biblioteca Napoletana et apparato agli huomini illustri in lettere di Napoli.* Naples, 1678.

Vesalius, Andreas. *De humani corporis fabrica.* Basel: Ex Officina Ioannis Oporini, 1543.

MODERN WORKS CITED

Agasse, J. M. "La bibliothèque d'un médecin humaniste: *L'Index librorum* de Girolamo Mercuriale." *Cahiers d'humanisme* 3–4 (2002–3): 201–53.

Agrimi, Jole. *Ingeniosa scientia nature: Studi sulla fisiognomica medievale.* Florence: Sismel, 2002.

Allacci, Leone. *Relazione sul trasporto della Bibliteca Palatina da Heidelberg a Roma.* Florence: Tipografia del Fratelli Bencini, 1882.

Ames, Christine Caldwell. "Does Inquisition Belong to Religious History?" *American Historical Review* 110, no. 1 (2005): 11–37.

Ames, Christine Caldwell. *Righteous Persecution.* Philadelphia: University of Pennsylvania Press, 2009.

Andretta, Elisa. "Dedicare libri di medicina: Medici e potenti nella Roma del XVI secolo." In Romano, *Rome et la science moderne,* 207–55.

Angeli, Giovanni, and Antonino Poppi, eds. *Lettere del Sant'Ufficio di Roma all'Inquisizione di Padova, 1567–1660: Con nuovi documenti sulla carcerazione padovana di Tommaso Campanella in appendice (1594)*. Padua: Centro studi antoniani, 2013.

Angeli, Luigi. *Sulla vita e su gli scritti di alcuni medici imolesi: Memorie storiche*. Imola, It.: per Gianbenedetto Filippini, 1808.

Arcangeli, Alessandro, and Vivian Nutton, eds. *Girolamo Mercuriale: Medicina e cultura nell'Europa del Cinquecento; Atti del convegno "Girolamo Mercuriale e lo spazio scientifico e culturale del Cinquecento" (Forlì, 8–11 novembre 2006)*. Florence: L. S. Olschki, 2008.

Ariosto, Lodovico. *Orlando Furioso*. Edited by Lanfranco Caretti. Turin: Einaudi, 1992.

Ariosto, Lodovico. *Orlando Furioso, The Frenzy of Orlando: A Romantic Epic*. Edited and translated by Barbara Reynolds. 2 vols. New York: Penguin, 1975.

Azzolini, Monica. *The Duke and the Stars: Astrology and Politics in Renaissance Milan*. Cambridge, MA: Harvard University Press, 2013.

Bacchi, Maria Christina. "Ulisse Aldrovandi e i suoi libri." *L'Archiginnasio: Bollettino della Biblioteca Comunale di Bologna* 100 (2005): 255–366.

Bainton, Roland. *Bernardino Ochino, esule e riformatore senese del Cinquecento, 1487–1563*. Florence: Sansoni, 1941.

Baldacci, A. "Contributo alla bibliografia delle opere di Ulisse Aldrovandi." In *Intorno alla vita e alle opere di Ulisse Aldrovandi*, edited by Ludovico Frati, 69–139. Bologna: Beltrami, 1907.

Baldassari, F. "Per Pietro Paolini: Un importante ritrovamento." *Paragone* 42, no. 26–27 (1991): 73–77.

Baldini, Ugo. "Bellarmino tra vecchia e nuova scienza." In *Legem impone subactis: Studi su filosofia e scienza dei Gesuiti in Italia, 1540–1632*, 305–31. Rome: Bulzoni, 1992.

Baldini, Ugo. "Cardano negli archive dell'inquisizione e dell'indice: Note su una ricerca." *Rivista di storia della filosofia*, no. 4 (1998): 761–66.

Baldini, Ugo. "Il pubblico della scienza nei permissi di lettura di libri proibiti delle Congregazioni del Sant'Ufficio e dell'Indice (secolo XVI): Verso una tipologia professionale e disciplinare." In *Censura ecclesiastica e cultura politica in Italia tra Cinquecento e Seicento: Atti del convegno 5 marzo 1999*, edited by Cristina Stango, 171–201. Florence: Olschki, 2001.

Baldini, Ugo. "L'astronomia del Cardinale," in *Legem impone subactis*, 285–303.

Baldini, Ugo. "The Roman Inquisition's Condemnation of Astrology: Reasons and Consequences." In Fragnito, *Church, Censorship, and Culture in Early Modern Italy*, 79–110.

Baldini, Ugo, and Leen Spruit. "Cardano e Aldrovandi nelle lettere del Sant'Uffizio romano all'Inquisitore di Bologna (1571–73)." *Bruniana & Campanelliana* 6 (2000): 145–63.

Ball, Philip. *The Devil's Doctor: Paracelsus and the World of Renaissance Magic and Science*. London: William Heinemann, 2006.

Balsamo, Luigi. *Bibliography: History of a Tradition*. Translated by William A. Pettas. Berkeley, CA: Bernard M. Rosenthal, 1990.

Balsamo, Luigi. "How to Doctor a Bibliography: Antonio Possevino's Practice." In Fragnito, *Church, Censorship, and Culture in Early Modern Italy*, edited by Gigliola 50–78.

Bamji, Alexandra, Geert H. Janssen, and Mary Laven, eds. *The Ashgate Research Companion to the Counter-Reformation*. Farnham, UK: Ashgate, 2013.

Barbierato, Federico. "Il medico e l'inquisitore: Note su medici e perizie mediche nel tribunale del Sant'Uffizio veneziano fra Sei e Settecento." In *Paolo Zacchia: Alle origini della medicina legale, 1584–1659*, edited by Alessandro Pastore and Giovanni Rossi, 266–85. Milan: Franco Angeli, 2008.

Barbierato, Federico. *The Inquisitor in the Hat Shop: Inquisition, Forbidden Books and Unbelief in Early Modern Venice*. Burlington, VT: Ashgate, 2012.

Barbierato, Federico. *La rovina di Venetia in materia de' libri prohibiti: Il libraio Salvatore de' Negri e l'Inquisizione veneziana (1628–1661)*. Venice: Marsilio Editori, 2007.

Barbierato, Federico. *Nella stanza dei circoli: Clavicula Salomonis e libri di magia a Venezia nei secoli XVII e XVIII*. Milan: Edizioni Sylvestre Bonnard, 2002.

Barnes, Robin Bruce. *Astrology and Reformation*. New York: Oxford University Press, 2016.

Barthes, Roland. "From Word to Text." In *Textual Strategies: Perspectives in Poststructural Criticism*, ed. J. V. Harari. Ithaca, NY: Cornell University Press, 1979.

Bartholin, Thomas. *Thomas Bartholin on the Burning of His Library and on Medical Travel*. Translated by Charles Donald O'Malley. Lawrence: University of Kansas Libraries, 1961.

Baudry, Hervé. *Livro medico e censura na primeira modernidade em Portugal*. Lisbon: Cham Ebooks, 2017.

Bauer, Stefan. *The Censorship and Fortuna of Platina's Lives of the Popes in the Sixteenth Century*. Turnhout, Belg.: Brepols, 2006.

Bepler, Jill. "Vicissitudo Temporum: Some Sidelights on Book Collecting in the Thirty Years' War." *Sixteenth Century Journal* 32, no. 4 (Winter 2001): 953–68.

Beretta, Francesco. "L'archivio della Congregazione del Sant'Ufficio: Bilancio provvisorio della storia e natura dei fondi d'antico regime." In *L'Inquisizione romana: Metodologia delle fonti e storia istituzionale*, edited by Andrea Del Col and Giovanna Paolin, 119–44. Trieste, It.: Edizioni Università di Trieste, 2000.

Berns, Andrew D. *The Bible and Natural Philosophy in Renaissance Italy: Jewish and Christian Physicians in Search of Truth*. New York: Cambridge University Press, 2015.

Berns, Andrew D. "The Crypto-Judaism of Amatus Lusitanus." Paper presented at meeting of the Society of Crypto-Judaic Studies, Phoenix, AZ, August 4, 2008.

Bethancourt, Francisco. *The Inquisition: A Global History, 1478–1834*. Translated by Jean Birrel. Cambridge: Cambridge University Press, 2009.

Biagioli, Mario. *Galileo Courtier: The Practice of Science in an Age of Absolutism*. Chicago: University of Chicago Press, 1993.

Biasiori, Lucio. *L'eresia di un umanista: Celio Secondo Curione nell'Europa del Cinquecento*, edited by Studi Storici Carocci. Rome: Carocci Editore, 2015.

Biblioteca Panizzi. "Libri proibiti: Stampa e censura nel Cinquecento." Accessed January 19, 2020. http://panizzi.comune.re.it/Sezione.jsp?idSezione=106. Online exhibition.

Bizzarini, Marco. *Luca Marenzio: The Career of a Musician between the Renaissance and the Counter-Reformation.* Translated by James Chater. Burlington, VT: Ashgate, 2003.

Black, Christopher. *The Italian Inquisition.* New Haven, CT: Yale University Press, 2009.

Black, Christopher. "Perugia and Post-Tridentine Church Reform." *Journal of Ecclesiastical History* 35 (1984): 429–51.

Blackwell, Richard J. *Galileo, Bellarmine, and the Bible.* Notre Dame, IN: University of Notre Dame Press, 1991.

Blair, Ann. "Conrad Gessner et la publicité: Un humaniste au carrefour des voies de circulation du savoir." In *L'Annonce faite au lecteur,* edited by Annie Charon, Sabine Juratic, and Isabelle Pantin, Collection L'Atelier d'Erasme, 21–55. Louvain, Belg.: Presses universitaires de Louvain, 2017.

Blair, Ann. "Conrad Gessner's Paratexts." *Gesnerus* 73, no. 1 (2016): 73–123.

Blair, Ann. "The Dedication Strategies of Conrad Gessner." In Manning and Klestinec, *Professors, Physicians and Practices,* 169–209.

Blair, Ann. "Humanist Methods in Natural Philosophy: The Commonplace Book." *Journal of the History of Ideas* 53, no. 4 (1992): 541–51.

Blair, Ann. "Mosaic Physics and the Search for a Pious Natural Philosophy in the Late Renaissance." *Isis* 91, no. 1 (2000): 32–58.

Blair, Ann. "The Rise of Note-Taking in Early Modern Europe." *Intellectual History Review* 20, no. 3 (September 2010): 303–16.

Blair, Ann. "Textbooks and Methods of Note-Taking in Early Modern Europe." In *Scholarly Knowledge: Textbooks in Early Modern Europe,* edited by Emidio Campi, Simone De Angelis, Anja-Silvia Goeing, and Anthony Grafton, 39–73. Geneva: Droz, 2008.

Blair, Ann. *Too Much to Know: Managing Scholarly Information before the Modern Age.* New Haven, CT: Yale University Press, 2010.

Blair, Ann, and Anja-Silvia Goeing, eds. *For the Sake of Learning: Essays in Honor of Anthony Grafton.* 2 vols. Leiden, Neth.: Brill, 2016.

Bonani, Vittoria, ed. *Dal torchio alle fiamme: Inquisizione e censura; Nuovi contributi della più antica biblioteca provinciale d'Italia; Atti del convegno nazionale di studi, Salerno, 5–6 novembre 2004.* Salerno: Biblioteca Provinciale, 2005.

Bonati, Maurizio Rippa, and José Pardo-Tomás, eds. *Il teatro dei corpi: Le pitture colorate d'anatomia di Girolamo Fabrici d'Acquapendente.* Milan: Mediamed, 2004.

Borraccini Verducci, Rosa Marisa, and Roberto Rusconi. *Libri, biblioteche e cultura degli ordini regolari nell'Italia moderna attraverso la documentazione della Congregazione dell'Indice: Atti del convegno internazionale; Macerata, 30 maggio–1 giugno 2006.* Vatican City: Biblioteca apostolica Vaticana, 2006.

Bots, Hans, and Françoise Waquet. *La république des lettres.* Paris: Belin, 1997.

Bouley, Bradford A. *Pious Postmortems: Anatomy, Sanctity, and the Catholic Church in Early Modern Europe.* Philadelphia: University of Pennsylvania Press, 2017.

Boutroue, Marie-Élisabeth. "'Ne dites plus qu'elle est amarante': Les problèmes de l'identification des plantes et de leurs noms dans la botanique de la Renaissance." *Nouvelle revue du XVIe siècle* 20, no. 1 (2002): 47–64.

Braccesi, Simonetta Adorni. "Il dissenso religioso nel contest urbano lucchese della Controriforma." In *Citta italiane del '500 tra Riforma e Controriforma: Atti del Convegno Internazionale di Studi Lucca, 13–15 ottobre 1983*, 225–39. Lucca, It.: Maria Pacini Fazzi editore, 1988.

Brevaglieri, Sabina. "Libri e circolazione della cultura medico-scientifica nella Roma del Seicento: La biblioteca di Johannes Faber." *Mélanges de L'Ecole française de Rome: Italie et Méditerranée* 120, no. 2 (2008): 425–44.

Brevaglieri, Sabina. *Natural desiderio di sapere: Roma barocca fra vecchi e nuovi mondi*. Rome: Viella, 2019.

Brevaglieri, Sabina. "Science, Books and Censorship in the Academy of the Lincei: Johannes Faber as Cultural Mediator." In Donato and Kraye, *Conflicting Duties*, 109–33.

Briggs, Robin. "The Académie Royale des Sciences and the Pursuit of Utility." *Past & Present* 131, no. 1 (May 1991): 38–88.

Brinkhus, Gerd, and Claudine Pachnicke. *Leonhart Fuchs (1501–1566): Mediziner und Botaniker*. Tübingen, Ger.: Kulturamt, 2001.

Brooke, John, and Ian Maclean, eds. *Heterodoxy in Early Modern Science and Religion*. Oxford: Oxford University Press, 2005.

Bruni, Flavia. *Erano di molti libri proibiti: Frate Lorenzo Lucchesi e la censura libraria a Lucca alla fine del Cinquecento*. Rome: Marianum, 2009.

Bucciantini, Massimo. *Contro Galileo: Alle origini dell'affaire*. Florence: Olschki, 1995.

Bucciantini, Massimo, Michele Camerota, and Franco Giudice. *Galileo's Telescope: A European Story*. Cambridge, MA: Harvard University Press, 2015.

Burmeister, Karl Heinz. *Achilles Pirmin Gasser, 1505–1577: Arzt und Naturforscher, Historiker und Humanist*. Wiesbaden, Ger.: G. Pressler, 1970.

Burt, Richard. "(Un)Censoring in Detail: The Fetish of Censorship in the Early Modern Past and the Postmodern Present." In *Censorship and Silencing: Practices of Cultural Regulation*, edited by Robert C. Post, 17–41. Los Angeles: Getty Research Institute for the History of Art and the Humanities, 1998.

Buzzi, Franco, and Roberta Ferro, eds. *La Biblioteca Ambrosiana: Tra Roma, Milano e l'Europa*. Rome: Bulzoni, 2005.

Bynum, Caroline Walker. *Christian Materiality: An Essay on Religion in Late Medieval Europe*. New York: Zone Books, 2011.

Camerota, Michele. *Galileo Galilei e la cultura scientifica nell'età della Controriforma*. Rome: Salerno, 2004.

Camerota, Michele, Franco Giudice, and Salvatore Ricciardo. "The Reappearance of Galileo's Original Letter to Benedetto Castelli." *Notes and Records: The Royal Society Journal of the History of Science* 73, no. 1 (2019): 11–28.

Caplan, Jay. *Postal Culture in Europe: 1500–1800*. Oxford: Voltaire Foundation, 2016.

Caramuel y Lobkowitz, Juan. *Syntagma de arte typographica*. Edited and translated by Pablo Andrés Escapa. Salamanca: Instituto de Historia del Libro y de la Lectura, 2004.

Caravale, Giorgio. *Forbidden Prayer: Church Censorship and Devotional Literature in Renaissance Italy*. Translated by Peter Dawson. Farnham, UK: Ashgate, 2011.

Cardano, Girolamo. *The Book of My Life (de Vita Propria Liber)*. Translated by Jean Stoner. New York: New York Review of Books, 2002.

Carlino, Andrea. *Books of the Body: Anatomical Ritual and Renaissance Learning*. Translated by John Tedeschi and Anne C. Tedeschi. Chicago: University of Chicago Press, 1999.

Carlsmith, Christopher. "'Cacciò fuori un bastone bianco': Conflicts Between the Ancarano College and the Episcopal Seminary in Bologna." In *The Culture of Violence in Renaissance Italy: Proceedings of the International Conference; Georgetown University at Villa Le Balze, 3–4 May, 2010*, edited by Samuel Kline Cohn and Fabrizio Ricciardelli. Florence: Le lettere, 2012.

Carlsmith, Christopher. "Struggling toward Success: Jesuit Education in Italy, 1540–1600." *History of Education Quarterly* 42, no. 2 (Summer 2002): 215–46.

Carrara, Daniela Mugnai. *La biblioteca di Nicolò Leoniceno: Tra Aristotele e Galeno; Cultura e libri di un medico umanista*. Florence: Leo S. Olschki Editore, 1991.

Carrara, Daniela Mugnai. "Le epistole prefatorie sull'ordine dei libri di Galeno di Giovan Battista da Monte: Esigenze di metodo e dilemmi editoriali." In *Vetustatis indagatur: Scritti offerti a Filippo di Benedetto*, edited by Vicenzo Fera and Agostino Guida, 207–34. Messina: Università degli Studi, 1999.

Carrara, Daniela Mugnai. "Profilo di Nicolò Leoniceno 1428–1524." *Interpres* 2 (1979): 169–212.

Carella, Candida. "Antonio Possevino e la biblioteca 'selecta' del principe cristiano." In *Bibliothecae selectae: Da Cusano a Leopardi*, Lessico Intellettuale Europeo 58, edited by Eugenio Canone, 507–16. Florence: L. S. Olschki, 1993.

Cavarzere, Marco. *La prassi della censura nell'Italia del Seicento: Tra repressione e mediazione*. Rome: Edizioni di storia e letteratura, 2011.

Celati, Alessandra. "*Contra medicos:* Physicians Facing the Inquisition in 16th-Century Venice." *Early Science and Medicine* 23 (July 2018): 72–91.

Celati, Alessandra. "Heresy, Medicine and Paracelsianism in Sixteenth-Century Italy." *Gesnerus* 71, no. 1 (2014): 5–37.

Celati, Alessandra. "Heretical Physicians in Sixteenth-Century Italy: The Fortunes of Girolamo Massari, Guglielmo Grataroli, and Teofilo Panarelli." *Societate si Politica* 12, no. 1 (2018): 11–31.

Celati, Alessandra. "Medici ed eresie nel Cinquecento italiano." PhD diss., Università degli studi di Pisa, 2016.

Ceresa, Massimo, ed. *La Biblioteca Vaticana tra riforma cattolica, crescita delle collezioni e nuovo edificio (1535–1590)*. Vol. 2 of *Storia della Biblioteca Apostolica Vaticana*. Vatican City: Biblioteca Apostolica Vaticana, 2012.

Chartier, Roger. "Foucault's Chiasmus: Authorship between Science and Literature in the Seventeenth and Eighteenth Centuries." In *Scientific Authorship: Credit and Intellectual Property in Science*, edited by Mario Biagioli and Peter Galison, 13–31. New York: Routledge, 2003.

Chartier, Roger. *Inscription and Erasure: Literature and Written Culture from the Elev-*

enth to the Eighteenth Century. Translated by Arthur Goldhammer. Philadelphia: University of Pennsylvania Press, 2007.

Chartier, Roger. *The Order of Books*. Translated by Lydia Cochrane. Cambridge: Polity Press, 1994.

Chartier, Roger. "Texts, Printings, Readings." In *The New Cultural History*, edited by Lynn Hunt, 154–75. Berkeley: University of California Press, 1989.

Chernow, Ron. *The Warburgs: The Twentieth-Century Odyssey of a Remarkable Jewish Family*. New York: Random House, 1993.

Chieppi, Sergio. *I servizi postali dei Medici dal 1500 al 1737*. San Giovanni Valdarno, It.: Servizio editoriale fiesolano, 1997.

Cicco, Giuseppe Gianluca. "La censura e le opere di argomento medico-scientifico." In *Dal torchio alle fiamme: Inquisizione e censura; Nuovi contributi della più antica biblioteca provinciale d'Italia; Atti del Convegno Nazionale di Studi, Salerno, 5–6 novembre 2004*, edited by Vittoria Bonani, 173–94. Salerno: Biblioteca Provinciale, 2005.

Cipolla, Carlo M. *Public Health and the Medical Profession in the Renaissance*. Cambridge: Cambridge University Press, 1976.

Clark, John Willis. *The Care of Books: An Essay on the Development of Libraries and Their Fittings, from the Earliest Times to the End of the Eighteenth Century*. Cambridge: Cambridge University Press, 1902.

Clericuzio, Antonio. "Chemical Medicines in Rome: Pietro Castelli and the Vitriol Debate (1616–1626)." In Donato and Kraye, *Conflicting Duties*, 281–302.

Clutton-Brock, Martin. "Copernicus's Path to His Cosmology: An Attempted Reconstruction." *Journal for the History of Astronomy* 36, no. 2 (May 2005): 197–216.

Cochrane, Eric W. *Historians and Historiography in the Italian Renaissance*. Chicago: University of Chicago Press, 1981.

Coltrini, Francesca. "Romolo Spezioli (1642–1723) medico, collezionista e committente d'arte fra Roma e Fermo." In Sani and Zurlini, *La formazione del medico in età moderna*, 183–228.

Cook, Harold J. "The History of Medicine and the Scientific Revolution." *Isis* 102, no. 1 (2011): 102–8.

Cook, Harold J. "Medicine." In Park and Daston, *Early Modern Science*, 407–34.

Copernicus, Nicolaus. *On the Revolutions*. Translated by Edward Rosen. Baltimore: Johns Hopkins University Press, 1978.

Coryate, Thomas, and George Coryate. *Coryate's Crudities: Hastily Gobled [sic] up in Five Months Travels in France, Savoy, Italy, Rhetia Commonly Called the Grisons Country, Helvetia Alias Switzerland, Some Parts of High Germany and the Netherlands; Newly Digested in the Hungry Aire of Odcombe in the County of Somerset, and Now Dispersed to the Nourishment of the Travelling Members of This Kingdome*. Glasgow: J. MacLehose, 1905.

Council of Trent. "Ten Rules Concerning Prohibited Books Drawn Up by the Fathers Chosen by the Council of Trent and Approved by Pope Pius." In *Canons and Decrees of the Council of Trent*, translated by H. J. Schroeder, 273–78. 1941. Reprint, Rockford, IL: Tan Books, 1978.

Crowther, Kathleen M. "Sacred Philosophy, Secular Theology: The Mosaic Physics of Levinus Lemnius (1505–1568) and Francisco Vallés (1524–1592)." In *Nature and Scripture in the Abrahamic Religions: Up to 1700*, edited by Jitse M. van der Meer, 407–38. Leiden, Neth.: Brill, 2008.

Cuomo, Serafina. *Pappus of Alexandria and the Mathematics of Late Antiquity.* Cambridge: Cambridge University Press, 2000.

Dall'Olio, Guido. *Eretici e inquisitori nella Bologna del Cinquecento.* Bologna: Istituto per la Storia di Bologna, 1999.

Damanti, Alfredo. *Libertas philosophandi: Teologia e filosofia nella Lettera alla granduchessa Cristina di Lorena di Galileo Galilei.* Rome: Edizioni di storia e letteratura, 2010.

Danielson, Dennis Richard. *The First Copernican: Georg Joachim Rheticus and the Rise of the Copernican Revolution.* New York: Walker, 2006.

Darnton, Robert. *Censors at Work: How States Shaped Literature.* New York: W. W. Norton, 2014.

De Collenberg, Wipertus Hugues Rudt. "Les 'custodi' de la Marciana Giovanni Sozomenos et Giovanni Matteo Bustron." *Miscellanea Marciana* 5 (1990): 9–76.

De Feller, Francesco Saverio. *Dizionario storico ossia storia compendiata degli uomini memorabili per ingegno, dottrina, virtù, errori, delitti, dal principio del mondo fino ai nostri giorni.* 10 vols. Venice, 1834.

Delisle, Candice. "The Letter: Private Text or Public Place? The Mattioli-Gesner Controversy about the *aconitum primum.*" *Gesnerus* 61, no. 3–4 (2004): 161–76.

De Luca, Francesco. *Biblioteche monastiche in Puglia nel Cinquecento.* Lecce, It.: Conte, 1996.

De Maio, Romeo. "La Biblioteca Vaticana nell'età della Controriforma." In *Riforme e miti nella Chiesa del Cinquecento*, 313–63. Naples: Guida, 1973.

De Pasquale, Andrea. "La biblioteca di Orlando Fresia di Moncalvo, medico del Duca di Savoia." *Il Platano: Rivista di cultura astigiana* 27 (2002): 23–40.

De Renzi, Silvia. "Medical Competence, Anatomy and the Polity in Seventeenth-Century Rome." *Renaissance Studies* 21, no. 4 (2007): 551–67.

De Santillana, Giorgio. *The Crime of Galileo.* Chicago: University of Chicago Press, 1955.

Deutsches Medizinhistorisches Museum. *Leonhart Fuchs Zum 500. Geburtstag: Philologe, Mediziner, Botaniker—Ein Universalgelehrter an Der Universität Ingolstadt; Ausstellung vom 13. Juli 2001 bis 7. Oktober 2001 im Deutschen Medizinhistorischen Museum Ingolstadt.* Ingolstadt, Ger.: Deutsches Medizinhistorisches Museum, 2001.

De Vivo, Filippo. "Pharmacies as Centres of Communication in Early Modern Venice." *Renaissance Studies* 21, no. 4 (September 2007): 505–21.

De Vivo, Filippo, Andrea Guidi, and Alessandro Silvestri. *Archivi e archivisti in Italia tra Medioevo ed età moderna.* Libri di Viella 203. Rome: Viella, 2015.

De Vries, Lyke, and Leen Spruit. "Paracelsus and Roman Censorship: Johannes Faber's 1616 Report in Context." *Intellectual History Review* 28, no. 2 (2018): 225–54.

De Waardt, Hans. "Justus Velsius Haganus: An Erudite but Rambling Prophet." In *Exile and Religious Identity, 1500–1800*, edited by Jesse Spohnholz and Gary K. Waite, 97–110. London: Pickering and Chatto, 2014.

Ditchfield, Simon. *Liturgy, Sanctity and History in Tridentine Italy: Pietro Maria Campi and the Preservation of the Particular.* Cambridge: Cambridge University Press, 1995.

Donati, Claudio. "A Project of 'Expurgation' by the Congregation of the Index: Treatises on Duelling." In Fragnito, *Church, Censorship and Culture in Early Modern Italy,* 134–62.

Donato, Maria Pia. "Les doutes de l'inquisiteur: Philosophie naturelle, censure et théologie à l'époque moderne." *Annales HSS* 64, no. 1 (2009): 15–43.

Donato, Maria Pia. "Scienze della natura." In Prosperi, Lavenia, and Tedeschi, *Dizionario storico dell'Inquisizione,* 1394–98.

Donato, Maria Pia. "Scienza e teologia nelle congregazioni romane: La questione atomista, 1626–1727." In Romano, *Rome et la science moderne,* 595–634.

Donato, Maria Pia, and Jill Kraye, eds. *Conflicting Duties: Science, Medicine, and Religion in Rome, 1550–1750.* London: Warburg Institute, 2009.

Drake, Stillman. *Galileo at Work.* Chicago: University of Chicago Press, 1978.

Driver, Martha W. *The Image in Print: Book Illustration in Late Medieval England and Its Sources.* London: British Library, 2004.

Durling, Richard J. "Conrad Gesner's Liber amicorum: 1555–1565." *Gesnerus* 22 (1965): 134–59.

Durling, Richard J. "Girolamo Mercuriale's De modo studendi." *Osiris* 6 (1990): 181–95.

Duroselle-Melish, Caroline, and David A. Lines. "Editor's Choice: The Library of Ulisse Aldrovandi (†1605): Acquiring and Organizing Books in Sixteenth-Century Bologna." *Library* 16, no. 2 (2015): 133–61.

Eamon, William. *Science and the Secrets of Nature: Books of Secrets in Medieval and Early Modern Culture.* Princeton, NJ: Princeton University Press, 1994.

Egmond, Florike. "Clusius and Friends: Cultures of Exchange in the Circles of European Naturalists." In *Carolus Clusius: Towards a Cultural History of a Renaissance Naturalist,* edited by Florike Egmond, Paul Hoftijzer, and Robert Visser. Amsterdam: Koninklijke Nederlandse Akademie van Wetenscappen, 2007.

Eire, Carlos M. N. "Calvin and Nicodemism: A Reappraisal." *Sixteenth Century Journal* 10, no. 1 (Spring 1979): 44–69.

Elliott, Dyan. "Violence against the Dead: The Negative Translation and *damnatio memoriae* in the Middle Ages." *Speculum* 92, no. 4 (October 2017): 1020–55.

Ernst, Germana, and Guido, Giglioni. *Il linguaggio dei cieli: Astri e simboli nel Rinascimento.* Rome: Carocci, 2012.

Fahy, Conor. *Printing a Book at Verona in 1622: The Account Book of Francesco Calzolari Junior.* Paris: Foundation Custodia, 1993.

Favaro, Antonio. *Atti della Nazione Germanica di Padova.* Venice: A spese della Società, 1911–12.

Favaro, Antonio. *Documenti inediti per la storia dei manoscritti Galileiani nella Biblioteca Nazionale di Firenze.* Rome, 1886.

Fedele, Clemente. *Per servizio di Nostro Signore: Strade, corrieri e poste dei papi dal Medioevo al 1870.* Modena, It.: E. Mucchi editore, 1988.

Feingold, Mordechai. "The Grounds for Conflict: Grienberger, Grassi, Galileo, and Posterity." In *The New Science and Jesuit Science: Seventeenth Century Perspec-*

tives, edited by Mordechai Feingold, 121–57. Dordrecht, Neth.: Kluwer Academic Publishers, 2003.

Feingold, Mordechai. "Jesuits: Savants." In *Jesuit Science and the Republic of Letters*, edited by Mordechai Feingold, 1–45. Cambridge: MIT Press, 2003.

Feingold, Mordechai. "Science as a Calling? The Early Modern Dilemma." *Science in Context* 15, no. 1 (March 2002): 79–119.

Feist-Hirsch, E. "The Strange Career of a Humanist: The Intellectual Development of Justus Velsius (1502–1582)." In *Aspects de la propagande religieuse*, edited by G. Berthoud, 308–24. Geneva: Droz, 1957.

Feldhay, Rivka. *Galileo and the Church: Political Inquisition or Critical Dialogue?* Cambridge: Cambridge University Press, 1995.

Ferrari, G. E. "Le opere a stampa del Guilandino: Per un paragrafo dell'editoria scientifica padovana del pieno Cinquecento." In *Libri e stampatori in Padova*, edited by Antonion Barzon, 377–463. Padua: Tipografia del Seminario, 1959.

Ferretto, Silvia. *Maestri per il metodo di trattar le cose: Bassiano Lando, Giovan Battista da Monte e la scienza della medicina nel XVI secolo*. Padua: Belzoni, 2012.

Ferro, Roberta. "Per la storia del Fondo Pinelli all'Ambrosiana: Notizie dalle lettere di Paolo Gualdo." In *Tra i fondi dell'Ambrosiana: Manoscritti italiani antichi e moderni*, edited by Marco Ballarini, Gennaro Barbarisi, Claudia Berra, and Giuseppe Frasso. Milan: Cisalpino, 2008.

Findlen, Paula. "Anatomy Theaters, Botanical Gardens, and Natural History Collections." In Park and Daston, *Early Modern Science*, 272–89.

Findlen, Paula. "Aristotle in the Pharmacy: The Ambitions of Camilla Erculiani in Sixteenth-Century Padua," foreword to *Letters on Natural Philosophy: The Scientific Correspondence of a Sixteenth-Century Pharmacist, with Related Texts*, by Camilla Erculiani. Edited by Eleonora Carinci. Translated by Hannah Marcus. Toronto: Iter Press, 2020.

Findlen, Paula. "The Death of a Naturalist: Knowledge and Community in Late Renaissance Italy." In Manning and Klestinec, *Professors, Physicians and Practices*, 155–95.

Findlen, Paula. "The Formation of a Scientific Community: Natural History in Sixteenth-Century Italy." In *Natural Particulars: Nature and the Disciplines in Renaissance Europe*, edited by Anthony Grafton and Nancy Siraisi, 369–400. Cambridge, MA: MIT Press, 1999.

Findlen, Paula. *Possessing Nature: Museums, Collecting, and Scientific Culture in Early Modern Italy*. Berkeley: University of California Press, 1994.

Findlen, Paula, and Tara Nummedal. "Words of Nature: Scientific Books in the 17th Century." In *Thornton and Tully's Scientific Books, Libraries and Collectors*, edited by Andrew Hunter, 216–57. Brookfield, VT: Ashgate, 2000.

Finocchiaro, Maurice A. "Science, Religion, and the Historiography of the Galileo Affair: On the Undesirability of Oversimplification." *Osiris* 16 (2001): 114–32.

Finocchiaro, Maurice A. *Defending Copernicus and Galileo*. Boston Studies in the Philosophy of Science 280. Dordrecht: Springer Netherlands, 2009.

Finocchiaro, Maurice A., ed. *The Galileo Affair: A Documentary History*. Berkeley: University of California Press, 1989.

Firpo, Luigi. "Correzioni d'autore coatte." *Studi e problemi di critica testuale.* Bologna: Commissione per i Testi di Lingua, 1961.

Firpo, Luigi. "The Flowering and Withering of Speculative Philosophy—Italian Philosophy and the Counter-Reformation: The Condemnation of Francesco Patrizi." In *The Late Italian Renaissance 1525–1630,* edited by Eric Cochrane, 266–84. London: Macmillan, 1970.

Fish, Stanley. *Is There a Text in This Class? The Authority of Interpretive Communities.* Cambridge, MA: Harvard University Press, 1980.

Flower, Harriet I. *The Art of Forgetting: Disgrace and Oblivion in Roman Political Culture.* Chapel Hill: University of North Carolina Press, 2006.

Forster, Marc R. *Catholic Revival in the Age of the Baroque: Religious Identity in Southwest Germany, 1550–1750.* Cambridge: Cambridge University Press, 2001.

Forster, Marc R. *The Counter-Reformation in the Villages: Religion and Reform in the Bishopric of Speyer, 1560–1720.* Ithaca, NY: Cornell University Press, 1992.

Foucault, Michel. "What Is an Author." In *The Foucault Reader,* edited by Paul Rabinow. New York: Pantheon Books, 1984.

Fragnito, Gigliola. "Aspetti e problemi della censura espurgatoria." In *L'inquisizione e gli storici: Un cantiere aperto; Tavola rotonda nell'ambito della conferenza annuale della ricerca; Roma, 24–25 giugno 1999.* Rome: Accademia nazionale dei lincei, 2000.

Fragnito, Gigliola. "The Central and Peripheral Organization of Censorship." In Fragnito, *Church, Censorship, and Culture in Early Modern Italy,* 13–49.

Fragnito, Gigliola, ed. *Church, Censorship, and Culture in Early Modern Italy.* Translated by Adrian Belton. Cambridge: Cambridge University Press, 2001.

Fragnito, Gigliola. "Girolamo Savonarola e la censura ecclesiastica." *Rivista di storia e letteratura religiosa* 35 (1999): 501–29.

Fragnito, Gigliola. *La Bibbia al rogo: La censura ecclesiastica e i volgarizzamenti della scrittura: 1471–1605.* Bologna: Il Mulino, 1997.

Fragnito, Gigliola. *Proibito capire: La Chiesa e il volgare nella prima età moderna.* Bologna: Il Mulino, 2005.

Fragnito, Gigliola. "Un archivio conteso: Le 'carte' dell'Indice tra Congregazione e Maestro del Sacro Palazzo." *Rivista storica italiana* 119 (2007): 1276–18.

Frajese, Vittorio. "Le licenze di lettura tra vescovi ed inquisitori: Aspetti della politica dell'Indice dopo il 1596." *Società e storia* 86 (October 1999): 766–818.

Frajese, Vittorio. *Nascita dell'Indice: La censura ecclesiastica dal Rinascimento alla Controriforma.* Brescia, It.: Morcelliana, 2006.

Frampton, Stephanie. "What to Do with Books in the *De finibus.*" *Transactions of the American Philological Association* 146, no. 1 (Spring 2016): 117–47.

Franza, Annarita. "Romolo Spezioli, Andrea Vesalio ed il *manuum munus,* 'Il dono delle mani' nella practica medica moderna." In Sani and Zurlini, *La formazione del medico in età moderna,* 139–48.

Frascarelli, Dalma. "The Salone Sistino: The Iconographic Plan." In *The Vatican Library,* edited by Ambrogio M. Piazzoni et al., 178–265. Milan: Jaca Book, 2012.

Friedman, Jerome. *Michael Servetus: A Case Study in Total Heresy.* Geneva: Droz, 1978.

Front, Dov. "The Expurgation of Medical Books in Sixteenth-Century Spain." *Bulletin of the History of Medicine* 75, no. 2 (Summer 2001): 290–96.

Front, Dov. "The Expurgation of the Books of Amatus Lusitanus: Censorship and the Bibliography of the Individual Book." *Book Collector* 47, no. 4 (1998): 520–36.

Galen. *On the Usefulness of the Parts of the Body*. Translated by Margaret Tallmadge May. Ithaca, NY: Cornell University Press, 1968.

Galilei, Galileo. *Dialogue Concerning the Two Chief World Systems*. Edited by Stephen Jay Gould. Translated by Stillman Drake. Paperback ed. New York: Modern Library, 2001.

Galison, Peter. "Removing Knowledge: The Logic of Modern Censorship." In Proctor and Schiebinger, *Agnotology*, 37–54.

Galluzzi, Paolo. "Motivi paracelsiani nella Toscana di Cosimo II e di Don Antonio dei Medici: Alchemia, medicina 'chimica' e riforma del sapere." In *Scienze, credenze occulte, livelli di cultura: Convegno internazionale di studi; Firenze, 26–30 giugno 1980*. Florence: L. S. Olschki, 1982.

Galluzzi, Paolo. "The Sepulchers of Galileo: The 'Living' Remains of a Hero of Science." In *The Cambridge Companion to Galileo*, edited by Peter Machamer, 417–48. Cambridge: Cambridge University Press, 1998.

Gaukroger, Stephen. *Francis Bacon and the Transformation of Early-Modern Philosophy*. Cambridge: Cambridge University Press, 2001.

Gentilcore, David. *From Bishop to Witch: The System of the Sacred in Early Modern Terra d'Otranto*. Manchester: Manchester University Press, 1992.

Gentilcore, David. *Healers and Healing in Early Modern Italy*. Manchester: Manchester University Press, 1998.

Gentilcore, David. *Medical Charlatanism in Early Modern Italy*. Oxford: Oxford University Press, 2006.

Gingerich, Owen. *The Book Nobody Read: Chasing the Revolutions of Nicolaus Copernicus*. New York: Penguin Books, 2005.

Gingerich, Owen. "The Censorship of Copernicus's De Revolutionibus." *Annali dell'Istituto e museo di storia della scienza di Firenze* 7 (1981): 45–61.

Ginzburg, Carlo. *The Cheese and the Worms: The Cosmos of a Sixteenth-Century Miller*. Translated by Anne and John Tedeschi. Baltimore: Johns Hopkins University Press, 1980.

Ginzburg, Carlo. *Il nicodemismo: Simulazione e dissimulazione religiosa nell'Europa del '500*. Turin: Einaudi, 1970.

Giovanni Battista Rossi: Carmelitano ravennate; atti del convegno organizzato a Ravenna il 14/15 dicembre 1979. Ravenna: Centro Studio G. Donati, 1980.

Goddu, André. *Copernicus and the Aristotelian Tradition: Education, Reading, and Philosophy in Copernicus's Path to Heliocentrism*. Leiden, Neth.: Brill, 2010.

Godman, Peter. *From Poliziano to Machiavelli: Florentine Humanism in the High Renaissance*. Princeton, NJ: Princeton University Press, 1998.

Godman, Peter. *The Saint as Censor: Robert Bellarmine between Inquisition and Index*. Leiden, Neth.: Brill, 2000.

González Echeverría, Francisco Javier. *El amor a la verdad: Vida y obra de Miguel Servet*. Tudela, Sp.: Gobierno de Navarra, 2011.

Grafton, Anthony. *Cardano's Cosmos: The Worlds and Works of a Renaissance Astrologer.* Cambridge, MA: Harvard University Press, 1999.

Grafton, Anthony. *The Culture of Correction in Renaissance Europe.* London: British Library, 2011.

Grafton, Anthony. *The Footnote: A Curious History.* Rev. ed. Cambridge, MA: Harvard University Press, 1997.

Grafton, Anthony. *Rome Reborn: The Vatican Library and Renaissance Culture.* Washington, DC: Library of Congress, 1993.

Grafton, Anthony. "Philological and Artisanal Knowledge Making in Renaissance Natural History: A Study in Cultures of Knowledge." *History of Humanities* 3, no. 1 (2018): 39–55.

Grafton, Anthony. "A Sketch Map of a Lost Continent: The Republic of Letters." In *Worlds Made by Words: Scholarship and Community in the Modern West.* Cambridge, MA: Harvard University Press, 2009.

Grafton, Anthony, and Nancy Siraisi. "Between the Election and My Hopes: Girolamo Cardano and Medical Astrology." In *Secrets of Nature: Astrology and Alchemy in Early Modern Europe,* edited by William R. Newman and Anthony Grafton, 69–131. Cambridge, MA: MIT Press, 2001.

Grell, Ole Peter, ed. *Paracelsus: The Man and His Reputation, His Ideas and Their Transformation.* Leiden, Neth.: Brill, 1998.

Grell, Ole Peter, and Andrew Cunningham, eds. *Medicine and the Reformation.* New York: Routledge, 2013.

Grell, Ole Peter, Andrew Cunningham, and Jon Arrizabalaga. *Centres of Medical Excellence? Medical Travel and Education in Europe, 1500–1789.* Farnham, UK: Ashgate, 2010.

Grendler, Marcella T. "Book-Collecting in Counter-Reformation Italy: The Library of Gian Vincenzo Pinelli, 1535–1601." *Journal of Library History* 16, no. 1 (1981): 143–51.

Grendler, Marcella T., and Paul F. Grendler. "The Survival of Erasmus in Italy." *Erasmus in English* 8 (1976): 2–22.

Grendler, Paul F. "The Conditions of Enquiry: Printing and Censorship." In *The Cambridge History of Renaissance Philosophy,* edited by Charles B. Schmitt, Quentin Skinner, Eckhard Kessler, and Jill Kraye, 22–54. Cambridge: Cambridge University Press, 1988.

Grendler, Paul F. *Critics of the Italian World, 1530–1560: Anton Francesco Doni, Nicolò Franco and Ortensio Lando.* Madison: University of Wisconsin Press, 1969.

Grendler, Paul F. *The Roman Inquisition and the Venetian Press, 1540–1605.* Princeton, NJ: Princeton University Press, 1977.

Grendler, Paul F. "The 'Tre Savii sopra Eresia' 1547–1605: A Prosopographical Study." *Studi veneziani* 3 (1979): 283–340.

Grendler, Paul F. *The Universities of the Italian Renaissance.* Baltimore: Johns Hopkins University Press, 2002.

Gunnoe, Charles D. *Thomas Erastus and the Palatinate: A Renaissance Physician in the Second Reformation.* Leiden, Neth.: Brill, 2011.

Gutwirth, Eleazar. "Amatus Lusitanus and the Location of Sixteenth-Century Cultures." In *Cultural Intermediaries: Jewish Intellectuals in Early Modern Italy*, edited by David B. Ruderman and Giuseppe Veltri, 216–38. Philadelphia: University of Pennsylvania Press, 2004.

Hall, Crystal. *Galileo's Reading*. Cambridge: Cambridge University Press, 2013.

Hankins, James. *Repertorium Brunianum: A Critical Guide to the Writings of Leonardo Bruni*. Rome: Istituto storico italiano per il medioevo, 1997–.

Hannaway, Owen. *The Chemists and the Word: The Didactic Origins of Chemistry*. Baltimore: Johns Hopkins University Press, 1975.

Hasse, Hans-Peter, and Günther Wartenberg, eds. *Caspar Peucer (1525–1602): Wissenschaft, Glaube und Politik im konfessionellen Zeitalter*. Leipzig: Evangelische Verlagsanstalt, 2004.

Hedrick, Charles. *History and Silence: Purge and Rehabilitation of Memory in Late Antiquity*. Austin: University of Texas Press, 2000.

Heesakkers, C. L., and Dirk van Miert, "An Inventory of the Correspondence of Hadrianus Junius (1511–1575)." *Lias: Journal of Early Modern Intellectual Culture and Its Sources* 37, no. 2 (2010): 108–268.

Heilbron, John. *Galileo*. Oxford: Oxford University Press, 2010.

Heller, Marvin J. *Printing the Talmud: A History of the Earliest Printed Editions of the Talmud*. Brooklyn: Im Hasefer, 1992.

Helm, Jennifer. "Literary Censorship: The Case of the Orlando Furioso." *Dimensioni e problemi della ricerca storica* 1 (2012): 193–214.

Hendrickson, Thomas, ed. and trans. *Ancient Libraries and Renaissance Humanism: The "De bibliothecis" of Justus Lipsius*. Leiden, Neth.: Brill, 2017.

Hieronymus, Frank. "Physicians and Publishers: The Translation of Medical Works in Sixteenth-Century Basle." In *The German Book 1450–1750: Studies Presented to David L. Paisey in His Retirement*, edited by John L. Flood and William A. Kelly, 97–101. London: British Library, 1995.

Hillar, Marian. *The Case of Michael Servetus (1511–1553): The Turning Point in the Struggle for Freedom of Conscience*. Lewiston, NY: Edwin Mellen Press, 1997.

Hobson, Anthony. "A Sale by Candle in 1608." *Library* 26, no. 3 (1971): 215–33.

Holzberg, Niklas. "Ein vergessener Schüler Philipp Melanchthons: Georg Aemilius (1517–1569)." *Archiv für Reformationsgeschichte—Archive for Reformation History* 73 (1982): 94–122.

Horace. *Satires; Epistles; The Art of Poetry*. Translated by H. Rushton Fairclough. Loeb Classical Library 194. Cambridge, MA: Harvard University Press, 1926.

Hsia, R. Po-Chia. *The World of Catholic Renewal, 1540–1770*. Cambridge: Cambridge University Press, 1998.

Infelise, Mario. "Deposito legale e censura a Venezia (1569–1650)." *La Bibliofilía* 109 (2007): 71–77.

Infelise, Mario. *I libri proibiti: Da Gutenberg all'Encyclopédie*. Rome: GLF editori Laterza, 1999.

Jardine, Lisa, and Anthony Grafton. "'Studied for Action': How Gabriel Harvey Read His Livy." *Past & Present* 129 (November 1990): 30–78.

Jardine, Nicholas. "Keeping Order in the School of Padua; Jacopo Zabarella and Francesco Piccolomini on the Offices of Philosophy." In *Method and Order in Renais-*

sance Philosophy of Nature: The Aristotelian Commentary Tradition. Aldershot, UK: Ashgate, 1997.

Johns, Adrian. *The Nature of the Book: Print and Knowledge in the Making.* Chicago: University of Chicago Press, 1998.

Jones, John. "The Censor Censored: The Case of Benito Arias Montano." *Romance Studies* 13 (1995): 19–29.

Jones, Pamela M. *Federico Borromeo and the Ambrosiana: Art, Patronage, and Reform in Seventeenth-Century Milan.* Cambridge: Cambridge University Press, 1993.

Jones, Peter Murray. "Book Ownership and the Lay Culture of Medicine in Tudor Cambridge." In *The Task of Healing: Medicine, Religion and Gender in England and the Netherlands 1450–1800,* edited by H. Marland and M. Pelling, 49–68. Rotterdam: Erasmus, 1996.

Jones, Peter Murray. "Medical Libraries and Medical Latin, 1400–1700." In *Medical Latin from the Late Middle Ages to the Eighteenth Century,* edited by Wouter Bracke and Herwig Deumens, 115–35. Brussels: Koninklijke Academie Voor Geneeskunde Van Belgie, 1999.

Jostock, Ingeborg. *La censure négociée: Le contrôle du livre a Genève, 1560–1625.* Geneva: Droz, 2007.

Kagan, R. "Universities in Italy, 1500–1700." In *Les Universités européennes du XVIe au XVIIIe siècle: Histoire sociale des populations étudiantes,* edited by Dominique Julia, Jacques Revel, and Roger Chartier, 153–86. Paris: Editions de l'Ecole des hautes études en sciences sociales, 1986.

Keller, Vera. *Knowledge and the Public Interest, 1575–1725.* Cambridge: Cambridge University Press, 2015.

Klestinec, Cynthia. *Theaters of Anatomy: Students, Teachers, and Traditions of Dissection in Renaissance Venice.* Baltimore: Johns Hopkins University Press, 2011.

Koestler, Arthur. *The Sleepwalkers.* New York: Macmillan, 1968.

Kostyło, Joanna. "Commonwealth of All Faiths: Republican Myth and the Italian Diaspora in Sixteenth-Century Poland-Lithuania." In *Citizenship and Identity in a Multinational Commonwealth: Poland-Lithuania in Context, 1550–1772,* edited by Karin Friedrich and Barbara M. Pendzich, 171–206. Leiden, Neth.: Brill, 2008.

Kusukawa, Sachiko. *Picturing the Book of Nature: Image, Text, and Argument in Sixteenth-Century Human Anatomy and Medical Botany.* Chicago: University of Chicago Press, 2012.

Labowsky, Lotte. *Bessarion's Library and the Biblioteca Marciana: Six Early Inventories.* Rome: Edizione di Storia e Letteratura, 1979.

Lando, Ortensio. *Paradossi: Cioè sentenze fuori del comun parere,* edited by Antonio Corsaro. Rome: Edizione di Storia e Letteratura, 2000.

L'Apertura degli archive del Sant'Uffizio romano: Giornata di studio; Roma, 22 gennaio 1998. Rome: Accademia nazionale dei Lincei, 1998.

Lebreton, M. M., and L. Fiorani, eds. *Codices Vaticani Latini: Codices 11266–11326 Inventari di biblioteche religiose italiane alla fine del Cinquecento.* Rome: Typis Polyglottis Vaticanis, 1985.

Leong, Elaine, and Alisha Michelle Rankin. *Secrets and Knowledge in Medicine and Science, 1500–1800.* Burlington, VT: Ashgate, 2011.

Leu, Urs B. *Conrad Gessner (1516–1565): Universalgelehrter und Naturforscher der Renaissance*. Zurich: Verlag Neue Zürcher Zeitung, 2016.

Lewis, Sinclair. *Arrowsmith*. New York: Harcourt, Brace, 1925.

Lindberg, David C. *The Beginnings of Western Science: The European Scientific Tradition in Philosophical, Religious, and Institutional Context, Prehistory to A.D. 1450*. Chicago: University of Chicago Press, 2007.

Lindberg, David C. "The Medieval Church Encounters the Classical Tradition: Saint Augustine, Roger Bacon, and the Handmaiden Metaphor." In *When Science and Christianity Meet*, edited by David C. Lindberg and Ronald L. Numbers, 7–32. Chicago: University of Chicago Press, 2008.

Lines, David A. "Latin and Vernacular in Francesco Piccolomini's Moral Philosophy." In *"Aristotele fatto volgare": Tradizione aristotelica e cultura volgare nel Rinascimento*, edited by David A. Lines and Eugenio Refini. Pisa: Edizioni ETS, 2014.

Long, Pamela. *Artisan/Practitioners and the Rise of the New Sciences, 1400–1600*. Corvallis: Oregon State University Press, 2011.

Lopez, Pasquale. *Inquisizione, stampa e censura nel Regno di Napoli tra '500 e '600*. Naples: Edizioni del delfino, 1974.

López Piñero, José María. *La traducción por Juan de Jarava de Leonhart Fuchs y la terminología botánica castellana del siglo XVI*. Valencia: Instituto de Estudios Documentales e Históricos sobre la Ciencia, 1994.

Lowry, Martin. *The World of Aldus Manutius: Business and Scholarship in Renaissance Venice*. Oxford: Blackwell, 1979.

Lynn, Kimberly. *Between Court and Confessional: The Politics of Spanish Inquisitors*. Cambridge: Cambridge University Press, 2013.

Maclean, Ian. "The Diffusion of Learned Medicine in the Sixteenth Century through the Printed Book." In Maclean, *Learning and the Market Place*, 59–86.

Maclean, Ian. "Heterodoxy in Natural Philosophy and Medicine: Pietro Pomponazzi, Guglielmo Gratarolo, Girolamo Cardano." In Brooke and Maclean, *Heterodoxy in Early Modern Science and Religion*, 1–30.

Maclean, Ian. *Logic, Signs, and the Order of Nature in the Renaissance*. Cambridge: Cambridge University Press, 2002.

Maclean, Ian. "Mediations of Zabarella in Northern Germany, 1586–1623." In Maclean, *Learning and the Market Place*, 39–55.

Maclean, Ian. *Learning and the Market Place: Essays in the History of the Book*. Edited by Andrew Pettegree. Leiden, Neth.: Brill, 2009.

Maclean, Ian. "The Medical Republic of Letters before the Thirty Years War." *Intellectual History Review* 18, no. 1 (2008): 15–30.

Maclean, Ian. *Scholarship, Commerce, Religion: The Learned Book in the Age of Confessions, 1560–1630*. Cambridge, MA: Harvard University Press, 1998.

Magnien, Michel. "Le 'Nomenclator' de Robert Constantin (1555): Première bibliographie française?" *Renaissance and Reformation / Renaissance et Réforme* 34, no. 3 (2011): 65–89.

Maifreda, Germano. *I denari dell'inquisitore: Affari e giustizia di fede nell'Italia moderna*. Einaudi Storia 52. Turin: Giulio Einaudi Editore, 2014.

Manfredi, Antonio, ed. *Le origini della Biblioteca Vaticana tra umanesimo e Rinasci-*

mento, 1447–1534. Vol. 1 of *Storia della Biblioteca Apostolica Vaticana.* Vatican City: Biblioteca Apostolica Vaticana, 2010.

Mangani, Lorella, and Giuseppe Martini. *La biblioteca di Francesco Redi e della sua famiglia, Catalogo.* Arezzo: Accademia Petrarca, 2006.

Manning, Gideon, and Cynthia Klestinec, eds. *Professors, Physicians and Practices in the History of Medicine: Essays in Honor of Nancy Siraisi.* Cham, Switz.: Springer, 2017.

Maragi, Mario. "Corrispondenze mediche di Ulisse Aldrovandi coi paesi germanici." *Pagine della storia della medicina* 13 (July–August 1969): 102–10.

Marcus, Hannah. "Bibliography and Book Bureaucracy: Reading Licenses and the Circulation of Prohibited Books in Counter-Reformation Italy." *Papers of the Bibliographical Society of America* 110, no. 4 (December 2016): 433–57.

Marcus, Hannah. "Expurgated Books as an Archive of Practice." *Archive Journal,* August 2017. http://www.archivejournal.net/essays/expurgated-books-as-an-archive -of-practice/.

Marcus, Hannah, and Paula Findlen. "Deciphering Galileo: Communication and Secrecy before and after the Trial." *Renaissance Quarterly* 72, no. 3 (2019): 953–95.

Margócsy, Dániel, Mark Somos, and Stephen N. Joffe. *The Fabrica of Andreas Vesalius: A Worldwide Descriptive Census, Ownership, and Annotations of the 1543 and 1555 Editions.* Medieval and Early Modern Philosophy and Science 28. Leiden, Neth.: Brill, 2018.

Marí, Antonio Beltrán. *Talento y poder: Historia de las relaciones entre Galileo y la Iglesia católica.* 2nd ed. Pamplona: Laetoli, 2007.

Martin, John Jeffries. *Venice's Hidden Enemies: Italian Heretics in a Renaissance City.* Berkeley: University of California Press, 1993.

Mayer, Thomas F. *The Roman Inquisition: A Papal Bureaucracy and Its Laws in the Age of Galileo.* Philadelphia: University of Pennsylvania Press, 2013.

Mayer, Thomas F. *The Roman Inquisition on the Stage of Italy.* Philadelphia: University of Pennsylvania Press, 2014.

Mayer, Thomas F. *The Roman Inquisition: Trying Galileo.* Haney Foundation Series. Philadelphia: University of Pennsylvania Press, 2015.

McClure, George W. *The Culture of Profession in Late Renaissance Italy.* Toronto: University of Toronto Press, 2004.

Mendelsohn, Andrew, and Hess Volker. "Case and Series: Medical Knowledge and Paper Technology, 1600–1900." *History of Science* 48, nos. 3–4 (2010): 287–314.

Metzler, Josef. "Francesco Ingoli, der erste Sekretär der Kongregation (1578–1649)." In *Sacrae Congregationis de Propaganda Fide Memoria Rerum,* 197–243. Rome: Herder, 1971.

Meyer, Frederick G., Emily Emmart Trueblood, and John Lewis Heller. *The Great Herbal of Leonhart Fuchs: De Historia Stirpium Commentarii Insignes, 1542 (notable Commentaries On the History of Plants).* Stanford, CA: Stanford University Press, 1999.

Miggiano, G. "Libri prohibiti: Qualche appunto dalle carte di Johannes Faber Lynceus Bambergensis." In *L'organizzazione del sapere: Studi in onore di A. Serrai,* ed. M. T. Biagetti, 245–73. Milan: S. Bonnard, 2004.

Miglietti, Sara. "The Censor as Reader: Censorial Responses to Bodin's Methodus in Counter-Reformation Italy (1587–1607)." *History of European Ideas* 42, no. 5 (2016): 707–21.

Minuzzi, Sabina. *Sul filo dei segreti: Farmacopea, libri e pratiche terapeutiche a Venezia in età moderna*. Milan: Edizioni unicopli, 2016.

Mittler, Elmar. *Bibliotheca Palatina: Catalogue to the Exhibition from July 8th to November 2nd 1986 in the Heiliggeist Church, Heidelberg*. Heidelberg: Edition Braus, 1986.

Montcher, Fabien. "Early Modern Bibliopolitics." *Pacific Coast Philology* 52, no. 2 (2017): 206–18.

Montuschi, Claudia, ed. *La Vaticana nel Seicento (1590–1700)*. Vol. 3 of *Storia della Biblioteca Apostolica Vaticana*. Vatican City: Biblioteca Apostolica Vaticana, 2014.

Moran, Bruce T. *Andreas Libavius and the Transformation of Alchemy: Separating Chemical Cultures with Polemical Fire*. Sagamore Beach, MA: Science History Publications, 2007.

Moran, Bruce T. *Distilling Knowledge: Alchemy, Chemistry, and the Scientific Revolution*. Cambridge, MA: Harvard University Press, 2005.

Moss, Ann. *Printed Commonplace-Books and the Structuring of Renaissance Thought*. Oxford: Clarendon Press, 1996.

Muir, Edward. *Culture Wars of the Late Renaissance: Skeptics, Libertines, and Opera*. Cambridge, MA: Harvard University Press, 2007.

Murphy, Hannah. "Common Places and Private Spaces: Libraries, Record-Keeping and Orders of Information in Sixteenth-Century Medicine." *Past & Present* 230 (November 2016): 253–68.

Naudé, Gabriel. *Advice on Establishing a Library*. Berkeley: University of California Press, 1950.

Nauert, Charles G. Jr. *Agrippa and the Crisis of Renaissance Thought*. Urbana: University of Illinois Press, 1995.

Navoni, Marco. "Gli uomini di Federico Borromeo: Gli oblati, i primi dottori e i primi conservatori." In Buzzi and Ferro, *La Biblioteca Ambrosiana*, 281–310.

Nesvig, Martin Austin. *Ideology and Inquisition: The World of the Censors in Early Mexico*. New Haven, CT: Yale University Press, 2009.

Nicholls, Matthew C. "Galen and Libraries in the Peri Alupias." *Journal of Roman Studies* 101 (2011): 123–42.

Nuovo, Angela. "A proposito del carteggio Pinelli-Dupuy." *Bibliotheca: Rivista di studi bibliografici* 2 (2002: 96–115.

Nuovo, Angela. *The Book Trade in the Italian Renaissance*. Leiden, Neth.: Brill, 2013.

Nuovo, Angela. "The Creation and Dispersal of the Library of Gian Vincenzo Pinelli." In *Books on the Move: Tracking Copies through Collections and the Book Trade*, edited by Robin Myers, Michael Harris, and Giles Mandelbrote, 39–68. New Castle, DE: Oak Knoll Press; London: British Library, 2007.

Nuovo, Angela. "Dispersione di una biblioteca privata: La biblioteca di Gian Vincenzo Pinelli dall'agosto 1601 all'ottobre 1604." In *Biblioteche private in età moderna e contemporanea: Atti del convegno internazionale Udine, 18–20 ottobre 2004*, edited by Angela Nuovo, 43–54. Milan: Ed. Bonnard, 2005.

Nussdorfer, Laurie. *Brokers of Public Trust: Notaries in Early Modern Rome*. Baltimore: Johns Hopkins University Press, 2009.

Nutton, Vivian. "Dr. James's Legacy: Dutch Printing and the History of Medicine." In *The Bookshop of the World: The Role of the Low Countries in the Book Trade, 1473–1941*, edited by Lotte Hellinga, Alastair Duke, Jacob Harskamp, and Theo Hermans, 207–18. Goy-Houten: Hes & De Graaf, 2001.

Nutton, Vivian. *John Caius and the Manuscripts of Galen*. Cambridge: Cambridge Philological Society, 1987.

Nutton, Vivian. "Greek Science in the Sixteenth-Century Renaissance." In *Renaissance and Revolution: Humanists, Scholars, Craftsmen, and Natural Philosophers in Early Modern Europe*, edited by J. V. Field and Frank A. J. L. James, 15–28. Cambridge: Cambridge University Press, 1993.

Nutton, Vivian. "Hellenism Postponed: Some Aspects of Renaissance Medicine, 1490–1530." *Sudhoffs Archiv* 81, no. 2 (1997): 158–70.

Nutton, Vivian. "The Reception of Fracastroro's Theory of Contagio: The Seed That Fell among Thorns?" *Osiris* 2, vol. 6 (1990): 196–234.

Ogilvie, Brian W. *The Science of Describing: Natural History in Renaissance Europe*. Chicago: University of Chicago Press, 2006.

Olds, Katrina. *Forging the Past: Invented Histories in Counter-Reformation Spain*. New Haven, CT: Yale University Press, 2015.

Olmi, Giuseppe. *L'Inventario del mondo: Catalogazione della natura e luoghi del sapere nella prima età moderna*. Bologna: Il Mulino, 1992.

O'Malley, John W. *Trent and All That: Renaming Catholicism in the Early Modern Era*. Cambridge, MA: Harvard University Press, 2000.

Omodeo, Pietro. "*Utilitas astronomiae* in the Renaissance: The Rhetoric and Epistemology of Astronomy." In *The Structures of Practical Knowledge*, edited by Matteo Valleriani, 307–31. Cham, Switz.: Springer, 2017.

Ordine, Nuccio. "Caravage et Bruno: Les relations de Gian Vincenzo Pinelli avec Della Rovere, Paolo Gualdo et les frères Del Monte." In *Une Traversée des Savoirs: Mélanges Offerts à Jackie Pigeaud*, edited by Philippe Heuzé, Yves Hersant, and Eric Van Der Schueren, 611–25.Québec: Presses de L'Université Laval, 2008.

Ottone, Andrea. "Pastoral Care and Cultural Accuracy: Book Collections of Secular Clergy in Three Southern Italian Dioceses." In *Documenting the Early Modern Book World: Inventories and Catalogues in Manuscript and Print*, ed. Malcolm Walsby and Natasha Constantinidou, 231–60. Leiden, Neth.: Brill, 2013.

Palmer, Ada. *Censorship and Information Control*. Chicago: Swift Impressions, 2018. Published in conjunction with an exhibition of the same title, presented at University of Chicago Library, September 17–December 14, 2018.

Palmer, Richard. "Medicine at the Papal Court in the Sixteenth Century." In *Medicine at the Courts of Europe*, edited by Vivian Nutton, 49–78. London: Routledge, 1990.

Palmer, Richard. "Pharmacy in the Republic of Venice in the Sixteenth Century." In *The Medical Renaissance in the Sixteenth Century*, edited by Andrew Wear, R. K. French, and Ian M. Lonie, 100–117. Cambridge: Cambridge University Press, 1985.

Palmer, Richard. "Physicians and the Inquisition in Sixteenth-Century Venice." In

Grell and Cunningham, *Medicine and the Reformation*, 118–33. New York: Routledge, 2013.

Palumbo, Margherita. *"Lexica malvagia et perniciosa*: The Case of Estienne's *Thesaurus graecae linguae." Lexicon Philosophicum: International Journal for the History of Texts and Ideas* 3 (2015): 1–22.

Pardo Tomás, José. *Ciencia y censura: La Inquisición Española y los libros científicos en los siglos XVI y XVII*. Estudios sobre la ciencia 13. Madrid: Consejo Superior de Investigaciones Científicas, 1991.

Park, Katharine. *Doctors and Medicine in Early Renaissance Florence*. Princeton, NJ: Princeton University Press, 1985.

Park, Katharine. *Secrets of Women: Gender, Generation, and the Origins of Human Dissection*. New York: Zone Books, 2006.

Park, Katharine, and Lorraine Daston, eds. *Early Modern Science*. Vol. 3 of *The Cambridge History of Science*. Cambridge: Cambridge University Press, 2006.

Pasini, Cesare. "Il progetto biblioteconomico di Federico." In Buzzi and Ferro, *La Biblioteca Ambrosiana*, 260–70.

Passanzini, Stefano. "Giovanni Battista Rossi Carmelitano: La famiglia, la patria, il personaggio." In *Giovanni Battista Rossi*, 7–38.

Pastor, Ludwig. *The History of the Popes, from the Close of the Middle Ages*. 29 vols. London: K. Paul, Trench, Trübner, 1906.

Patterson, Annabel M. *Censorship and Interpretation: The Conditions of Writing and Reading in Early Modern England*. Madison: University of Wisconsin Press, 1984.

Paviolo, Maria Gemma. *I testamenti dei cardinali: Carlo Gaudenzio Madruzzo (1562–1629)*. Self-published, 2015.

Pavord, Anna. *The Naming of Names: The Search for Order in the World of Plants*. London: Bloomsbury, 2005.

Pelling, Margaret. *The Common Lot: Sickness, Medical Occupations, and the Urban Poor in Early Modern England; Essays*. London: Longman, 1998.

Pender, Stephen, and Nancy S. Struever, eds. *Rhetoric and Medicine in Early Modern Europe*. Burlington, VT: Ashgate, 2012.

Pennuto, Conchetta. "Francisco Vallés' *De Sacra Philosophia*: A Medical Reading of the Bible," in *Lay Readings of the Bible in Early Modern Europe*, ed. Erminia Ardissino and Élise Boillet, 235. Leiden, Neth.: Brill, 2019.

Perini, Leandro. *La vita e i tempi di Pietro Perna*. Rome: Edizioni di storia e letteratura, 2002.

Peters, Edward. *Inquisition*. New York: Free Press, 1988.

Petrarca, Francesco, and Robert M. Durling. *Petrarch's Lyric Poems: The Rime Sparse and Other Lyrics*. Cambridge, MA: Harvard University Press, 1976.

Petrarca, Francesco, and David Marsh. *Invectives*. I Tatti Renaissance Library 11. Cambridge, MA: Harvard University Press, 2003.

Pierpaoli, Mario. "Girolamo Rossi medico e storico ravennate." In *Storie Ravennati*, translated and edited by Mario Pierpaoli. Ravenna: Longo Editore, 1996.

Pine, Martin L. *Pietro Pomponazzi: Radical Philosopher of the Renaissance*. Padua: Editrice Antinore, 1986.

Pinon, Laurent. "La culture scientifique à Rome à la Renaissance—Clématite bleue contre poissons séchés: Sept lettres inédites d'Ippolito Salviani à Ulisse Aldrovandi." *MEFRIM: Mélanges de l'École française de Rome: Italie et mediterranée* 114, no. 2 (2002): 477–92.

Pinto, Virgilio. "Censorship: A System of Control and an Instrument of Action." In *The Spanish Inquisition and the Inquisitorial Mind*, edited by Angel Alcala, 303–20. Boulder, CO: Social Science Monographs, 1987.

Pizzorno, Patrizia Grimaldi. *The Ways of Paradox from Lando to Donne.* Florence: Olschki, 2007.

Pomata, Gianna. "Sharing Cases: The Observationes in Early Modern Medicine." *Early Science and Medicine* 15, no. 3 (2010): 193–236.

Pomata, Gianna, and Nancy Siraisi, eds. *Historia: Empiricism and Erudition in Early Modern Europe.* Cambridge, MA: MIT Press, 2005.

Pomian, Krzysztof. *Collectors and Curiosities: Paris and Venice, 1500–1800.* Translated by Elizabeth Wiles Portier. London: Polity, 1990.

Poppi, Antonino. *Cremonini e Galilei inquisiti a Padova nel 1604: Nuovi documenti d'archivio.* Padua: Antenore, 1992.

Poppi, Antonino. *Cremonini, Galilei e gli inquisitori del Santo a Padova.* Padua: Centro studi antoniani, 1993.

Pormann, Peter E. "La querelle des médecins arabistes et hellénistes et l'héritage oublié." In *Lire les médecins grecs à la Renaissance: Aux origines de l'édition médicale; Actes du colloque international de Paris (19–20 septembre 2003)*, edited by Véronique Boudon Millot and Guy Cobolet, 113–41. Paris: De Boccard Edition-Diffusion, 2004.

Proctor, Robert, and Londa Schiebinger, eds. *Agnotology: The Making and Unmaking of Ignorance.* Stanford, CA: Stanford University Press, 2008.

Prodi, Paolo. *Il Cardinale Gabriele Paleotti (1522–1597).* 2 vols. Rome: Edizioni di storia e letteratura, 1959–67.

Prodi, Paolo. *The Papal Prince: One Body and Two Souls; The Papal Monarchy in Early Modern Europe.* Cambridge: Cambridge University Press, 1987.

Prodi, Paolo, and Wolfgang Reinhard, eds. *Il concilio di Trento e il moderno.* Bologna: Il Mulino, 1996.

Prosperi, Adriano. "Anime in trappola: Confessione e censura ecclesiastica all'Università di Pisa fra '500 e '600." *Belfagor* 54, no. 321 (1999): 257–87.

Prosperi, Adriano. "'Damnatio memoriae': Nomi e libri in una proposta della Controriforma." In *L'Inquisizione romana: Letture e ricerche*, 385–411. Rome: Edizioni di storia e letteratura, 2003.

Prosperi, Adriano. "Presentazione." In *La Biblioteca antica dell'Osservatorio Ximeniano, Catalogo*, edited by Mauro Guerrini. Florence: Regione Toscana, 1994.

Prosperi, Adriano. *Tribunali della coscienza: Inquisitori, confessori, missionari.* Turin: Einaudi, 1996.

Prosperi, Adriano, Vincenzo Lavenia, and John A. Tedeschi, eds. *Dizionario storico dell'Inquisizione.* 4 vols. Pisa: Scuola Normale Superiore, 2010.

Proverbio, Edoardo. "Francesco Giuntini e l'utilizzo delle tavole copernicane in Italia nel XVI secolo." In *La diffusione del copernicanesimo in Italia, 1543–1610*, edited by Massimo Bucciantini and Maurizio Torrini, 37–56. Florence: Olschki, 1997.

Pugliese, Silvia. "Melchiorre Guilandino, 'Bazzaro Venetoteutonico' alla guida dell'Orto botanico di Padova: Studi su una biblioteca scientifica del Cinquecento." PhD diss., Università degli Studi di Udine, 2014.

Rabelais, François, and Andrew Brown. *Pantagruel: King of the Dipsodes Restored to His Natural State with His Dreadful Deeds and Exploits Written by the Late M. Alcofribas, Abstractor of the Quintessence*. London: Hesperus, 2003.

Raffaelli, Filippo. *La Biblioteca Comunale di Fermo*. Recanati, It.: R. Simboli, 1890.

Ragland, Evan R. "'Making Trials' in Sixteenth- and Early Seventeenth-Century European Academic Medicine." *Isis* 108, no. 3 (2017): 503–28.

Raphael, Renée Jennifer. *Reading Galileo: Scribal Technologies and the Two New Sciences*. Baltimore: Johns Hopkins University Press, 2017.

Raz-Krakotzkin, Amnon. *The Censor, the Editor, and the Text: The Catholic Church and the Shaping of the Jewish Canon in the Sixteenth Century*. Philadelphia: University of Pennsylvania Press, 2007.

Reeves, Eileen Adair. *Galileo's Glassworks: The Telescope and the Mirror*. Cambridge, MA: Harvard University Press, 2008.

Reeves, Eileen Adair. *Painting the Heavens: Art and Science in the Age of Galileo*. Princeton, NJ: Princeton University Press, 1997.

Regier, Jonathan. "Reading Cardano with the Roman Inquisition: Astrology, Celestial Physics and the Force of Heresy." *Isis* 10, no. 4 (December 2019): 661–79.

Ricci, Saverio. *Inquisitori, censori, filosofi sullo scenario della contrariforma*. Rome: Salerno Editrice, 2008.

Rietbergen, Peter. *Power and Religion in Baroque Rome: Barberini Cultural Policies*. Leiden, Neth.: Brill, 2006.

Riondato, Ezio, and Antonino Poppi, eds. *Cesare Cremonini: Aspetti del pensiero e scritti; Atti del convegno di studio (Padova, 26–27 febbraio 1999)*. Padua: Accademia galileiana di scienze, lettere ed arti in Padova, 2000.

Robey, Tracy E. "Damnatio memoriae: The Rebirth of Condemnation of Memory in Renaissance Florence." *Renaissance and Reformation* 36, no. 3 (2013): 5–32.

Rodella, Massimo. "Fortuna e sfortuna della biblioteca di Gian Vincenzo Pinelli: La vendita a Federico Borromeo." *Biblioteca: Revista di studi bibliografici* 2 (2003): 87–125.

Roebel, Martin. *Humanistische Medizin und Kryptocalvinismus: Leben und medizinisches Werk des Wittenberger Medizinprofessors Caspar Peucer (1525–1602)*. Herbolzheim, Ger.: Centaurus Verlag & Media, 2012.

Romano, Antonella, ed. *Rome et la science moderne*. Rome: École française de Rome, 2009.

Rose, Paul. "A Venetian Patron and Mathematician of the Sixteenth Century: Francesco Barozzi (1537–1604)." *Studi Veneziani* 1 (1977): 119–80.

Rosen, Edward. *Copernicus and His Successors*. London: Hambledon Press, 1995.

Rosen, Edward. "Was Copernicus' *Revolutions* Approved by the Pope?" *Journal of the History of Ideas* 36, no. 3 (July–September 1975): 531–42.

Rosetti, Lucia, ed. *Acta Nationis Germanicae Artistarum (1616–1636).* Padua: Antenore, 1967.

Rosetti, Lucia. "Le biblioteche delle nationes nello studio di Padova." *Quaderni per la storia dell'Università di Padova* 2 (1969): 53–67.

Ross, Sarah Gwyneth. *Everyday Renaissances: The Quest for Cultural Legitimacy in Venice.* Cambridge, MA: Harvard University Press, 2016.

Ross, Tricia M. "Anthropologia: An (Almost) Forgotten Early Modern History." *Journal of the History of Ideas* 79, no. 1 (January 2018): 1–22.

Ross, Tricia M. "Care of Bodies, Cure of Souls: Religion and Medicine in Early Modern Germany." PhD diss., Duke University, 2017.

Ross, Tricia M. "Sacred Medicine and the Bible: Thomas Bartholin's *On Biblical Diseases* (1672)." *Early Science and Medicine* 24, no. 1 (May 2019): 90–116.

Rossi, Girolamo. *Storie Ravennati.* Edited and translated by Mario Pierpaoli. Ravenna: Longo Editore, 1996.

Roth, F. W. E. "Otto Brunfels, 1489–1534: Ein deutscher Botaniker." *Botanische Zeitung* 58 (1900): 191–232.

Rotondò, Antonio. "La censura ecclesiastica e la cultura." In *Storia d'Italia,* vol. 5, *I documenti,* edited by Ruggero Romano and Corrado Vivanti, pt. 2, pp. 1397–1492. Turin: G. Einaudi, 1973.

Rotondò, Antonio. *Studi e ricerche di storia ereticale italiana del Cinquecento.* Turin: Giappichelli, 1974.

Rozzo, Ugo. "Biblioteche e censura: Da Conrad Gesner a Gabriel Naudé." *Bibliotheca: Revista di studi bibliografici* 2 (2003): 33–72.

Rozzo, Ugo. *Biblioteche italiane del Cinquecento tra Riforma e Controriforma.* Udine, It.: Arti Grafiche Friulane, 1994.

Rozzo, Ugo. "Gli 'Hecatommithi' all'Indice." *La Bibliofilia* 92 (1991): 21–51.

Rozzo, Ugo. "Italian Literature on the Index." In Fragnito, *Church, Censorship, and Culture in Early Modern Italy,* 194–222.

Ruderman, David B. *Jewish Thought and Scientific Discovery in Early Modern Europe.* New Haven, CT: Yale University Press, 1995.

Rummel, Erika. *The Correspondence of Wolfgang Capito.* 3 vols. Toronto: University of Toronto Press, 2005–9.

Sabato, Milena. *Il sapere che brucia: Libri, censure e rapporti Stato-Chiesa nel Regno di Napoli fra '500 e '600.* Galatina, It.: Congedo, 2009.

Sabba, Fiammetta. *La Bibliotheca universalis di Conrad Gesner: Monumento della cultura europea.* Rome: Bulzoni, 2012.

Salzburg Museum. "Der Spottofen im SMCA." *Kunstwerk des,* February 1989.

Sanfilippo, Isa Lori, and Antonio Rigoni, eds. *Condannare all'oblio: Pratiche della damnatio memoriae nel medioevo; Atti del convegno di studio svoltosi in occasione della XX edizione del premio internazionale Ascoli Piceno; Ascoli Piceno, Palazzo dei Capitani, 27–29 novembre 2008.* Ascoli Piceno, It.: Istituto superiore di studi medievali Cecco d'Ascoli, 2010.

Sani, Roberto, and Fabiolo Zurlini, eds. *La formazione del medico in età moderna (secc. XVI–XVIII).* Macerata, It.: Eum, 2012.

Sarton, George. "The Strange Fame of Demetrio Canevari, Philosopher and Physician, Genoese Patrician (1559–1625)." *Journal of the History of Medicine and Allied Sciences* 1, no. 3 (July 1946): 398–418.

Savelli, Rodolfo. *Censori e giuristi: Storie di libri, di idee e di costumi (secoli XVI–XVII)*. Per la storia del pensiero giuridico moderno 94. Milan: A. Giuffrè, 2011.

Savelli, Rodolfo. "The Censoring of Law Books." In Fragnito, *Church, Censorship, and Culture in Early Modern Italy*, 223–53.

Savelli, Rodolfo. "La biblioteca disciplinata: Una 'libraria' cinque-seicentesca tra censura e dissimulazione." In *Tra diritto e storia: Studi in onore di Luigi Berlinguer promossi dalle Università di Siena e di Sassari*, vol. 2, edited by Luigi Berlinguer, 856–944. Soveria Mannelli, It.: Rubbettino, 2008.

Savelli, Rodolfo. "La critica roditrice dei censori." In *Saperi e meraviglie: Tradizione e nuove scienze nella libraria del medico genovese Demetrio Canevari*, edited by L. Malfatto and E. Ferro, 41–76. Genoa: Sagep, 2004.

Schaich-Klose, Wiebke. *D. Hieronymus Schürpf: Leben Und Werk Des Wittenberger Reformationsjuristen, 1481–1554*. Trogen, Ger.: Buchdruckerei Meili, 1967.

Schauer, Frederick. "The Ontology of Censorship." In *Censorship and Silencing: Practices of Cultural Regulation*, edited by Robert C. Post, 147–68. Los Angeles: Getty Research Institute for the History of Art and the Humanities, 1998.

Schmitt, Charles B. "Philosophy and Science in Sixteenth-Century Universities: Some Preliminary Comments." In *Studies in Renaissance Philosophy and Science*, 485–537. London: Variorum Reprints, 1981.

Schmitz-Esser, Romedio. "The Cursed and the Holy Body: Burning Corpses in the Middle Ages." *Journal of Medieval and Early Modern Studies* 45, no. 1 (January 2015): 131–57.

Schutte, Anne Jacobson. "Palazzo del Sant'Uffizio: The Opening of the Roman Inquisition's Central Archive." *Perspectives on History* 37, no. 5 (May 1999): 25–28.

Schwarze. "Cornerus, Christoph." In *Allgemeine Deutsche Biographie* 4 (1876): 499.

Schwedt, Herman H. "Gli inquisitori generali di Siena, 1560–1782." In *Le lettere della Congregazione del Sant'Ufficio all'inquisitore di Siena*, edited by Oscar di Simplicio. Trieste, It.: Università di Trieste, 2009.

Schwedt, Herman H., Jyri Hasecker, Dominik Höink, Judith Schepers, and Hubert Wolf, eds. *Prosopographie von Römischer Inquisition und Indexkongregation 1701–1813*. Paderborn, Ger.: Ferdinand Schöningh, 2010.

Secret, François. "De Mésué à Hieronymus Rubeus en passant par Giovanni Mainardi et Jacques Dubois." *Chrysopoeia* 5 (1996): 453–66.

Seidel Menchi, Silvana. *Erasmo in Italia, 1520–1580*. Turin: Bollati Boringhieri, 1987.

Seidel Menchi, Silvana. "Sette modi di censurare Erasmo." In *La censura libraria nell'Europa del secolo 16.: Convegno internazionale di studi; Cividale del Friuli, 9–10 novembre 1995*, edited by Ugo Rozzo, 177–206. Udine, It.: Forum, 1997.

Seitz, Jonathan. *Witchcraft and Inquisition in Early Modern Venice*. Cambridge: Cambridge University Press, 2011.

Serrai, Alfredo. *Conrad Gesner*. Rome: Bulzoni, 1990.

Sforza, Giovanni. *F. M. Fiorentini ed i suoi contemporanei lucchesi: Saggio di storia letteraria del secolo xvii*. Florence: F. Menozzi, 1879.

Sgarbi, Marco, ed. *Pietro Pomponazzi: Tradizione e dissenso; Atti del Congresso internazionale di studi su Pietro Pomponazzi, Mantova 23–24 ottobre 2008*. Florence: Leo S. Olschki Editore, 2010.

Shank, Michael H. "Setting up Copernicus? Astronomy and Natural Philosophy in Giambattista Capuano da Manfredonia's 'Expositio' on the 'Sphere.'" *Early Science and Medicine* 14, nos. 1–3 (2009): 290–315.

Shaw, James E., and Evelyn S. Welch. *Making and Marketing Medicine in Renaissance Florence*. Amsterdam: Rodopi, 2011.

Shea, William, and Mariano Artigas. *Galileo in Rome*. Oxford: Oxford University Press, 2003.

Sherman, William H. *Used Books: Marking Readers in Renaissance England*. Philadelphia: University of Pennsylvania Press, 2008.

Shuger, Deborah. *Censorship and Cultural Sensibility: The Regulation of Language in Tudor-Stuart England*. Philadelphia: University of Pennsylvania Press, 2006.

Siraisi, Nancy G. *Avicenna in Renaissance Italy: The "Canon" and Medical Teaching in Italian Universities after 1500*. Princeton, NJ: Princeton University Press, 1987.

Siraisi, Nancy G. *The Clock and the Mirror: Girolamo Cardano and Renaissance Medicine*. Princeton, NJ: Princeton University Press, 1997.

Siraisi, Nancy G. *Communities of Learned Experience: Epistolary Medicine in the Renaissance*. Baltimore: Johns Hopkins University Press, 2013.

Siraisi, Nancy G. "The Fielding H. Garrison Lecture: Medicine and the Renaissance World of Learning." *Bulletin of the History of Medicine* 78, no. 1 (Spring 2004): 1–36.

Siraisi, Nancy G. "Giovanni Argenterio: Medical Innovation, Princely Patronage and Academic Controversy." In *Medicine and the Italian Universities*, 328–55.

Siraisi, Nancy G. *History, Medicine, and the Traditions of Renaissance Learning*. Ann Arbor: University of Michigan Press, 2007.

Siraisi, Nancy G. *Medicine and the Italian Universities: 1250–1600*. Leiden, Neth.: Brill, 2001.

Siraisi, Nancy G. "Medicine, 1450–1620, and the History of Science." *Isis* 103, no. 3 (2012): 491–514.

Siraisi, Nancy G. *Medieval and Early Renaissance Medicine: An Introduction to Knowledge and Practice*. Chicago: University of Chicago Press, 1990.

Siraisi, Nancy G. "Mercuriale's Letters to Zwinger and Humanist Medicine." In Arcangeli and Nutton, *Girolamo Mercuriale*, 77–95.

Siraisi, Nancy G. "Oratory and Rhetoric in Renaissance Medicine." *Journal of the History of Ideas* 65, no. 2 (April 2004): 191–211.

Smith, Pamela. *The Body of the Artisan: Art and Experience in the Scientific Revolution*. Chicago: University of Chicago Press, 2004.

Spruit, Leen. "Cremonini nelle carte del Sant'Uffizio Romano." In Riondato and Poppi, *Cesare Cremonini*, 193–205.

Spruit, Leen. "Roman Reading Permits for the Works of Bruno and Campanella." *Bruniana & Campanelliana* 18, no. 2 (2012): 571–78.

Stallybrass, Peter. "Petrarch and Babylon: Censoring and Uncensoring the Rime, 1559–1651." In Blair and Goeing, *For the Sake of Learning*, 2:581–601.

Stella, Aldo. *Anabattismo e antitrinitarismo in Italia nel XVI secolo.* Padua: Liviani, 1969.

Stella, Aldo. "L'età postridentina." In *Diocesi di Padova,* edited by Pierantonio Gios, 215–44. Padua: Gregoriana libreria editrice, 1996.

Steuart, Francis. "The Scottish 'Nation' at the University of Padua." *Scottish Historical Review* 3, no. 9 (October 1905): 53–62.

Stoddard, Roger E. *Marks in Books, Illustrated and Explained.* Cambridge, MA: Houghton Library, Harvard University, 1985.

Stolzenberg, Daniel. "A Spanner and His Works: Books, Letters, and Scholarly Communication Networks in Early Modern Europe." In Blair and Goeing, *For the Sake of Learning,* 1:157–172.

Stolzenberg, Daniel. "Utility, Edification, and Superstition: Jesuit Censorship and Athanasius Kircher's *Oedipus Aegyptiacus.*" In *The Jesuits II: Cultures, Sciences, and the Arts 1540–1773,* edited by John W. O'Malley, Gauvin Alexander Bailey, Steven J. Harris, and T. Frank Kennedy, 336–54. Toronto: University of Toronto Press, 2006.

Stow, Kenneth R. "The Burning of the Talmud in 1553, in the Light of Sixteenth Century Catholic Attitudes toward the Talmud." *Bibliothèque d'Humanisme et Renaissance* 34, no. 3 (1972): 435–59.

Strocchia, Sharon. *Forgotten Healers: Women and the Pursuit of Health in Late Renaissance Italy.* Cambridge, MA: Harvard University Press, 2019.

Struhal, Eva. "Pittura e poesia a Lucca nel Seicento: Il caso di Pietro Paolini." In *Lucca città d'arte e i suoi archivi,* edited by Max Seidel and Romano Silva, 389–404. Venice: Marsilio, 2001.

Suarez, Michael J. *The Reach of Bibliography.* Lyell Lectures in Bibliography at Oxford University. Oxford: Oxford University Press, 2017.

Suitner, Riccarda. "Radical Reformation and Medicine in the Late Renaissance: The Case of the University of Padua." *Nuncius* 31, no. 1 (January 2016): 11–31.

Swerdlow, Noel M., and Otto Neugebauer. *Mathematical Astronomy in Copernicus's De Revolutionibus.* New York: Springer-Verlag, 1984.

Tarrant, Neil James. "Censoring Science in Sixteenth-Century Italy: Recent (And Not-So-Recent) Research." *History of Science* 52, no. 1 (2014): 1–27.

Tedeschi, John. "The Dispersed Archives of the Roman Inquisition." In *The Prosecution of Heresy: Collected Studies on the Inquisition in Early Modern Italy,* 23–45. Binghamton, NY: Medieval and Renaissance Texts and Studies, 1991.

Tedeschi, John. "Italian Reformers and the Diffusion of Renaissance Culture." *Sixteenth Century Journal* 5, no. 2 (October 1974): 79–94.

Thibodeau, Kenneth F. "Science and the Reformation: The Case of Strasbourg." *Sixteenth Century Journal* 7, no. 1 (April 1976): 35–50.

Tosi, Alessandro, ed. *Ulisse Aldrovandi e la Toscana: Carteggio e testimonianze documentarie.* Florence: Olschki, 1989.

Tutino, Stefania. *Empire of Souls: Robert Bellarmine and the Christian Commonwealth.* Oxford: Oxford University Press, 2010.

Tworek, Michael. "Learning Ennobles: Study Abroad, Renaissance Humanism, and the

Transformation of the Polish Nation in the Republic of Letters, 1517–1605." PhD diss., Harvard University, 2014.

Uccellini, Primo. *Dizionario storico di Ravenna e di altri luoghi di Romagna*. Ravenna: Tipografia del Ven. Seminario Arcivescovile, 1855.

Vacalebre, Natale. *Come le armadure e l'armi: Per una storia delle antiche biblioteche della Compagnia di Gesù; con il caso di Perugia*. Florence: Leo S. Olschki editore, 2016.

Valente, Michaela. *Bodin in Italia: La "Démonomanie des sorcier" e le vincende della sua traduzione*. Florence: Centro Editoriale Toscano, 1999.

Valente, Michaela. *"Correzioni d'autore" e censure dell'opera di Cardano*. In *Cardano e la tradizione dei saperi*, edited by M. Baldi and G. Canziani, 437–56. Milan: FrancoAngeli, 2003.

Van der Poel, M. *Cornelius Agrippa, the Humanist Theologian and His Declamations*. Leiden, Neth.: Brill, 1997.

Van Helden, Albert. Introduction to *Sidereus Nuncius, or the Starry Messenger*, 2nd ed., by Galileo Galilei, 1–26. Chicago: University of Chicago Press, 2016.

Van Liere, Katherine, Simon Ditchfield, and Howard Louthan, eds. *Sacred History: Uses of the Christian Past in the Renaissance World*. Oxford: Oxford University Press, 2012.

Van Miert, Dirk, ed. *The Kaleidoscopic Scholarship of Hadrianus Junius (1511–1575): Northern Humanism at the Dawn of the Dutch Golden Age*. Leiden, Neth.: Brill, 2011.

Van Miert, Dirk. "The Religious Beliefs of Hadrianus Junius (1511–1575)." In *Acta Conventus Neo-Latini Cantabrigiensis: Proceedings of the Eleventh International Congress of Neo-Latin Studies, Cambridge 30 July–5 August 2000*, edited by Rhoda Schnur, Jean-Louis Charlet, Lucia Gualda Rosa, Heinz Hofmann, Brenda Hosington, Elena Rodriguez Peregrina, and Ronald Truman 583–94. Tempe: Arizona Center for Medieval and Renaissance Studies, 2003.

Varner, Eric R. *Mutilation and Transformation: Damnatio Memoriae and Roman Imperial Portraiture*. Leiden, Neth.: Brill, 2004.

Visser, Arnould. "Irreverent Reading: Martin Luther as Annotator of Erasmus." *Sixteenth Century Journal* 48, no. 1 (2017): 87–109.

Vittinghoff, Friedrich. *Der Staatsfeind in der römischen Kaiserzeit: Untersuchungen zur "damnatio memoriae."* Vol. 2 of *Neue deutsche Forschungen Abteilung Alte Geschicte*. Berlin: Juncker & Dünnhaupt, 1936.

Walsham, Alexandra. "The Social History of the Archive: Record-Keeping in Early Modern Europe." *Past & Present* 230, suppl. 11 (November 2016): 9–48.

Walter, Tilman. "New Light on Antiparacelsianism (c. 1570–1610): The Medical Republic of Letters and the Idea of Progress in Science." *Sixteenth Century Journal* 43, no. 3 (Fall 2012): 701–25.

Watts, Ian. "Philosophical Intelligence: Letters, Print and Experiment during Napoleon's Continental Blockade." *Isis* 106, no. 4 (December 2015): 749–70.

Wear, A., R. K. French, and Iain M. Lonie, eds. *The Medical Renaissance of the Sixteenth Century*. Cambridge: Cambridge University Press, 1985.

Weber, Domizia. *Sanare e maleficiare: Guaritrici, streghe e medicina a Modena nel XVI secolo*. Rome: Carocci, 2011.

Webster, Charles. *The Great Instauration: Science, Medicine, and Reform, 1626–1660*. London: Duckworth, 1975.

Webster, Charles. *Health, Medicine, and Mortality in the Sixteenth Century*. Cambridge: Cambridge University Press, 1979.

Webster, Charles. *Paracelsus: Medicine, Magic and Mission at the End of Time*. New Haven, CT: Yale University Press, 2008.

Weisz, Leo. *Leo Jud: Ulrich Zwinglis Kampfgenosse, 1482–1542*. Zwingli-bücherei 27. Zurich: Zwingli-verlag, 1942.

Wenz, Andrea Beth. "Bernardino Ochino of Siena: The Composition of the Italian Reform at Home and Abroad." PhD diss., Boston College, 2017.

Werner, Sarah. "Correcting with Cancel Slips." *The Collation: Research and Exploration at the Folger* (blog). Folger Shakespeare Library. April 14, 2015. https://collation.folger.edu/2015/04/correcting-with-cancel-slips/.

Westfall, Richard S. "Patronage and the Publication of the *Dialogue*." In *Essays on the Trial of Galileo*. Notre Dame, IN: University of Notre Dame Press and Vatican Observatory Publication, 1989.

Westman, Robert S. *The Copernican Question: Prognostication, Skepticism, and Celestial Order*. Berkeley: University of California Press, 2011.

Westman, Robert S. "The Melanchthon Circle, Rheticus, and the Wittenberg Interpretation of the Copernican Theory." *Isis* 66, no. 2 (1975): 164–93.

White, Eric Marshall. *"Heresy and Error": The Ecclesiastical Censorship of Books 1400–1800*. Dallas: Bridwell Library, Perkins School of Theology, Southern Methodist University, 2010. Exhibition catalog.

Wilding, Nick. *Galileo's Idol: Gianfrancesco Sagredo and the Politics of Knowledge*. Chicago: University of Chicago Press, 2014.

Wilding, Nick. "Manuscripts in Motion: The Diffusion of Galilean Copernicanism." *Italian Studies* 66, no. 2 (July 2011): 221–33.

Williams, Gerhild Scholz, and Charles D. Gunnoe, eds. *Paracelsian Moments: Science, Medicine, and Astrology in Early Modern Europe*. Kirksville, MO: Truman State University Press, 2002.

Woolfson, Jonathan. *Padua and the Tudors: English Students in Italy, 1485–1603*. Toronto: University of Toronto Press, 1998.

Wootton, David. *Galileo: Watcher of the Skies*. New Haven, CT: Yale University Press, 2010.

Yale, Elizabeth. *Sociable Knowledge: Natural History and the Nation in Early Modern Britain*. Philadelphia: University of Pennsylvania Press, 2016.

Zambelli, Paola. *Astrologi Hallucinati: Stars and the End of the World in Luther's Time*. Berlin: De Gruyter, 1986.

Zedelmaier, Helmut. *Bibliotheca universalis und bibliotheca selecta: Das Problem der Ordnung des gelehrten Wissens in der frühen Neuzeit*. Cologne: Böhlau, 1992.

Zen Benetti, Francesca. "La libreria di Girolamo Fabrici d'Acquapendente." *Quaderni per la storia dell'Università di Padova* 9–10 (1976–77): 161–71.

Zorzi, Marino. *La libreria di San Marco: Libri, lettori, società nella Venezia dei Dogi.* Milan: A. Mondadori, 1987.

Zorzi, Marino. "La produzione e la circolazione del libro." In *Storia di Venezia dalle origini alla caduta della Serenissima,* vol. 7, edited by Gino Benzoni and Gaetano Cozzi, 921–85. Rome: Istituto della Enciclopedia Italiana, 1998.

Zurlini, Fabiola. *Cultura scientifica, formazione e professione medica tra la Marca e Roma nel Seicento: Il caso di Romolo Speziolo.* Fermo, It.: Litografica Com, 2009.

Zurlini, Fabiola. *Romolo Spezioli (Fermo, 1642—Roma, 1723): Un medico fermano nel XVII secolo a Roma.* Rome: Vecchiarelli Editore, 2000.

INDEX

Page references to tables are marked in bold. Page references to images are marked in italics.